Topological Methods, Variational Methods

and their applications

ICM 2002 Satellite Conference on
Nonlinear Functional Analysis

Topological Methods, Variational Methods

and their applications

Taiyuan, Shan Xi, P.R.China / August 14 – 18, 2002

edited by

H. Brezis
University of Paris VI, France and Rutgers University, USA

K. C. Chang
Peking University, P.R.China

S. J. Li
Chinese Academy of Science, P.R.China

P. Rabinowitz
University of Wisconsin,USA

World Scientific
New Jersey • London • Singapore • Hong Kong

Published by

World Scientific Publishing Co. Pte. Ltd.
5 Toh Tuck Link, Singapore 596224
USA office: Suite 202, 1060 Main Street, River Edge, NJ 07661
UK office: 57 Shelton Street, Covent Garden, London WC2H 9HE

British Library Cataloguing-in-Publication Data
A catalogue record for this book is available from the British Library.

TOPOLOGICAL METHODS, VARIATIONAL METHODS AND THEIR APPLICATIONS
Proceedings of the ICM2002 Satellite Conference on Nonlinear Functional Analysis

ISBN 981-238-262-3

Printed in Singapore by World Scientific Printers (S) Pte Ltd

Preface

ICM 2002 Satellite Conference on Nonlinear Analysis was held in the period: August 14-18, 2002 at Taiyuan, Shanxi Province, China. This conference was organized by Mathematical school of Peking University, Academy of Mathematics and System Sciences of Chinese Academy of Sciences, Mathematical school of Nankai University, and Department of Mathematics of Shanxi University, and was sponsored by Shanxi Province Education Committee, Tian Yuan Mathematics Foundation, and the Shanxi University.

166 mathematicians from 21 countries and areas in the world attended the conference. 53 invited speakers and 30 contributors presented their lectures. This conference aims at an overview of the recent development in nonlinear analysis. It covers the following topics: variational methods, topological methods, fixed point theory, bifurcations, nonlinear spectral theory, nonlinear Schrodinger equations, semilinear elliptic equations, Hamiltonian systems, central configuraton in N-body problems and variational problems arising in geometry and physics.

Many young scholars and graduate students from different universities and institutes in China joined the conference. The lectures presented in the conference reflect from different angles the broad and rapid development of nonlinear analysis. This volume contains most of these lectures.

We wish to thank all the following institutions: Shanxi Province, Shanxi Province Education Committee, Shanxi University, National Natural Science Foundation of China and Tien Yuan Math Foundation, for their generous financial support which made the conference possible, and all the related personnel in Shanxi who greatly contributed to the success of the conference.

We extend our thanks to all speakers of the conference and the World Scientific Publisher, specially the editor Dr. Q. Ye , who greatly helped us on the publication of this proceedings volume.

Haïm Brezis
Kung-Ching Chang
Shujie Li
Paul H. Rabinowitz

November 2002

Contents

List of Academic Committees

Honorary Chairman:

F. Browder, V. Lakshmikantham, Tian Fangzong, Peng Kunchi, Guo Guichun

Chairman:

K. C. Chang, H. Brezis, P. Rabinowitz

Members of Committee:

J. Mawhin, Chen Wenyuan, Guo Dajun, Liang Zhandong

List of Organizing Committees

Honorary Chairman:

Wang Xin, Xiang Congzhi, Liu Diansheng, Jia Suotang

Chairman:

Li Shujie, Long Yiming, Li Shengjia

Members of Committee:

Zhang Shisheng, Yan Jurang, Wu Xian, Li Yongqing, Zhao Lizhu, Su Jiabao, Zhang Zhitao, Zhong Chengkui

Organizor

School of Mathematical Sciences, Peking University
Academy of Mathematics and System Sciences, Chinese Academy of Sciences
College of Mathematics, Nankai University
Department of Mathematics, Shanxi University

List of Invited Speakers

N. Ackermann	Universitaet Giessen, Germany
A. Albouy	IMCCE, France
J. Appell	University of Wurzburg, Germany
T. Bartsch	Universitaet Giessen, Germany
B. Buffoni	Swiss Federal Institute of Technology, Switzerland
J.Y. Byeon	Pohang University of Science and Technology, Korea
D.M. Cao	AMSS, China
K.C. Chang	Peking Univ., China
A. Chenciner	Universite Paris VII, France
M. Clapp	UNAM, Mexico
D. Costa	University of Nevada - Las Vegas, USA
Y.H. Ding	AMSS, China
Z.H. Ding	Univ. of Nevada, USA
Y.H. Du	University of New England, Australia
M. Esteban	Universite Paris IX-Dauphine, France
X.L. Fan	Lanzhou Univ., China
C.F. Gui	University of Connecticut, USA
Y.X. Guo	Tsinghua Univ., China
B.N. Hirano	Yokohama National University, Japan
A. Ioffe	Haifa, Israel
M.Y. Jiang	Peking Univ., China
A. Kartsatos	University of South Florida, USA
F.Y. Li	Shanxi Univ., China
J.Q. Liu	Peking Univ., China
Z.L. Liu	Shandong Univ., China
Y.M. Long	Nankai University, China
R.Y Ma	Tsinghua Univ., China
G. Mancini	Universite degli Studi "Roma Tre", Italy
W. Marzantowicz	University of Poznañ, Poland
J. Mawhin	Universite Cathelique de Louvain, Belgium
V. Moroz	University of Bristol, UK
N. Papageorgiou	National Technical University, Greece
Y.W. Qi	University of California, Santa Barbara, USA
M. Ramos	Universidade de Lisboa, Portugal

T. Ricciardi	Federico II University of Naples, Italy
S. Rybicki	Nicolaus Copernicus University, Poland
J.P. Shi	College of William & Mary, USA
P.E. Sobolevski	Institute of Mathematics Givat Ram, Israel
C.A. Stuart	Ecole Polytechnique Fdrate Lausanne, Switzerland
A. Szulkin	Stockholm University, Sweden
K. Tanaka	Waseda University, Japan
G. Tian	MIT, USA; Peking Univ., China
H.Y. Wang	Yangzhou Univ., China
L.H. Wang	University of Iowa, USA
S.H. Wang	National Tsing Hua University, Taiwan, China
Z.Q. Wang	Utah State Univ., USA
T. Weth	Universitaet Giessen, Germany
J.G. You	Nanjing Univ, China
S.Q. Zhang	Chongqing Univ.,China
A.M. Zhao	Shanxi Univ., China
C.K. Zhong	Lanzhou Univ., China
J.X. Zhou	Texas A&M University, USA
C.F. Zhu	Nankai Univ. China

On the multiplicity of sign changing solutions to nonlinear periodic Schrödinger equations

Nils Ackermann

Mathematisches Institut, Universität Giessen Arndtstr. 2, 35392 Giessen Germany

ABSTRACT. Under weak non degeneracy assumptions we show that for certain superlinear, stationary, periodic Schrödinger equations there exist infinitely many geometrically distinct two-bump solutions that change sign exactly once. We give examples of potentials that satisfy the assumptions.

1. Introduction

We are concerned with the periodic nonlinear Schrödinger equation

$$\begin{cases} -\Delta u + V(x)u = f(x,u), & x \in \mathbb{R}^N \\ u \in H^1(\mathbb{R}^N) \end{cases} \tag{P}$$

where $N \in \mathbb{N}$, f has superlinear and subcritical growth in u, and f and V are periodic in x. Our intention is to prove existence of multibump solutions using a weak non degeneracy condition. There is a vast literature devoted to the existence of multibumps, using different flavors of non degeneracy assumptions on the solution set. In the pioneering works [**5, 12, 11, 1**] it was required that the solution set near the lowest energy level is finite or at most countable, modulo translations. The survey article [**9**] contains many more references of this type of results.

There have been some papers where this assumption could be relaxed, but only in the case of one space dimension [**2, 10**]. We will use a condition that arises from translating the conditions used in these particular references to our setting (indeed, it can be shown that in our setting these conditions are equivalent).

The main novelty in the present paper is therefore to prove existence of multibumps in a degenerate setting in higher space dimension. It turns out that, using some new results on pseudo gradient vector fields whose flows preserve the order

---

email: `nils.ackermann@math.uni-giessen.de`. This is joint work with Tobias Weth.

structure in $H^1(\mathbb{R}^N)$ (compare with [**3**]), it is much easier to prove the existence of *sign changing* two-bump solutions than of positive multibump solutions.

In the second half of our work we give examples where the non degeneracy condition can be verified. We do this for certain parameter-dependent problems that are inspired by the results in [**4, 6**].

To be more specific, let 2^* denote the critical Sobolev exponent, i.e. $2^* = 2N/(N-2)$ for $N \geq 3$ and $2^* = \infty$ for $N = 1, 2$. Assume

(A_1) $V \in L^\infty(\mathbb{R}^N)$, $\operatorname{essinf} V > 0$.

(A_2) $f\colon \mathbb{R}^N \times \mathbb{R} \to \mathbb{R}$ is a Caratheodory function, and there are constants $C > 0$ and $p \in (2, 2^*)$ such that

$$|f(x,u)| \leq C(1 + |u|^{p-1}) \quad \text{for } x \in \mathbb{R}^N,\ u \in \mathbb{R}.$$

Moreover, $f(x,u) = o(|u|)$ as $u \to 0$ uniformly in x.

(A_3) There is $\eta > 2$ such that

$$0 < \eta F(x,u) \leq f(x,u)u \quad \text{for } x \in \mathbb{R}^N,\ u \in \mathbb{R} \setminus \{0\},$$

where $F(x,u) := \int_0^u f(x,s)\,ds$.

(A_4) V and f are periodic with respect to some real invertible $N \times N$ matrix B, i.e. there holds $V(x + Bl) = V(x)$ and $f(x + Bl, u) = f(x,u)$ for all $l \in \mathbb{Z}^N$, $x \in \mathbb{R}^N$, $u \in \mathbb{R}$.

(A_5) f is odd in the second argument.

(A_6) The function $u \mapsto f(x,u)/|u|$ is strictly increasing on $\mathbb{R} \setminus \{0\}$ for a.e. $x \in \mathbb{R}^N$.

In the following we put $E := H^1(\mathbb{R}^N)$, and for matters of convenience we endow E with the scalar product

$$(u,v) := \int_{\mathbb{R}^N} (\nabla u \nabla v + V(x)uv)\,dx \qquad \text{for } u, v \in E.$$

By (A_1) the induced norm $\|\cdot\|$ is equivalent to the standard norm of $H^1(\mathbb{R}^N)$. Consider the functional

$$\Phi\colon E \to \mathbb{R}, \qquad \Phi(u) = \frac{1}{2}\|u\|^2 - \int_{\mathbb{R}^N} F(x,u)\,dx\,.$$

Standard arguments show that $\Phi \in C^1(E)$, and that critical points of Φ are weak solutions of (P). As a consequence of (A_4), Φ remains invariant under the action of the group $\mathbb{Z}^N$ given by

$$\mathbb{Z}^N \times E \to E, \qquad (k,u) \mapsto k * u := u(\cdot - Bk).$$

Due to this translation invariance, Φ does not satisfy the Palais-Smale condition. Nevertheless it has been shown that (A_1)–(A_4) imply the existence of at least one positive solution of (P). For a more precise statement, if $A \subset E$ let us introduce the notations

$$A_+ := \{\, u \in A \mid u(x) \geq 0 \quad \text{for } x \in \mathbb{R}^N \,\},$$
$$A_- := \{\, u \in A \mid u(x) \leq 0 \quad \text{for } x \in \mathbb{R}^N \,\},$$

$$A^s := \{ u \in A \mid \Phi(u) \le s \}, \qquad s \in \mathbb{R}$$

and

$$A^s_\pm := A_\pm \cap A^s \,.$$

Moreover consider

$$K := \{ u \in E \setminus \{0\} \mid \Phi'(u) = 0 \}$$

and

$$c_0 := \inf \Phi(K) \,.$$

Then $c_0 > 0$ and $K_+^{c_0} \neq \varnothing$, i.e., the value c_0 is attained by a positive solution. Furthermore we set

$$a_0 := \inf\{ \|u\| \mid u \in K \}$$

and remark that $a_0 > 0$.

We now state our final assumption. Put $\mathbb{Z}^N_* := \mathbb{Z}^N \setminus \{0\}$.

(A_7) There is $\alpha \in (0, c_0)$ and a compact set $\mathcal{K} \subseteq K^{c_0+\alpha}$ such that
  (a) $K^{c_0+\alpha} = \mathbb{Z}^N * \mathcal{K}$
  (b) $\operatorname{dist}(\mathcal{K}, \mathbb{Z}^N_* * \mathcal{K}) > 0$.

Our main result is the following

THEOREM 1.1. *Suppose that* (A_1)–(A_7) *hold and let* $\varepsilon > 0$. *Then there exists* $L = L(\varepsilon)$ *such that the following holds:*

For $l_1, l_2 \in \mathbb{Z}^N$ *with* $|l_1 - l_2| \ge L$ *there exist a sign changing solution* u *of* (P) *and* $v_1, v_2 \in \mathcal{K}_+$ *such that*

(i) $\Phi(u) \in [2c_0, 2c_0 + \varepsilon)$
(ii) $\|u - (l_1 * v_1 - l_2 * v_2)\| \le \varepsilon$
(iii) *u has precisely two nodal domains.*

COROLLARY 1.2. *Suppose that* (A_1)–(A_7) *hold. Then* (P) *has infinitely many geometrically distinct solutions with precisely two nodal domains.*

REMARK 1.3.

(a) Note that for every sign changing solution u of (P) there holds $\Phi(u) \ge 2c_0$. Indeed, denoting by u^+ resp. u^- the positive resp. negative part of u (that is $u^+ = \max\{u, 0\}$, $u^- = \min\{u, 0\}$), we have $\Phi'(u_+)u_+ = 0 = \Phi'(u_-)u_-$. Hence $\Phi(u) = \Phi(u_+) + \Phi(u_-) \geq 2c_0$.
(b) Similarly we see that every sign changing solution u with $\Phi(u) < 3c_0$ has precisely two nodal domains. Suppose in contradiction that u has at least three nodal domains Ω_1, Ω_2, Ω_3. Since u is continuous, it has been shown in [**8**, Lemma 1] that $u_i := u\chi_{\Omega_i} \in E$ for $i = 1, 2, 3$. Moreover we have $\Phi'(u_i)u_i = \Phi'(u)u_i = 0$, hence $\Phi(u) \ge \sum_{i=1}^3 \Phi(u_i) \ge 3c_0$, a contradiction.

REMARK 1.4. The assumption (A_5) is not crucial for our results and can be relaxed. Its only purpose is to assure that the energy levels of the positive and negative solutions with minimal energy are equal.

For our examples the following is useful:

Proposition 1.5. *Condition* (A_7) *is equivalent to the following property: There is* $\alpha \in (0, c_0)$ *and* $\mathcal{U} \subseteq K^{c_0+\alpha}$*, relatively open, such that*

$$\{\, l * \mathcal{U} \mid l \in \mathbb{Z}^N \,\}$$

forms a disjoint cover of $K^{c_0+\alpha}$*. Indeed, one can set* $\mathcal{K} = \mathcal{U}$.

In the present contribution for a proceedings volume we just sketch the main line of argumentation and state our lemmata without proof. There is a paper with full details in preparation.

1.1. Notation. Let X be a metric space with metric d. If $A \subseteq X$ and $r \geq 0$ we put $U_r(A) := \{\, x \in X \mid \operatorname{dist}(x, A) < r \,\}$. If $r \geq 0$ and $x \in X$ we set $B_r(x) := \{\, y \in X \mid d(x, y) \leq r \,\}$.

2. Existence of sign changing two-bump solutions

2.1. Analysis of two-bump functions with sign change. Here we work out consequences of (A_7) for sign changing functions. Consider the sets

$$D^1_\tau := \{\, u \in E \mid \operatorname{dist}(u, E_-) < \tau, \}, \qquad D^2_\tau := \{\, u \in E \mid \operatorname{dist}(u, E_+) < \tau \,\}$$

as well as $D_\tau := \overline{D^1_\tau} \cup \overline{D^2_\tau}$ for $\tau > 0$. Note that D^1_τ and D^2_τ are open convex subsets of E, whereas D_τ is closed in E. Moreover, $E \setminus D_\tau$ contains only sign changing functions.

Lemma 2.1. *For some* $\tau > 0$ *suppose that* $(u_n) \subseteq E \setminus D_\tau$ *and* $\Phi'(u_n) \to 0$. *Then* $\liminf_{n\to\infty} \Phi(u_n) \geq 2c_0$.

From now on we consider pairs of integers which we write as $\bar{l} = (l^1, l^2) \in \mathbb{Z}^N \times \mathbb{Z}^N$. We define $d(\bar{l}) := |l^1 - l^2|$. Also we put $\mathcal{N}(\bar{l}) = l^1 * \mathcal{K}_+ + l^2 * \mathcal{K}_-$. If $L \geq 0$ we set $Z(L) = \{\, \bar{l} \in \mathbb{Z}^N \mid d(\bar{l}) \geq L \,\}$.

Lemma 2.2. *There is* $b_0 > 0$ *such that for any* $\varepsilon \in (0, b_0)$ *there are* $L > 0$ *and* $\delta > 0$ *such that for any* $\bar{l}, \bar{m} \in Z(L)$:

(i) $U_{b_0}(\mathcal{N}(\bar{l})) \subseteq E \setminus D_{a_0/4}$
(ii) $U_{b_0}(\mathcal{N}(\bar{l})) \cap U_{b_0}(\mathcal{N}(\bar{m})) = \varnothing$ *if* $\bar{l} \neq \bar{m}$
(iii) $\inf\{\, \|\Phi'(u)\| \mid u \in U_{b_0}(\mathcal{N}(\bar{l})) \setminus U_\varepsilon(\mathcal{N}(\bar{l})),\ \Phi(u) \leq 2c_0 + \alpha \,\} \geq \delta$
(iv) *Any* $(\mathrm{PS})_c$*-sequence* $(u_n) \subseteq U_{b_0}(\mathcal{N}(\bar{l}))$ *with* $c \leq 2c_0 + \alpha$ *is relatively compact. Its cluster points lie in* $U_\varepsilon(\mathcal{N}(\bar{l}))$.

We define for $v_1, v_2 \in E_+$ the cone

$$\mathcal{C}(v_1, v_2) := \{\, s_1 v_1 - s_2 v_2 \mid s_1, s_2 \geq 0 \,\}.$$

Lemma 2.3. *Let* $v \in \mathcal{K}^{c_0}_+$. *Then for every* $\varepsilon_1, \varepsilon_2 > 0$ *there is* L *such that for* $\bar{l} \in Z(L)$ *the following properties are satisfied:*

(i) $\sup\{\, \Phi(u) \mid u \in \mathcal{C}(l_1 * v, l_2 * v) \,\} < 2c_0 + \varepsilon_1$.
(ii) *If* $\Phi(u) \geq 2c_0$ *for some* $u \in \mathcal{C}(l_1 * v, l_2 * v)$, *then* $u \in U_{\varepsilon_2}(\mathcal{N}(\bar{l}))$.

2.2. The descending flow. We remark that the gradient of Φ has the form $\nabla\Phi(u) = u - G(u)$ where $G\colon E \to E$ is uniquely determined by the relation

$$(G(u), v) = \int_{\mathbb{R}^N} f(x,u)v\,dx \quad \text{for all } v \in E.$$

The main ingredient for proving existence of order-preserving flows is the following

LEMMA 2.4. *There is $\tau_0 > 0$ such that for $0 < \tau \le \tau_0$ there holds*

(i) *$G(\partial D^1_\tau) \subset D^1_\tau$, and every nontrivial solution $u \in D^1_\tau$ of (P) is negative;*
(ii) *$G(\partial D^2_\tau) \subset D^2_\tau$, and every nontrivial solution $u \in D^2_\tau$ of (P) is positive.*

From now on we fix $\tau \in (0, \min\{a_0/4, \tau_0\})$ such that

$$\Phi \ge 0 \qquad \text{on } B_{\sqrt{2}\tau}(0) \supset D^1_\tau \cap D^2_\tau,$$

this is possible by (A_1) and (A_2).

Put $E_0 = \{u \in E \mid \Phi'(u) \ne 0\}$. Suppose we are given a (local) flow φ on E_0. If $A \subseteq E$ we say that *A is positive invariant for the flow φ* if $\varphi(t,u) \in A$ for every $u \in A \cap E_0$ and $t \ge 0$ such that $\varphi(t,u)$ exists. If $A \subseteq E$ is positive invariant, denote by

$$\mathcal{A}(A) := A \cup \{u \in E_0 \mid \varphi(t,u) \in A \text{ for some } t \ge 0\}$$

the *set of attraction of A*.

In view of Lemma 2.4 the next proposition just follows from [**7**, Lemma 3.2].

PROPOSITION 2.5. *There is a pseudo gradient vector field $\mathcal{V}$ for Φ in E_0 such that D^1_τ and D^2_τ are positive invariant for the associated descending flow. Moreover, $\partial D^i_\tau \subset \mathcal{A}(D^i_\tau)$ for $i = 1, 2$.*

In the following, φ denotes a flow on E_0 with the properties listed in Proposition 2.5. For any $u \in E_0$ let $T(u) \in (0, \infty]$ denote the maximal existence time for the trajectory $\varphi(t,u)$.

2.3. Proof of the main theorem. This section is devoted to the proof of Theorem 1.1. Recall the definition of b_0 in Lemma 2.2. We fix $\varepsilon \in (0, b_0)$ and choose $v \subset \mathcal{K}^{co}_+$. Let $L > 0$ and $\delta > 0$ be given by Lemma 2.2 for this choice of ε. Set $\varepsilon_0 = \varepsilon\delta/4$. Take L large enough so that the assertions of Lemma 2.3 hold true for this choice of v, with $\varepsilon_1 = \min\{\varepsilon/2, \varepsilon_0, \alpha\}$ and $\varepsilon_2 = \varepsilon$. Moreover, fix $\bar{l} \in Z(L)$.

The next Proposition reflects the fact that, if we look for sign changing solutions, there is a topological obstruction that can replace minimax arguments.

PROPOSITION 2.6. *There exists $u_0 \in \mathcal{C}(l^1 * v, l^2 * v) \setminus D_\tau$ such that either $\Phi'(u_0) = 0$ or the following properties hold.*

(i) *$\varphi(t, u_0) \in E \setminus D_\tau$ for $t \in [0, T(u_0))$.*
(ii) *For some sequence $(t_n)_n \subset [0, T(u_0))$ with $t_n \to T(u_0)$ it holds that $\Phi'(\varphi(t_n, u_0)) \to 0$.*

If $\Phi'(u_0) = 0$ put $u = u_0$. Then Remark 1.3 implies $\Phi(u) \geq 2c_0$, and thus $u \in U_\varepsilon(\mathcal{N}(\bar{l}))$ by Lemma 2.3(ii). By Lemma 2.3(i) we have that $\Phi(u) \in [2c_0, 2c_0 + \min\{\varepsilon, \alpha\}]$.

To handle the case $\Phi'(u_0) \neq 0$ we need the

LEMMA 2.7. *Suppose that $u \in U_\varepsilon(\mathcal{N}(\bar{l}))$ satisfies $\Phi(u) \leq 2c_0 + \alpha$ and $\varphi(t,u) \notin U_{b_0}(\mathcal{N}(\bar{l}))$ for some $t > 0$. Then $\Phi(u) - \Phi(\varphi(t,u)) \geq \varepsilon_0$.*

Put $u_n = \varphi(t_n, u_0)$. By Lemma 2.3(i), for all n we have $\Phi(u_n) < 2c_0 + \min\{\varepsilon/2, \varepsilon_0, \alpha\}$. Using Lemma 2.1 we see that $\liminf_{n\to\infty} \Phi(u_n) \geq 2c_0$. Now by Lemma 2.3(ii) $u_0 \in U_\varepsilon(\mathcal{N}(\bar{l}))$, and Lemma 2.7 gives that $u_n \in U_{b_0}(\mathcal{N}(\bar{l}))$ for all n. By Lemma 2.2(iv), after extraction of a subsequence, there is $u \in U_\varepsilon(\mathcal{N}(\bar{l}))$ such that $u_n \to u$ as $n \to \infty$. Therefore $\Phi(u) = \lim_{n\to\infty} \Phi(u_n) \in [2c_0, 2c_0 + \min\{\varepsilon/2, \alpha\}]$.

In any case, the solution u we have found must have precisely two nodal domains by Lemma 2.2(i) and Remark 1.3. We have seen that $\Phi(u) \in [2c_0, 2c_0 + \varepsilon)$. The remaining assertion follows from $u \in U_\varepsilon(\mathcal{N}(\bar{l}))$. The proof is finished.

3. Two examples

In this section we try to illustrate how condition (A_7) can be verified in particular cases. To simplify notation we restrict ourselves to the case $N \geq 3$, even though the results are true for $N = 1, 2$ also.

In contrast to the former notation, in this section let $\|\cdot\|$ denote the standard norm on E, and for $V \in L^\infty(\mathbb{R}^N)$ we define $\|\cdot\|_V$ by

$$\|u\|_V^2 := \int_{\mathbb{R}^N} (|\nabla u|^2 + Vu^2)\, dx .$$

Similarly we define the corresponding scalar products $(\cdot,\cdot)$ and $(\cdot,\cdot)_V$.

3.1. The center of mass. For any $\delta \in (0,1]$ and $C \geq 0$ denote by $\Gamma(\delta, C)$ the set of weak solutions in $E \setminus \{0\}$ of equations of the form

$$-\Delta u + Vu = f(x,u) ,$$

where $V \in L^\infty(\mathbb{R}^N)$ with $\operatorname{essinf} V \geq \delta$ and $f\colon \mathbb{R}^N \times \mathbb{R} \to \mathbb{R}$ is a Caratheodory function such that for $u \in \mathbb{R}$ and a.e. $x \in \mathbb{R}^N$

$$|f(x,u)| \leq \frac{\delta}{2}|u| + C|u|^{2^*-1} .$$

From

$$\delta\|u\|^2 \leq \|u\|_V^2 = \int f(x,u)u\, dx \leq \frac{\delta}{2}\|u\|^2 + \widetilde{C}C\|u\|^{2^*}$$

for $u \in \Gamma(\delta, C)$ and some constant $\widetilde{C}$ independent of u, δ and C, it follows that $\|u\|$ is bounded away from 0 in $\Gamma(\delta, C)$, so that

$$\overline{\Gamma(\delta, C)} \subseteq E \setminus \{0\} .$$

PROPOSITION 3.1. *Given $\delta \in (0,1]$ and $C \geq 0$, the map $\beta\colon \Gamma(\delta, C) \to \mathbb{R}^N$ with*

$$\beta(u) = \frac{1}{\|u\|^2} \int_{\mathbb{R}^N} (|\nabla u|^2 + u^2) x \, dx$$

is well defined. Furthermore, β extends to a continuous map on $\overline{\Gamma(\delta, C)}$.

3.2. Generalities. For the rest of Sect. 3 we fix a function f that satisfies (A_2) and (A_3). In addition we assume that there is a real invertible $N \times N$-matrix B such that f is periodic in the first variable with respect to B. Define F and Ψ as in the Introduction.

Now suppose that we are given $V \in L^\infty(\mathbb{R}^N)$ with $\operatorname{essinf} V > 0$ and let

$$\Phi_V(u) = \frac{1}{2}\|u\|_V^2 - \Psi(u)$$

for $u \in E$. Furthermore, we put

$$K_V := \{\, u \in E \setminus \{0\} \mid \Phi_V'(u) = 0 \,\}$$

and

$$c_V := \inf \Phi_V(K_V) \, .$$

We have by (A_3) that $c_V > 0$.

3.3. Steep potential wells. For our first example, suppose that $g_0, g \in L^\infty(\mathbb{R}^N)$ are periodic with respect to B, that $\operatorname{essinf} g_0 > 0$, and $g \geq 0$. Moreover, suppose that there is a compact subset $M \subseteq \mathbb{R}^N$ with

$$\operatorname{conv}(M) \cap (B\mathbb{Z}_*^N + \operatorname{conv}(M)) = \varnothing$$

and such that there is $u \in E \setminus \{0\}$ with $u(x) = 0$ for $x \in \mathbb{R}^N \setminus M$. The latter condition is satisfied for example if $\mathring{M} \neq \varnothing$. Lastly, setting $\mathcal{M} = B\mathbb{Z}^N + M$, we assume that $g = 0$ in $\mathcal{M}$, and $g > 0$ a.e. in $\mathbb{R}^N \setminus \mathcal{M}$. For $\mu \geq 0$ we set

$$V_\mu = g_0 + \mu g \, .$$

Also put $\|\cdot\|_\mu = \|\cdot\|_{V_\mu}$ and define Φ_μ, K_μ, and c_μ accordingly as in Sect. 3.2. It turns out that

$$\inf_{\mu \geq 0} c_\mu > 0 \, .$$

THEOREM 3.2. *There is $\alpha > 0$ such that condition (A_7) is satisfied for the functional Φ_μ provided that μ is large enough.*

SKETCH OF PROOF: First we show that

$$c_\infty := \lim_{\mu \to \infty} c_\mu > 0 \tag{3.1}$$

exists. Setting

$$\varepsilon := \operatorname{dist}(\operatorname{conv}(M), B\mathbb{Z}_*^N + \operatorname{conv}(M)) > 0$$

there is μ_0 so that for any $\mu \geq \mu_0$

$$\beta(K_\mu^{c_\mu + c_\infty/2}) \subseteq U_{\varepsilon/3}(\mathcal{M}) \, .$$

Fix $\mu \geq \mu_0$ and put $\mathcal{U} := \beta^{-1}(U_{\varepsilon/3}(M)) \subseteq K_\mu$. By Prop. 3.1 β is continuous on K_μ, so that $\mathcal{U}$ is open in K_μ. Clearly $\{ l * \mathcal{U} \mid l \in \mathbb{Z}^N \}$ is a disjoint cover of $K_\mu^{c_\mu + c_\infty/2}$. By (3.1) we can also assume that $c_\mu > c_\infty/2$, so that with $\alpha = c_\infty/2$ condition (A_7) is satisfied for the functional Φ_μ by Prop. 1.5. ∎

3.4. Separated potential wells. Let $g_0 \in L^\infty(\mathbb{R}^N)$ be B-periodic. Suppose that $M \subseteq \mathbb{R}^N$ is compact with $\mathrm{mes}(M) > 0$, and that $g \in L^\infty(M)$ satisfies $g \geq 0$ and $\|g\|_{L^\infty(M)} > 0$. Put $\mathcal{M}_n := nB\mathbb{Z}^N + M$ for $n \in \mathbb{N}$. Suppose also that for

$$g_1 = g_0 - \chi_M g$$

it holds that $\delta := \operatorname{essinf} g_1 > 0$. Lastly, for $n \in \mathbb{N}$ put

$$V_n(x) = \begin{cases} g_0(x) - g(x) & x \in \mathcal{M}_n \\ g_0(x) & x \in \mathbb{R}^N \setminus \mathcal{M}_n . \end{cases}$$

Then V_n is nB-periodic and $\operatorname{essinf} V_n = \delta$ for large n. For convenience we set $\|\cdot\|_n = \|\cdot\|_{V_n}$, $\Phi_n = \Phi_{V_n}$ and so forth.

THEOREM 3.3. *There is $\alpha > 0$ such that condition* (A_7) *is satisfied for the functional Φ_n provided that n is large enough.*

SKETCH OF PROOF: First we prove

$$c_\infty := \min\{ \inf_n c_n, c_{g_0}, c_{g_1} \} > 0 .$$

We show that there are $r > 0$, $n_0 \in \mathbb{N}$ such that for any $n \geq n_0$

$$\beta(K_n^{c_n+\alpha}) \subseteq U_r(\mathcal{M}_n)$$

and $\{ nBl + U_r(M) \mid l \in \mathbb{Z}^N \}$ is a disjoint cover of $\mathcal{M}_n$. As in Sect. 3.3 put $\mathcal{U} := \beta^{-1}(U_r(M))$ so that $\mathcal{U}$ is open in $K_n^{c_n+\alpha}$ and $\{ (nl) * \mathcal{U} \mid l \in \mathbb{Z}^N \}$ is a disjoint cover of $K_n^{c_n+\alpha}$. Hence (A_7) is satisfied for the functional Φ_n if $n \geq n_0$, where the periodicity is given by the matrix nB. ∎

References

[1] Alama, S. and Li, Y.Y.: On "multibump" bound states for certain semilinear elliptic equations. Indiana Univ. Math. J. **41**, 983–1026 (1992)

[2] Arioli, G., Gazzola, F., and Terracini, S.: Multibump periodic motions of an infinite lattice of particles. Math. Z. **223**, 627–642 (1996)

[3] Bartsch, T., Liu, Z., and Weth, T.: Sign changing solutions to superlinear schrödinger equations. Preprint

[4] Bartsch, T., Pankov, A., and Wang, Z.Q.: Nonlinear Schrödinger equations with steep potential well. Commun. Contemp. Math. **3**, 549–569 (2001)

[5] Coti Zelati, V. and Rabinowitz, P.H.: Homoclinic type solutions for a semilinear elliptic PDE on $\mathbf{R}^n$. Comm. Pure Appl. Math. **45**, 1217–1269 (1992)

[6] Kabeya, Y. and Tanaka, K.: Uniqueness of positive radial solutions of semilinear elliptic equations in $\mathbf{R}^N$ and Séré's non-degeneracy condition. Comm. Partial Differential Equations **24**, 563–598 (1999)

[7] Liu, Z. and Sun, J.: Invariant sets of descending flow in critical point theory with applications to nonlinear differential equations. J. Differential Equations **172**, 257–299 (2001)

[8] Müller-Pfeiffer, E.: On the existence of nodal domains for elliptic differential operators. Proc. Roy. Soc. Edinburgh Sect. A **94**, 287–299 (1983)
[9] Rabinowitz, P.H.: A variational approach to multibump solutions of differential equations. In: Hamiltonian dynamics and celestial mechanics (Seattle, WA, 1995), vol. 198 of Contemp. Math., pp. 31–43, Providence, RI: Amer. Math. Soc. (1996)
[10] Rabinowitz, P.H.: A multibump construction in a degenerate setting. Calc. Var. Partial Differential Equations **5**, 159–182 (1997)
[11] Séré, É.: Existence of infinitely many homoclinic orbits in Hamiltonian systems. Math. Z. **209**, 27–42 (1992)
[12] Séré, É.: Looking for the Bernoulli shift. Ann. Inst. H. Poincaré Anal. Non Linéaire **10**, 561–590 (1993)

The underlying geometry of the fixed centers problems

Alain Albouy

IMCCE
77, av. Denfert-Rochereau, 75014 Paris, France

ABSTRACT. We present a sequence of results about some natural generalizations of the one fixed center problem and the two fixed centers problem. The main original point is that we do not feel obliged to stay in the traditional Lagrangian or Hamiltonian frameworks.

Introduction

The Variational Method supposes that the space is endowed with a structure that defines a concept similar to "length": one wishes to minimize or extremize this length and produce a solution to the Variational Problem. The basic examples are stated in the framework of Riemannian geometry. Finsler manifolds, pseudo-Riemannian manifolds, metric spaces are also frameworks where something similar to a "length" can be extremized. But the usual vector and affine spaces are examples of spaces where nothing similar to a metric is defined[1].

The simplest variational problem is often used to define a straight line in Euclidean space: it is the shortest path between two points. If one is convinced of the existence of an "underlying affine space", or of a "subjacent affine structure", below the Euclidean space, this definition has to be rejected, and just considered as a property of a straight line that appears in the context of Euclidean geometry[2].

---

e-mail: `albouy@imcce.fr`.

[1]There is indeed a classical variational problem where one starts from a space with very poor local structure, namely the problem of finding an Einstein metric. The question is to extremize the metric, which in a certain way measures itself.

[2]Note that for other reasons Poincaré criticized this variational definition: "Peut-on définir la ligne droite ? La définition connue, le plus court chemin d'un point à un autre, ne me satisfait guère. Je partirais tout simplement de la *règle* [...] Qu'est-ce que la géométrie pour le philosophe ?

Affine geometry was developed mainly in the 19th century. The "vector space", also called (see [ScK], p. 1) "centred affine space", and the "affine space" are the basic objects of affine geometry. The words affine and affinity were introduced (see [Wey], p. xiii) to designate linear transformations which are not the familiar isometries of Euclidean space. An accepted terminology is such that: a Euclidean space is an affine space with a Euclidean structure (a positive definite form defined on the vector space of the translations); a vector space is an affine space with a preferred point called zero; a Euclidean vector space is a vector space with a positive definite form or a Euclidean space with a preferred point called zero.

The starting point of my research project is a guess: in Celestial Mechanics, also called Gravitational Dynamics, any property of the systems which is "unexpected" in the framework of Lagrangian mechanics, i.e. does not find its natural generalization in this framework, does find generalizations in a strange but somewhat elementary mechanics where no variational principle can be defined, and concepts such as action, kinetic energy, potential, Hamiltonian, symplectic form disappear. The space may be an affine space, a generalized (curved) affine space, or a locally projective space. It is not endowed with a pseudo-Riemannian structure or with anything that could define the "length" of a path.

What are examples of these "unexpected properties"? I am especially interested in dynamical facts. It is a standard remark among specialists, that "nothing is generic in Celestial Mechanics". The curves described by the particles in the two-body problem are conic sections: they are algebraic curves; they are closed when they are bounded. These two properties are unexpected when one looks at the differential system, even with the eyes of an expert in symplectic geometry. An unexpected regular behavior appears again in one of the oldest integrable problems: Euler's two-fixed center problem. I was more recently interested in the old theory of the attractions of homogeneously filled ellipsoids, starting from a result of [Vel] on the attraction of spheres. More generally potential theory also possesses unexpected features; some of them may be expressed considering only the field of force, not the potential. Finally, I have to mention Herman's resonance [AbA], a strange fact that decided me to really study this subject.

This kind of striking facts in classical mechanics, unexpected regularity in the behavior of a gravitational system, or unexpected formulas for certain fields of force or potentials, seems to extend in the direction of general relativity, with the Schwarzschild's and Kerr's universes. Maybe is there a very deep, but understandable theory responsible for this also.

All the results stated in this text were proved, and most of them were obtained by an easy computation. Due to limitations of size and time, I do not present the proofs here, and I stop the sequence of results quite arbitrarily.

I wish to thank A. Chenciner and T. Stuchi for their precious comments.

C'est l'étude d'un groupe, et quel groupe? de celui des mouvements des corps solides. Comment alors définir ce groupe sans faire mouvoir quelques corps solides?" (see [Poi], pp. 143–145)

1. One fixed center

1.0. Kepler problem. It is convenient to number some known facts for easy reference in each generalization. The notation in Property 5 is explained in [Alb], p. 93 and also briefly in 1.1.3. We consider the motion of a small particle in the field of a point O with mass m.

1) The trajectories $q(t)$ are the solutions of the differential system $\ddot{q} = -m\|q\|^{-3}q$.

2) Equal areas are swept from O in equal times.

3) The bivector $C = q \wedge \dot{q}$ is a first integral.

4) A curve described by a trajectory, when rescaled, is also a curve described by a trajectory. The state $(q, \dot{q})$ becomes $(\lambda q, \lambda^{-1/2}\dot{q})$ after rescaling. If the orbit was closed with period T, the new period is $\lambda^{3/2}T$.

5) The eccentricity vector $L = m^{-1}\dot{q} \lrcorner C + \|q\|^{-1}q$ is a first integral.

6) The particle describes conic sections. O is a focus.

7) An isometry transforms a trajectory into a trajectory.

8) The period T of a closed trajectory is a function of the semi-major axis a only (it is independent of the eccentricity e if a is fixed).

9) The energy $H = \|\dot{q}\|^2/2 - m\|q\|^{-1}$ is a first integral.

10) The period T is a function of the energy H only.

1.1. Jacobi-Darboux attractors. We start from a general case and add successively some hypotheses. The differential equation is:

$$\ddot{q} = -mf(q)q. \tag{C}$$

The space is a vector space U and $f : U \setminus \{O\} \to \mathbb{R}$. Without other hypothesis on f (it could be a function of $\dot{q}$ and t also), Properties 2 and 3 are unchanged. Number 2 is theorem 1 of Newton's Principia. Any unit of area may be chosen in the plane of motion. Number 3 is synonymous; its proof is simpler if we know how to compute with the wedge product $\wedge$. To get Property 4, we suppose that

$$f \text{ is positively homogeneous of degree } -3. \tag{J}$$

A surprising result by Jacobi is that (C) is now integrable by quadratures. The solutions possess the following property: to each trajectory is associated a line in the plane of motion, passing through O. The trajectory can cut this line only at two distinguished points M and N. The origin O is on the segment MN.

1.1.1. Planar case. Let us suppose first $\dim U = 2$. Suppose that (C) admits a closed trajectory, or alternatively suppose that f is even, i.e. $f(q) = f(-q)$. Then every ray from O has the property that any trajectory cuts it in at most one point. In particular, *every trajectory is closed or goes to infinity after less than one turn.* And the motion has to be periodic if the curve is closed.

There exists indeed a weaker condition on f that gives the same particular property. To state it we must introduce some notation. One way is to choose linear coordinates x and y on U and to call $\sigma = xdy - ydx$. Then the hypothesis may be stated in two equivalent ways:

$$\text{the forms } xf\sigma \text{ and } yf\sigma \text{ are exact on } U \setminus \{O\} \tag{A}$$

or

$$\left.\begin{array}{r}\text{there exists } F : U \setminus \{O\} \to \mathbb{R}, \text{ positively homogeneous of degree 1,} \\ \text{such that } \dfrac{\partial^2 F}{\partial x^2} = y^2 f, \quad \dfrac{\partial^2 F}{\partial x \partial y} = -xyf, \quad \dfrac{\partial^2 F}{\partial y^2} = x^2 f.\end{array}\right\} \quad (A)$$

Hypothesis (A) can also be stated in a more intrinsic language. It is convenient to choose a non-zero unit of area $\nu \in \bigwedge^2 U^*$. We put $\sigma = q \lrcorner \nu$. Hypothesis (A) reads:

$$\text{the vector-valued 1-form } f(q)\sigma \otimes q \text{ is exact on } U \setminus \{O\} \quad (A)$$

or

$$\left.\begin{array}{r}\text{there exists } F : U \setminus \{O\} \to \mathbb{R}, \text{ positively homogeneous of deg. 1,} \\ \text{with } \partial^2 F = f\sigma \otimes \sigma.\end{array}\right\} \quad (A)$$

To write Property 5 assuming Hypothesis (A), note that $\langle C|\nu\rangle = \langle \dot{q}, \sigma\rangle$. A choice of F is necessary: dF appears in the formula; instead, we can use the "symplectic gradient" X_F of F with respect to the conventional two-form ν. It satisfies $X_F \lrcorner \nu = dF$. Using the symbol d for differentiation of a vector valued function on $U \setminus \{O\}$, we get $dX_F = f\sigma \otimes q$.

Definition 1. We call a Jacobi-Darboux problem a problem defined by system (C) with Condition (A). Note that Condition (J) is implied by (A).

5) The covector

$$\mathcal{L} = (-\dot{y}C + m\frac{\partial F}{dx})dx + (\dot{x}C + m\frac{\partial F}{dy})dy = \langle C|\nu\rangle \dot{q} \lrcorner \nu + m dF$$

is a first integral. The vector $w = \dot{q} + m\langle C|\nu\rangle^{-1} X_F$ is a first integral.

Remark 1. The expression for w depends on the choice of F, but not on the choice of ν. We can choose the unique ν such that $\langle \nu | C\rangle = 1$. We denote it by $1/C$. The expression for the vector becomes $w = \dot{q} + mX_F$.

The trajectories are not conic sections, but they are described by a property similar to the Pappus' Focus-Directrix definition of a conic section. From 5 we obtain $\langle \mathcal{L}, q\rangle = -\langle C|\nu\rangle^2 + mF$. This means:

6) The trajectories are such that the "anisotropic distance" F from the particle to the "focus" O is proportional to the value of an affine function ("distance to a line").

Remark 2. Projective transformations and generalizations. Let K be a positively homogeneous function of degree 1. Consider the transformation

$$\begin{array}{ccc} U \setminus \{O\} & \longrightarrow & U \setminus \{O\} \\ q & \longmapsto & Q = \dfrac{q}{1 + K(q)} \end{array}$$

One may assume $K > 0$ to guarantee that the transformation is defined everywhere. If we introduce a new time parameterization and denote by $'$ the derivation with respect to the new time, and if we denote by k the function associated to K as f is associated to F in (A), we get

$$Q' = \dot{q} + dK \lrcorner C, \qquad C = q \wedge \dot{q} = Q \wedge Q', \qquad Q'' = -m\Big(f(Q) + \langle \nu | C\rangle^2 k(Q)\Big)Q.$$

Thus this transformation maps any orbit of a Jacobi-Darboux problem to an orbit of another such problem. If K is a linear form, $k = 0$ and the transformation sends an orbit to an orbit in the same Jacobi-Darboux problem. This projective symmetry was noticed by Halphen (see [App]) in the usual Kepler problem, where it coexists with Properties 4 and 7.

1.1.2. The multidimensional case. We suppose now $\dim U > 2$. The orbits are always planar. A first question arise: how can Condition (A) be satisfied for every plane passing through O?

The answer is: *the function f must be even, i.e. $f(q) = f(-q)$ for any $q \in U \setminus \{O\}$*. And this result is an interesting variation of a famous result due to Funk. It is sufficient to study the dimension three case. The results in higher dimensions follow.

Consider the unit sphere $\mathcal{S}$ of a three-dimensional Euclidean vector space. Then the space $\mathcal{S}^*$ of the oriented great circles of $\mathcal{S}$ has the same geometry as $\mathcal{S}$. It is possible to identify $\mathcal{S}^*$ to $\mathcal{S}$, considering a great circle as an "equator" and associating to it a "north pole".

The *Radon transform* of a function $\phi : \mathcal{S} \to \mathbb{R}$ is the function $\phi^\circ : \mathcal{S}^* \to \mathbb{R}$, which associate to a great circle the integral $\oint \phi$ of the function ϕ on it.

Funk's theorem states that the Radon transform of ϕ is identically zero if and only if ϕ is odd (i.e. $\phi(q) = -\phi(-q)$ for any $q \in \mathcal{S}$).

Our condition of exactness (A) is the cancellation of an integral on a cycle, and the cycles we choose are on a plane passing through O. But our question is distinct from Funk's question. We consider a vector valued integral instead of a scalar one, and no canonical metric is defined on the "sphere" where the integration paths are drawn.

But it happens that our result can be deduced from the following proposition, applying it to the restriction ϕ to a unit sphere of $f : U \to \mathbb{R}$.

Proposition 1. *Choose an orientation on $\mathcal{S}$. To a function $\phi : \mathcal{S} \to \mathbb{R}$ associate $\phi^\bullet : S^* \to \mathbb{R}$, which associate to an oriented great circle of $\mathcal{S}$ the integral of ϕ on the hemisphere bordered by ϕ, on the "left bank" of ϕ. The function $\phi^\bullet$ is constant if and only if ϕ is even.*

The vanishing of the derivative of $\phi^\bullet$ at a great circle $\mathcal{C}$ corresponds to the exactness hypothesis (A) for the plane carrying $\mathcal{C}$. A proof of the proposition is deduced from the fact that the spherical harmonics of a given odd degree form an eigenspace for the transformation $\phi \to \phi^\bullet$.

In the multidimensional case, it is thus natural to work with the hypothesis that f is even. In this case, if moreover $f > 0$ (attractive case), some trajectories play the role of circular trajectories in the classical problem: the orbits that are symmetric with respect to O.

The main change in the list of properties is at Property 5: we shall give a new expression for the eccentricity vector. The good candidate for generalization is the vector w, rather than the vector $\mathcal{L}$. This vector does not depend on the choice of ν in each plane, and we suggested in Remark 1 to choose $\nu = 1/C$, where $C = q \wedge \dot{q}$.

We needed to choose a F corresponding to f. In the case of an even f, there is a canonical choice: the unique even F. We will write our formula with these two choices. We call

$$W(q, C) = \frac{1}{2} \int_{Q=-q}^{q} f(Q) Q \Big(Q \lrcorner \frac{1}{C} \Big).$$

This is analogous to the path integral of $f(q)q\sigma$ in (A). The integral is taken on any path from $-q$ to q that stays on the plane defined by $C = q \wedge \dot{q} \in \bigwedge^2 U$. The vector W is only defined if $q \wedge C = 0$. Our claim is:

5) The vector $w = \dot{q} + mW(q, q \wedge \dot{q})$ is a first integral.

1.1.3. The energy. We are not able to give any equivalent to the Properties 7 to 10. One can argue that the transformations given in Remark 2 could carry the energy integral from the classical case, or even from the equation $\ddot{q} = 0$, to the general Jacobi-Darboux case. But this is not possible since C appears in the formula for the transformed attractor.

For a general Jacobi-Darboux attractor in the plane, a first integral that is a polynomial of degree 2 in the velocities must be a linear function of $\langle \nu | C \rangle^2$ and $\mathcal{L}$. There is an exception only in the classical case and easy generalizations of it: it is the energy integral, which is quadratic in the velocity. It is related to L and $||C||^2$ but the relation is nonlinear: $m^2(||L||^2 - 1) = 2H||C||^2$.

These easy generalizations of the classical case are defined by the attraction function $f = \langle q, q \rangle^{-3/2}$, where $\langle ., . \rangle$ is a non-degenerate bilinear form on the plane of motion U. Then $W = -\langle q, q \rangle^{-1/2} \langle C | C \rangle^{-1} q \lrcorner C$. We identify U and U^* using the scalar product in this formula. Note that $W \lrcorner C = \langle q, q \rangle^{-1/2} q$, so $L = m^{-1} w \lrcorner C$ gives the classical formula for the eccentricity vector, which has the advantage of being defined also for a degenerate $\langle ., . \rangle$ on the plane U. Of course we must always restrict the domain with the inequality $\langle q, q \rangle > 0$.

degenere: le calcul de la primitive F.

1.2. Jacobi-Darboux attractors on a curved screen

We consider a vector space V of dimension $n+1$. We give us a "screen function" $K : V \to \mathbb{R}$, positively homogeneous of degree 1, somewhere positive. Our aim is to define and study the Kepler problem and the Jacobi-Darboux generalization on the hypersurface of equation $K = 1$ which we call the "screen". The previous study corresponds to a "flat screen", i.e. the case where K is a linear form.

The attractor is now given by a vector $\mathrm{a} \in V$, and a positively homogeneous function f of degree -3 defined outside the line[3] $[\mathrm{a}] \subset V$ generated by a. The function f must satisfy $\langle df, \mathrm{a} \rangle = 0$ everywhere. It is thus defined on the quotient space $V/[\mathrm{a}]$. We assume moreover that it satisfies Hypothesis (A) on this quotient space if $n = 2$. If $n \geq 3$, we simply assume that f is even. Note that the quotient

[3]It is of course possible to study cases where f is not defined in all this domain.

space $V/[\mathrm{a}]$ is also the space of the $b \in \bigwedge^2 V$ such that $\mathrm{a} \wedge b = 0$: we can find a function g such $f(q) = g(\mathrm{a} \wedge q)$. This may become an important expression when continuing the theory.

1) The differential equation is $\ddot{q} = -f(q) dK \lrcorner(\mathrm{a} \wedge q) - \langle \partial^2 K, \dot{q} \otimes \dot{q} \rangle q$. It may be written also $\dot{\varpi} = -f(q)\mathrm{a} \wedge q$, where $\varpi = q \wedge \dot{q}$.

The first term is a "force" tangent to the screen $K = 1$. The second term is a kind of reaction. But usually reactions are normal to the surface. Ours is "radial". The particle is maintained on the screen by this reaction.

2) Equal volumes are swept in V by the triangle $(\mathrm{O}, \mathrm{a}, q)$ in equal times.

3) The trivector $\mathcal{C} = \mathrm{a} \wedge q \wedge \dot{q}$ is a first integral.

Property 4 seems less interesting in this context. To write Property 5, we define the bivector-valued integral

$$\mathcal{W}(q, \mathcal{C}) = \frac{1}{2} \int_{Q=-q}^{q} f(Q) \mathrm{a} \wedge Q \Big(\mathrm{a} \wedge Q \lrcorner \frac{1}{\mathcal{C}} \Big),$$

as we did in the previous flat case: the trivector $\mathcal{C}$ defines by its support the subspace where the integration path lives. We assume that f is even.

5) The bivector $q \wedge \dot{q} + \mathcal{W}(q, \mathcal{C})$ is a first integral.

Note that the wedge product of this bivector with the constant vector a is $\mathcal{C}$. It is possible to state something similar to Property 6 in the case $n = 2$. But we will rather insist on some particular cases.

1.3. Quadratic screens

If the screen is given by the screen function $K(q) = \langle q, q \rangle^{1/2}$, where $\langle .,. \rangle$ is a non degenerate bilinear form, it is canonically endowed with a pseudo-Riemannian structure, and it makes sense to ask if our system is defined by a force field which is derived from a potential. This property is satisfied in the following examples.

1) The differential equation is $\ddot{q} = -\|\mathrm{a} \wedge q\|^{-3} q \lrcorner(\mathrm{a} \wedge q) - \|\dot{q}\|^2 q$, or $\dot{\varpi} = -\|\mathrm{a} \wedge q\|^{-3} \mathrm{a} \wedge q$.

The "reaction" term is in this case normal to the surface, the normality being defined by the possibly indefinite bilinear form $\langle .,. \rangle$. This kind of attraction has been first considered, according to [Voz], by Lobachevsky.

2) Equal volumes are swept in V by the triangle $(\mathrm{O}, \mathrm{a}, q)$ in equal times.

3) The trivector $\mathcal{C} = \mathrm{a} \wedge q \wedge \dot{q}$ is a first integral.

5) The vector $\Lambda = \|\mathrm{a} \wedge q\|^{-1} \mathrm{a} \lrcorner(\mathrm{a} \wedge q) + \|\mathrm{a}\|^2 \varpi \lrcorner \mathcal{C}$ is a first integral.

6) The particle describes generalized conic sections. The point $\|\mathrm{a}\|^{-1}\mathrm{a}$ is a focus.

This curves are the intersections of a quadratic cone with the screen. They are the so-called "sphero-conic sections" in the case of a spherical screen. When a point moves on the curve, the sum of its geodesic distances to both foci is constant. Property 6 has been discovered by Paul Serret [Ser]. Note the formula $\langle \Lambda, q \rangle = \|\mathrm{a} \wedge q\|$.

7) An isometry transforms a trajectory into a trajectory.

8) The period T of a closed trajectory is a function of the geodesic length of the major axis.

9) The energy $H = ||\dot{q}||^2/2 - ||\mathrm{a}||^{-2}||\mathrm{a} \wedge q||^{-1}\langle \mathrm{a}, q\rangle$ is a first integral. We have $||\mathrm{a}||^{-2}||\Lambda||^2 - 1 = 2H||\mathrm{a}||^2||\mathcal{C}||^2$.

10) The period T is a function of the energy only.

1.4. General discussion

From the observation of the various cases of one center problems it appears that some dynamical properties may survive when the screen is changed. *In 1.2 if the function f is given, the unparameterized trajectories are defined independently of the screen function K, in the following sense: if $t \mapsto q(t)$ is the solution on a screen, the solution on another screen is of the form $s \mapsto t(s) \mapsto \lambda(t)q(t)$, the function λ being the adequate factor making λq a point of the new screen.*

In other words, the solutions on distinct screens correspond one to each other by central projection, except for the parameterization. This is why the conic sections of 1.0 became generalized conic sections in 1.3.

Let us concentrate on the case $n = 2$, the attraction being defined in 1.3, the screen being a sphere. We can take as a second screen a plane tangent to the sphere at a point $\tau \in V$. Elliptic trajectories on the sphere are sent by central projection to elliptic, parabolic or hyperbolic trajectories on the plane. If $\tau = \lambda\mathrm{a}$ then this point is a common focus of all these conic sections. If τ and a are independent vectors, a still points toward the common focus, but we have to be careful: the Euclidean structure on the plane is not defined by its embedding in the Euclidean space V, but by the formula $||v|| := ||\mathrm{a}||^{-1}||\mathrm{a} \wedge v||$, where v is a vector tangent to the plane. In particular the second focus of an ellipse on the sphere will not coincide by central projection with the second focus of the corresponding ellipse, parabola or hyperbola on the plane.

Also the energy in the spherical model does not correspond to the energy in the planar model. And of course the period in both models does not coincide. Last Properties 8 and 10 are rather unexpected. But 10 follows from a general property of Hamiltonian systems (see [KoH], [Mil], [Mos]). Property 8 has no explanation. It was discovered by C. Velpry in an addendum to [Vel], with the date July 2000.

2. Two fixed centers

The fundamental fact just discussed, that one can pass from a screen to another by central projection and change of time, may be generalized to any force field. It is enough to understand the correspondence between fields of forces tangent to different screens. This correspondence is very simple but we will describe it in another work. We refer for the moment to [App].

Let us choose two independent vectors $\mathrm{a} \in V$ and $\mathrm{b} \in V$. Suppose these vectors are two "fixed centers" defining an attraction as in 1.2. We need two functions $f_{\mathrm{a}} : V \setminus [\mathrm{a}] \to \mathbb{R}$ and $f_{\mathrm{b}} : V \setminus [\mathrm{b}] \to \mathbb{R}$, positively homogeneous of degree -3, such that $\langle df_{\mathrm{a}}, \mathrm{a}\rangle = 0$, $\langle df_{\mathrm{b}}, \mathrm{b}\rangle = 0$, satisfying Condition (A) as in 1.2. Define the bivector field

$$\gamma = f_{\mathrm{a}}(q)\mathrm{a} \wedge q + f_{\mathrm{b}}(q)\mathrm{b} \wedge q.$$

Consider now a screen function $K : V \to \mathbb{R}$. We state and study the generalized two fixed centers problem on this screen.

1) The differential system is $\ddot{q} = -dK|\gamma - \langle \partial^2 K, \dot{q} \otimes \dot{q} \rangle q$. Making $\varpi = q \wedge \dot{q}$, it reads $\dot{\varpi} = -\gamma$.

2) Equal 4-volumes are swept in V by the tetrahedron $(\mathrm{O}, \mathrm{a}, \mathrm{b}, q)$ in equal times.

3) The 4-vector $\mathrm{a} \wedge \mathrm{b} \wedge q \wedge \dot{q}$ is a first integral.

We stop the list here, because our curiosity first asks for information in the case of a flat screen.

2.1. Flat screen. We take for K a linear form. We call S the flat screen, an affine space of dimension n. In 1.1, we could identify S to its associated vector space U, which is also the subspace of equation $K = 0$. This was possible because of the obvious choice of putting the origin at the attracting point we called O. Now we have two attracting points $A = \mathrm{a}/K(\mathrm{a})$ and $B = \mathrm{b}/K(\mathrm{b})$. We avoid to choose an origin. We call $m_A = K(\mathrm{a})$, $m_B = K(\mathrm{b})$, $q_A = q - A$, $q_B = q - B$, $f_A(q_A) = f_\mathrm{a}(q)$ and $f_B(q_B) = f_\mathrm{b}(q)$. Last Property 1 becomes:

1) the differential system is $\ddot{q} = -m_A f_A(q_A) q_A - m_B f_B(q_B) q_B$.

2) Equal 3-volumes are swept in S by the triangle (A, B, q) in equal times.

3) The trivector $q_A \wedge q_B \wedge \dot{q}$ is a first integral.

There is a strong relation, noticed in [Per], §2.10, between the eccentricity vector and the unexpected Euler's integral of the two fixed centers problem. The following generalization makes this clearer.

2.1.1. Planar case. We choose a non-zero $\nu \in \bigwedge^2 U^*$. We denote F_A and F_B the functions associated to f_A and f_B as in 1.1.1. We make $u = q_B - q_A = A - B$.

5) The function $G = \langle q_A \wedge \dot{q} | \nu \rangle \langle q_B \wedge \dot{q} | \nu \rangle - \langle m_A dF_A - m_B dF_B, u \rangle$ is a first integral.

Of course we could write G in different ways, for example avoiding a choice of ν, or choosing coordinates. But it seems really difficult to guess a generalization to the 3D case.

2.1.2. Multidimensional case. We must assume that both attractors are defined by two possibly indefinite, possibly distinct, quadratic forms. We call $\langle .,. \rangle_A$ and $\langle .,. \rangle_B$ the scalar products and $||.||_A$, $||.||_B$ the "norms".

1) The equation is $\ddot{q} = -m_A ||q_A||_A^{-3} q_A - m_B ||q_B||_B^{-3} q_B$.

3) The trivector $q_A \wedge q_B \wedge \dot{q}$ is a first integral.

5')

$$\begin{aligned} &||q_B||_B^2 ||\dot{q}||_A^2 + ||q_A||_A^2 ||\dot{q}||_B^2 - 2\langle q_A, \dot{q} \rangle_A \langle q_B, \dot{q} \rangle_B \\ &\qquad -2m_A ||q_A||_A^{-1} \langle q_B, u \rangle_B + 2m_B ||q_B||_B^{-1} \langle q_A, u \rangle_A \end{aligned}$$

is a first integral.

2.1.3. The energy. As far as we know, the energy integral exists only in the classical Euler problem, the particular case of last System 1 where both scalar products coincide; and in the problems obtained from it by changing the flat screen, which are also particular cases of System 1, with distinct, but not arbitrarily distinct, quadratic forms. In these cases, the problem is integrable.

With the assumptions of 2.1.1 and 2.1.2 together, the first integrals 5 and 5' are independent. This indicates that one should be able to integrate, even when the energy does not exist. This interesting remark will be the object of a forthcoming work. The numerical experiments [AlS] in the general case 2.1.1, showing a quasi-integrable behavior, are also of interest.

2.2. Quadratic screens. The equation $\dot{\varpi} = -\gamma$ for the two fixed centers problem, with the choices $K = \langle q, q\rangle^{1/2}$ for the screen function, $f_{\rm a}(q) = ||{\rm a} \wedge q||^{-3}$, $f_{\rm b}(q) = ||{\rm b} \wedge q||^{-3}$ for the attraction functions, gives the problem of two fixed centers on a space of constant curvature, first studied and integrated in [KoH]. The integrability could have been predicted and proved as a corollary of the discussions of 1.4 and 2. Let us restrict our attention to the spherical screen, corresponding to a positive definite $\langle .,. \rangle$. The central projection on a flat screen tangent to the sphere at the middle point of both attractors sends trajectories of the Koslov-Harin problem to trajectories of the Euler problem. We must be careful again that the Euclidean structure on the flat screen is not the one induced by the embedding in V.

References

[AbA] K. Abdullah, A. Albouy, *On a strange resonance noticed by M. Herman*, Regular and Chaotic Dynamics, 6 (2001) pp. 421–432

[Alb] A. Albouy, *Lectures on the two-body problem*, "Classical and Celestial Mechanics. The Recife Lectures." edited by H. Cabral and F. Diacu, Princeton Un. Press (2002) pp. 63–116

[AlS] A. Albouy, T.J. Stuchi, *Quasi-integrability in a class of systems generalizing the problem of two fixed centers*, preprint (2002) http://arXiv.org/ab/nlin.CD/0210020

[App] P. Appell, *Sur les lois de forces centrales faisant décrire à leur point d'application une conique quelles que soient les conditions initiales*, American Journal of Mathematics 13 (1891) pp. 153–158

[Dar] M. Despeyrous, *Cours de Mécanique, avec des notes par M. G. Darboux*, Tome premier, A. Hermann, Paris (1884) note 11

[Jac] C.G.J. Jacobi, *De motu puncti singularis*, Crelle Journal für die reine und angewandte Mathematik, Bd. 24 (1842) pp. 5–27; Werke Bd. 4, pp. 263–288; cf. p. 282

[KoH] V.V. Koslov, A.O. Harin, *Kepler's problem in constant curvature spaces*, Celestial Mechanics and Dynamical Astronomy 54 (1992) pp. 393–399

[Mil] J. Milnor, *On the geometry of the Kepler problem*, American Math. Monthly, 90 (1983) pp. 353–365

[Mos] J. Moser, *Regularization of Kepler's Problem and the Averaging Method on a Manifold*, Communications on Pure and Applied Mathematics 23 (1970) pp. 609–636

[Per] A.M. Perelomov, *Integrable systems of classical mechanics and Lie algebras, vol. I*, Translated by A.G. Reyman, Birkhauser (1990)

[Poi] H. Poincaré, *Science et Méthode*, chap. II, *Les définitions mathématiques et l'Enseignement*, Flammarion, Paris

[ScK] J.A. Schouten, W. Kulk, *Pfaff's problem and its generalizations*, Oxford University Press (1949), Chelsea (1969)

[Ser] P. Serret, *Théorie nouvelle géométrique et mécanique des lignes à double courbure*, Mallet-Bachelier, Paris (1860) p. 204

[Vel] C. Velpry, *Kepler Laws and Gravitation in non-Euclidean (classical) Mechanics*, Heavy ion physics 11/1-2 (2000)

[Voz] T.G. Vozmischeva, *Classification of motions for generalization of the two centers problem on a sphere*, Celestial Mechanics & Dyn. Ast. 77 (2000) pp. 37–48

[Wey] H. Weyl, *Space-time-matter*, translated by H.L. Brose, Dover (1952)

Nonlinear spectral theory and applications: some current issues

Jürgen Appell

Department of Mathematics, University of Würzburg, Am Hubland, D-97074 Würzburg, Germany

The purpose of this talk is to discuss three particularly important spectra for nonlinear operators, viz. the "asymptotic" Furi-Martelli-Vignoli spectrum, the "global" Feng spectrum, and the "local" Väth phantom. Moreover, we show that these spectra coincide for certain classes of operators, and indicate how this may be used to prove some kind of "nonlinear Fredholm alternative" with applications to eigenvalue problems for the p-Laplace operator.

To put things in the right framework, let us first recall some properties of the "linear" spectrum. Given a Banach space X over $\mathbb{K} = \mathbb{R}$ or $\mathbb{K} = \mathbb{C}$ and an operator $L \in \mathfrak{L}(X)$, the algebra of all bounded linear operators on X, the *resolvent set* of L is defined by

$$\rho(L) = \{\lambda \in \mathbb{K} : (\lambda I - L)^{-1} \in \mathfrak{L}(X)\}, \tag{1}$$

and the *spectrum* of L by

$$\sigma(L) = \mathbb{K} \setminus \rho(L). \tag{2}$$

As is well known, this spectrum has some remarkable properties. It is compact and (in case $\mathbb{K} = \mathbb{C}$) nonempty, and it is at most countable (and so has empty interior) for L compact. Moreover, it is bounded by the spectral radius $r(L) = \lim_{n\to\infty} \sqrt[n]{||L^n||}$ (Gel'fand formula), and it commutes with any polynomial p, i.e. $\sigma(p(L)) = p(\sigma(L))$ (spectral mapping theorem). Finally, the map $\rho(L) \ni \lambda \mapsto (\lambda I - L)^{-1} \in \mathfrak{L}(X)$ is analytic, and the multivalued map $\mathfrak{L}(X) \ni L \mapsto \sigma(L) \in 2^{\mathbb{K}}$ is upper semicontinuous.

So if one wants to build a spectral theory in the nonlinear case, one is led to some "natural" requirements. So, given a continuous nonlinear operator $F : X \to X$, one should try to define a *spectrum* $\sigma(F)$ for F such that

---

e-mail: appell@mathematik.uni-wuerzburg.de.

- $\sigma(F)$ is the familiar spectrum for F bounded linear,
- $\sigma(F)$ has the usual properties (compact, nonempty etc.),
- $\sigma(F)$ contains the point spectrum $\sigma_p(F)$ (eigenvalues) of F, and
- $\sigma(F)$ has applications (e.g., existence, uniqueness, bifurcation problems).

We point out that the *point spectrum* of F in the third requirement is defined precisely as in the linear case, i.e.

$$\sigma_p(F) = \{\lambda \in \mathbb{K} : F(u) = \lambda u \text{ for some } u \neq 0\}. \tag{3}$$

Some "naive" definitions in this spirit have been given in the last 30 years. The philosophy is simply to replace the algebra $\mathfrak{L}(X)$ in (1) by other classes of continuous (nonlinear) operators. For the set $\mathfrak{C}(X)$ of all continuous operators $F : X \to X$ this leads to the *Rhodius resolvent spectrum* [19]. Another possible choice is that of *Lipschitz continuous* maps $F : X \to X$, which leads to the *Kachurovskij spectrum* [14]. Finally, considering the class of *continuously differentiable* operators, one is led to the *Neuberger spectrum* [17]. All these spectra, however, have very "bad" properties; for instance, they do not satisfy the above minimal requirements (see [1,2] for a comparative study of these spectra). The reason for this is that they are too much modelled on the linear case; therefore, the question arises how to define other spectra which are more adapted to nonlinear problems.

A promising approach to define "reasonable" spectra consists in decomposing the spectrum into several *subspectra*. Recall that a classical (not necessarily disjoint) decomposition of the spectrum in the linear case is

$$\sigma(L) = \sigma_\delta(L) \cup \sigma_p(L) \cup \sigma_{co}(L), \tag{4}$$

where $\lambda \in \sigma_\delta(L)$ if $\lambda I - L$ is not onto ("defect spectrum"), $\lambda \in \sigma_p(L)$ if $\lambda I - L$ is not 1-1 ("point spectrum", see (3)), and $\lambda \in \sigma_{co}(L)$ if $\lambda I - L$ is not proper ("compression spectrum"). Unfortunately, there is no canonical analogue to this in the nonlinear case. However, one may "imitate" these properties in different ways.

To this end we have to recall some definitions. The (Kuratowski) *measure of noncompactness* $\alpha(M)$ of a bounded set $M \subset X$ is defined as infimum of all $\delta > 0$ such that M may be covered by finitely many sets of diameter $\leq \delta$. So we have $\alpha(M) = 0$ if and only if M has compact closure (which motivates the name). Now, given a nonlinear operator F between two Banach spaces X and Y, the two conditions

$$\alpha(F(M)) \leq k\alpha(M), \quad \alpha(F(M)) \geq \ell\alpha(M) \qquad (M \subset X \text{ bounded}) \tag{5}$$

are crucial. The smallest constant k in (5) will be denoted by $[F]_A$, and the largest constant ℓ in (5) by $[F]_a$ in what follows. So $[F]_A = 0$ if and only if F is compact, and $[F]_a > 0$ implies that F is proper on closed bounded sets.

For $F : X \to Y$ we call the two characteristics

$$[F]_Q = \limsup_{||u||\to\infty} \frac{||F(u)||}{||u||}, \qquad [F]_q = \liminf_{||u||\to\infty} \frac{||F(u)||}{||u||} \tag{6}$$

the *upper and lower quasinorm* of F, respectively. Following Furi, Martelli and Vignoli [11] we call a continuous operator $F : X \to Y$ *stably solvable* if, for any compact operator $G : X \to Y$ satisfying $[G]_Q = 0$, the coincidence equation

$$F(u) = G(u) \tag{7}$$

has a solution $u \in X$. Taking $G(u) \equiv v$ for fixed $v \in Y$, one readily sees that stable solvability implies surjectivity. The converse is not true: the map $F(u) = u/\sqrt{1+|u|}$ is a homeomorphism in $X = \mathbb{R}$, but not stably solvable (consider $G(u) = F(u) + 1$). On the other hand, one may show [11] that for $L \in \mathfrak{L}(X,Y)$ stable solvability is *equivalent* to surjectivity of L.

Before introducing the first "non-standard" spectrum, we make a general remark. The examples in the preceding section show that the definition (3) of the point spectrum is not suitable for nonlinear operators F. As a matter of fact, there is absolutely no reason to compare a nonlinear operator F with the *identity* I, because this is too much modelled on the linear case. Instead, it is a useful device to replace the identity by some other "well-behaved" *nonlinear* operator J which takes into account the analytical and topological properties of the given operator F. So we will from now on define spectra for *pairs of operators* (F, J) between two Banach spaces X and Y, although the original definitions have been given only in case $X = Y$ and $J = I$.

The *Furi-Martelli-Vignoli spectrum* of $F, J : X \to Y$ [12] is defined as union

$$\sigma_{FMV}(F,J) = \sigma_{ss}(F,J) \cup \sigma_q(F,J) \cup \sigma_a(F,J), \tag{8}$$

where $\lambda \in \sigma_{ss}(F,J)$ if $\lambda J - F$ is not stably solvable, $\lambda \in \sigma_q(F,J)$ if $[\lambda J - F]_q = 0$, and $\lambda \in \sigma_a(F,J)$ if $[\lambda J - F]_a = 0$. By what we have observed before, for $L \in \mathfrak{L}(X)$ we get the relations $\sigma_{ss}(L,I) = \sigma_\delta(L)$, $\sigma_q(L,I) \supseteq \sigma_p(L)$, and $\sigma_a(L,I) \subseteq \sigma_{co}(L)$.

We collect some important properties of the spectrum (8) in the following

Theorem 1. *Suppose that $J : X \to Y$ is stably solvable with $[J]_Q < \infty$. Then the Furi-Martelli-Vignoli spectrum (8) is closed and upper semicontinuous. Moreover, the inclusion*

$$\sigma_{FMV}(F,J) \subseteq \left\{ \lambda \in \mathbb{K} : |\lambda| \leq \max \left\{ \frac{[F]_A}{[J]_a}, \frac{[F]_Q}{[J]_q} \right\} \right\} \tag{9}$$

holds, and so $\sigma_{FMV}(F,J)$ is compact if $[F]_A < \infty$, $[F]_Q < \infty$, $[J]_a > 0$, and $[J]_q > 0$. Finally, $\sigma_{FMV}(L,I)$ coincides with the familiar spectrum (2) for $L \in \mathfrak{L}(X)$.

We remark that Theorem 1 was proved for $X = Y$ and $J = I$ in [12]. In this case we have $[I]_Q = [I]_q = [I]_a = 1$, and so $\sigma_{FMV}(F,I)$ is compact if just $[F]_A < \infty$ and $[F]_Q < \infty$.

So we see that the Furi-Martelli-Vignoli spectrum (8) has many good properties in common with the usual linear spectrum. However, it has the series flaw of *not containing the point spectrum* (3). In fact, for the map $F(u) = \sqrt{|u|}$ in $\mathbb{R}$ we obviously have $\sigma_p(F) = \mathbb{R} \setminus \{0\}$. On the other hand, from (9) and the trivial relations $[F]_A = [F]_Q = 0$ and $[I]_a = [I]_q = 1$ we get $\sigma_{FMV}(F, I) = \{0\}$. Thus, in this example the spectrum and the point spectrum are even *disjoint*, and this is of course in sharp contrast to what one expects.

A scrutiny of the situation shows that this unpleasant phenomenon is due to the "bad" definition of the point spectrum (3) which is simply inadequate for the Furi-Martelli-Vignoli spectrum. Indeed, the definition (3) is "global", while the definition (8) is "asymptotic". Consequently, one has to replace the inappropriate definition (3) by the *asymptotic point spectrum* $\sigma_q(F, J)$ which, by (8), is always a part of the spectrum $\sigma_{FMV}(F)$. For instance, for the scalar map $F(u) = \sqrt{|u|}$ we have $\sigma_q(F, I) = \{0\}$, because $[\lambda I - F]_q = |\lambda|$.

Let us show how to calculate the Furi-Martelli-Vignoli spectrum by means of a more interesting example in an infinite dimensional space.

Example 1. Let X be an arbitrary infinite dimensional real Banach space, and let $F : X \to X$ be defined by $F(u) = ||u||u$. It is not hard to see that F is a proper homeomorphism with inverse $F^{-1}(v) = v/\sqrt{||v||}$. We claim that $\sigma_{FMV}(F, I) = [0, \infty)$. First of all, the eigenvalue equation $F(u) = \lambda u$ has a nontrivial solution u with $\lambda = ||u||$ if and only if $\lambda > 0$, and so $\sigma_p(F, I) = (0, \infty)$. On the other hand, it is easy to see that F has no asymptotic eigenvalue at all, i.e. $\sigma_q(F, I) = \emptyset$.

Taking as M the closed ball $B_{1/n} = \{u \in X : ||u|| \leq 1/n\}$ in (5) shows that $[F]_a = 0$. Likewise, the fact that $\lambda I - F$ maps, for $\lambda > 0$, the sphere $\{u \in X : ||u|| = \lambda\}$ into $\{0\}$ implies that $[\lambda I - F]_a = 0$ for such λ. Now fix $\lambda < 0$ and consider the function $f : (0, \infty) \to \mathbb{R}$ defined by

$$f(t) = \frac{\sqrt{\lambda^2 + 4t} - \lambda}{2t} \qquad (0 < t < \infty).$$

L'Hospital's rule shows that f admits a continuous extension to 0 by putting $f(0) = 1/|\lambda|$. A straightforward computation shows that

$$(\lambda I - F)^{-1}(v) = f(||v||)v$$

for every $v \in X$; for $\lambda = 0$ this coincides of course with the formula for F^{-1} above. It is not hard to se that f is bounded on $[0, \infty)$, say $|f(t)| \leq M_0$. Moreover, the derivative of f has the property that $t|f'(t)|$ remains bounded as $t \to 0$ or $t \to \infty$. For $||u|| \leq ||v||$ we have

$$||f(||u||)u - f(||v||)v|| = ||[f(||u||) - f(||v||)]u + f(||v||)(u - v)||$$

$$\leq |f'(\tau)(||u|| - ||v||)| \, ||u|| + M_0||u - v||,$$

where $||u|| \leq \tau \leq ||v||$. Since the map $t \mapsto t|f'(t)|$ is bounded, say $t|f'(t)| \leq M_1$, we conclude that

$$||(\lambda I - F)^{-1}(u) - (\lambda I - F)^{-1}(v)|| \leq (M_0 + M_1)||u - v||.$$

Interchanging the role of u and v we get the same estimates for $||v|| \leq ||u||$, and so we have proved that $(\lambda I - F)^{-1}$ is Lipschitz continuous on X for $\lambda < 0$.

Consequently, $[\lambda I - F]_a = [(\lambda I - F)^{-1}]_A^{-1} \geq [(\lambda I - F)^{-1}]_{Lip}^{-1} > 0$, and so $\sigma_a(F, I) = [0, \infty)$.

It remains to calculate the subspectrum $\sigma_{ss}(F, I)$. Suppose that $G : X \to X$ is compact with quasinorm $[G]_Q = 0$; in particular, the last condition implies that G maps a closed ball B_R for sufficiently large $R > 0$ into itself. Since $||(\lambda I - F)(u)|| \geq ||u||$, for any λ, hence $||(\lambda I - F)^{-1}(v)|| \leq ||v||$, the (compact!) operator $(\lambda I - F)^{-1}G$ also maps the ball B_R into itself, and so has a fixed point $\hat{u} \in B_R$, by Schauder's fixed point theorem. But then $\lambda\hat{u} - F(\hat{u}) = G(\hat{u})$ which shows that $\lambda I - F$ is stably solvable. We conclude that $\sigma_{ss}(F, I) = \emptyset$, and so $\sigma_{FMV}(F, I) = [0, \infty)$ as claimed. □

The drawback of the spectrum $\sigma_{FMV}(F, I)$ not to contain the eigenvalues of F in the sense of (3) has motivated Wenying Feng [8] to define another spectrum $\sigma_F(F, J)$ (for $J = I$) which is quite similar to the spectrum $\sigma_{FMV}(F, J)$ but contains the classical point spectrum

$$\sigma_p(F, J) = \{\lambda \in \mathbb{K} : F(u) = \lambda J(u) \text{ for some } u \neq 0\}. \tag{10}$$

To describe this spectrum, again some definitions are in order. First we have to replace the "asymptotic" characteristics (6) by the "global" characteristics

$$[F]_B = \sup_{u \neq 0} \frac{||F(u)||}{||u||}, \qquad [F]_b = \inf_{u \neq 0} \frac{||F(u)||}{||u||}. \tag{11}$$

It is clear that $[F]_b \leq [F]_q \leq [F]_Q \leq [F]_B$, and so $\sigma_b(F, J) \supseteq \sigma_q(F, J)$, where $\lambda \in \sigma_b(F, J)$ if $[\lambda J - F]_b = 0$.

In what follows, we denote by $\mathfrak{O}(X)$ the family of all open, bounded, connected subsets Ω of a Banach space X with $0 \in \Omega$; as a model example one may think of the open ball $\Omega = B_r^o$ for $r > 0$. Given $\Omega \in \mathfrak{O}(X)$, a continuous map $F : \overline{\Omega} \to Y$ is called *epi* [13] if, first, $F(u) \neq 0$ for $u \in \partial\Omega$ and, second, the coincidence equation (7) has a solution $u \in \Omega$ for every compact operator $G : \overline{\Omega} \to Y$ satisfying $G(u) \equiv 0$ on $\partial\Omega$. Thus, epi operators have a similar meaning on small balls as stably solvable operators on large spheres. It is not hard to see that, if an operator $F : X \to Y$ is stably solvable on X with $[F]_b > 0$, then F is epi on every $\Omega \in \mathfrak{O}(X)$. In fact, the condition $[F]_b > 0$ implies that $F(u) \neq 0$ for $u \neq 0$. Moreover, if $G : \overline{\Omega} \to Y$ is compact and satisfies $G(u) \equiv 0$ on $\partial\Omega$, we may extend G by continuity to the whole space X putting $G(u) \equiv 0$ outside $\overline{\Omega}$. Then trivially $[G]_Q = 0$, and so the equation (7) has a solution $\hat{u} \in X$, by assumption. Clearly, we must have $\hat{u} \in \Omega$, since $F(u) \neq 0 = G(u)$ outside Ω, and so we see that F is epi on Ω.

Now, the *Feng spectrum* of $F, J : X \to Y$ [8] is defined as union

$$\sigma_F(F, J) = \sigma_e(F, J) \cup \sigma_b(F, J) \cup \sigma_a(F, J), \tag{12}$$

where $\lambda \in \sigma_e(F, J)$ if $\lambda J - F$ is not epi on *some* $\Omega \in \mathfrak{O}(X)$. Our previous discussion shows that the inclusion $\sigma_{FMV}(F, J) \subseteq \sigma_F(F, J)$ is true. Moreover, we have the following result which is parallel to Theorem 1.

Theorem 2. *Suppose that $J : X \to Y$ is epi on every $\Omega \in \mathfrak{O}(X)$ with $[I]_B <$ ∞. Then the Feng spectrum* (12) *is closed and upper semicontinuous. Moreover, the inclusion*

$$\sigma_F(F, J) \subseteq \left\{ \lambda \in \mathbb{K} : |\lambda| \leq \max \left\{ \frac{[F]_A}{[J]_a}, \frac{[F]_B}{[J]_b} \right\} \right\} \tag{13}$$

holds, and so $\sigma_F(F, J)$ is compact if $[F]_A < \infty$, $[F]_B < \infty$, $[J]_a > 0$, and $[J]_b > 0$. Finally, $\sigma_F(L, I)$ coincides with the familiar spectrum (2) *for $L \in \mathfrak{L}(X)$.*

Again, Theorem 2 essentially simplifies for $X = Y$ and $J = I$. In this case we have $[I]_B = [I]_b = [I]_a = 1$, and so $\sigma_F(F, I)$ is compact if just $[F]_A < \infty$ and $[F]_B < \infty$.

Consider, for example, again the map $F(u) = \sqrt{|u|}$ (and $J = I$) in $\mathbb{R}$. We have seen that $\sigma_{FMV}(F, I) = \{0\}$ in this case. On the other hand, from the relation $\sigma_p(F, I) = \mathbb{R} \setminus \{0\}$ and the closedness of the Feng spectrum in case $J = I$ we deduce that $\sigma_F(F, J) = \mathbb{R}$. This shows that the Feng spectrum takes into account the global behaviour of F, while the Furi-Martelli-Vignoli spectrum reflects only the asymptotic properties of F.

On the other hand, for the operator F from Example 1 (and $J = I$) the Feng spectrum and the Furi-Martelli-Vignoli spectrum are the same. To see this, fix $\lambda < 0$ and $\Omega \in \mathfrak{O}(X)$, and let $G : \overline{\Omega} \to X$ be compact with $G(u) \equiv 0$ on $\partial\Omega$. The operator $(\lambda I - F)^{-1}G : \Omega \to (\lambda I - F)^{-1}G(\Omega)$ is then a compact homeomorphism which vanishes on the boundary $\partial\Omega$, and so has a fixed point in Ω. We conclude that $\lambda I - F$ is epi for $\lambda < 0$, and so $\sigma_F(F, I) = [0, \infty)$.

Yet another approach was proposed by Väth [20-22] which goes as follows. Let $\Omega \in \mathfrak{O}(X)$, and let $F : \overline{\Omega} \to Y$ be continuous with

$$\inf \{||F(u)|| : u \in \partial\Omega\} > 0. \tag{14}$$

Then F is called *strictly epi* if there exists some $k > 0$ such that, for any operator $G : \overline{\Omega} \to Y$ satisfying $G(u) \equiv 0$ on $\partial\Omega$ and $[G]_A \leq k$, the coincidence equation (7) has a solution in Ω. Thus, strictly epi maps are stronger than epi maps in two ways: first, the condition $0 \notin F(\partial\Omega)$ is replaced by the stronger condition (14), and instead of compact perturbations G in (7) one allows also "slightly noncompact" right-hand sides.

The *Väth phantom* of $F, J : X \to Y$ [22] is defined by

$$\phi(F, J) = \bigcap_{\Omega \in \mathfrak{O}(X)} \phi(F, J; \Omega), \tag{15}$$

where $\lambda \in \phi(F, J; \Omega)$ if $\lambda J - F$ is not strictly epi on Ω, or $F(u) = \lambda J(u)$ for some $u \in \partial\Omega$. There is also an analogue of Theorems 1 and 2 for the phantom (15): if J is strictly epi on some $\Omega \in \mathfrak{O}(X)$, then $\phi(F, J)$ is closed and upper semicontinuous (with respect to a suitable topology), and even compact if

$$[F|_{\overline{\Omega}}]_A < \infty, \quad [J|_{\overline{\Omega}}]_a > 0, \quad \sup_{u \in \partial\Omega} ||F(u)|| < \infty, \quad \inf_{u \in \partial\Omega} ||J(u)|| > 0.$$

In case $J = I$ these conditions are automatically satisfied on each $\Omega \in \mathfrak{O}(X)$. We remark that $\phi(L, I)$ also coincides with the usual spectrum (2) in the linear case.

As for the other spectra, we still have to introduce an appropriate notion of point spectrum associated to the phantom (15). Let us call $\lambda \in \mathbb{K}$ a *connected eigenvalue* [20] of the pair (F, J) if the solution set of the generalized eigenvalue equation

$$F(u) = \lambda J(u) \tag{16}$$

contains an unbounded connected subset C with $0 \in C$. We write $\phi_p(F, J)$ for the set of all connected eigenvalues and call this set the *point phantom* of (F, J). Of course, for $L \in \mathfrak{L}(X)$ and $J = I$ we simply have $\phi_p(L, I) = \sigma_p(L)$, since the eigenvectors of $\lambda \in \sigma_p(L)$ form a linear space.

To show how to calculate the phantom and point phantom in the nonlinear case consider again $F : X \to X$ from Example 1. From $\phi_p(F, I) \subseteq \sigma_q(F, I) = \emptyset$ it follows that F has no connected eigenvalues (which may also be verified directly). We claim that $\phi(F, I) = \{0\}$. Indeed, for $\lambda \neq 0$ and $\Omega = \{u \in X : ||u|| < |\lambda|/2\}$ the restriction $(\lambda I - F)|_{\overline{\Omega}} : \overline{\Omega} \to X$ is open and injective with $0 \in (\lambda I - F)(\Omega)$ and $[(\lambda I - F)|_{\overline{\Omega}}]_a > 0$. From this we may conclude that $\lambda I - F$ is strictly epi on Ω, and so $\lambda \notin \phi(F, I)$. On the other hand, it was proved by Furi [10] quite recently that the operator F is epi, but *not* strictly epi, and so $0 \in \phi(F, I)$.

The following table gives a general comparison of the three spectra introduced above.

Author	*Spectrum*	*Point spectrum*	*Character*
Furi-Martelli-Vignoli (1978)	FMV-spectrum $\sigma_{FMV}(F, J)$	asymptotic eigenvalues $\sigma_q(F, J)$	asymptotic $(\lVert u\rVert \to \infty)$
Feng (1997)	Feng spectrum $\sigma_F(F, J)$	classical eigenvalues $\sigma_p(F, J)$	global $(u \in X)$
Väth (2000)	phantom $\phi(F, J)$	connected eigenvalues $\phi_p(F, J)$	local $(u \in \overline{\Omega})$

To close this section, we give some relations between all the spectra and point spectra. First of all, we remark that the Väth phantom $\phi(F, J)$ is always contained in the Furi-Martelli-Vignoli spectrum $\sigma_{FMV}(F, J)$. In fact, for $\lambda \notin \sigma_{FMV}(F, J)$ we have, in particular, $[\lambda J - F]_q > 0$, and this implies that $||\lambda J(u) - F(u)|| \geq [\lambda J - F]_q \, ||u||$ for $||u|| \geq R$ with $R > 0$ sufficiently large. Extending then $G : B_R \to Y$ with $G(u) \equiv 0$ to be zero outside B_R and arguing as before one sees that the equation $\lambda J(u) - F(u) = G(u)$ has a solution in the interior of B_R.

So for general operators $F, J : X \to Y$ we get the following relations.

$$\begin{array}{ccccc}
\phi(F,J) & \subseteq & \sigma_{FMV}(F,J) & \subseteq & \sigma_F(F,J) \\
\cup\!| & & \cup\!| & & \cup\!| \\
\phi_p(F,J) & \subseteq & \sigma_q(F,J) & & \sigma_p(F,J)
\end{array}$$

In the linear case $L \in \mathfrak{L}(X)$ (and $J = I$) this table simplifies. Here all the spectra in the first row coincide with the usual spectrum $\sigma(L)$, and both the point spectrum $\sigma_p(L, I)$ and point phantom $\phi_p(L, I)$ coincide with the usual point spectrum $\sigma_p(L)$.

As one could expect, more can be said about these spectra if one restricts the class of operators in consideration. We restrict ourselves to the case of τ-*homogeneous operators* F and J, i.e.

$$F(tu) = t^\tau F(u), \quad J(tu) = t^\tau J(u) \qquad (t > 0,\, u \in X). \tag{17}$$

The following two theorems have been proved in [3].

Theorem 3 (Coincidence theorem). *Let X and Y be infinite dimensional Banach spaces, and suppose that $F, J : X \to Y$ satisfy* (17) *for some $\tau > 0$. Then*

$$\sigma_{FMV}(F,J) = \sigma_F(F,J) = \phi(F,J), \quad \sigma_q(F,J) \supseteq \sigma_p(F,J) = \phi_p(F,J).$$

Theorem 4 (Discreteness theorem). *Let X and Y be infinite dimensional Banach spaces, and suppose that $F, J : X \to Y$ are odd, $[F]_A = 0$ (i.e., F is compact), and $[J]_a > 0$. Then*

$$\sigma_{FMV}(F,J) \setminus \{0\} \subseteq \sigma_q(F,J), \quad \sigma_F(F,J) \setminus \{0\} \subseteq \sigma_p(F,J),$$

and

$$\phi(F,J) \setminus \{0\} \subseteq \phi_p(F,J).$$

Theorem 4 shows that, for F compact and odd, and J "regular" and odd, each nonzero spectral value is actually an eigenvalue (in a sense to be made precise). For F compact and linear and $J = I$ this is a classical fact.

To illustrate how these theorems apply to nonlinear problems, we consider the *eigenvalue problem for the p-Laplacian* which consists in finding solutions $u \not\equiv 0$ of

$$\begin{cases} -\operatorname{div}(|\nabla u|^{p-2}\nabla u)(x) = \mu |u(x)|^{p-2}u(x) & \text{in } G, \\ u(x) \equiv 0 & \text{on } \partial G, \end{cases} \tag{18}$$

where $G \subset \mathbb{R}^n$ is a bounded domain. Although this problem makes sense for $1 < p < \infty$, we restrict ourselves to the case $2 \le p < \infty$. The problem (18) may be reformulated as equivalent operator equation in weak form

$$F_p(u) = \lambda J_p(u), \tag{19}$$

where $\lambda = 1/\mu$, and $F_p, J_p : W_0^{1,p}(G) \to W^{-1,p'}(G)$ ($p' = p/(p-1)$) are defined by $F_p(u) = |u|^{p-2}u$ and

$$\langle J_p(u), v\rangle = -\int_G (|\nabla u(x)|^{p-2}\nabla u(x), \nabla v(x))\, dx \quad (u, v \in W_0^{1,p}(G)), \tag{20}$$

respectively. Equation (19) is of the form (16) and has been studied by many authors, e.g. by Drábek et al. in [4-7]. Interestingly, the eigenvalue theory for the problem (18) has many features in common with the classical *linear* eigenvalue problem $-\Delta u(x) = \mu u(x)$, which is a special case of (18) for $p = 2$. For instance, the first eigenvalue μ_1 of (18) is always positive and simple and may be "calculated" as Rayleigh quotient

$$\mu_1 = \inf_{\substack{u \in W_0^{1,p}(G) \\ u \neq 0}} \frac{\displaystyle\int_G |\nabla u(x)|^p\, dx}{\displaystyle\int_G |u(x)|^p\, dx}.$$

Moreover, the corresponding eigenfunction $u_1 \in W_0^{1,p}(G)$ is positive on G and simple (in the sense that any other eigenfunction is a scalar multiple of u_1). This function has the same "variational characterization" as in the linear case $p = 2$: it minimizes the functional $\Psi_p : W_0^{1,p}(G) \to \mathbb{R}$ defined by $\Psi_p(u) = \frac{1}{p}\langle J_p(u), u\rangle$, subject to the constraint

$$\frac{1}{p}\int_G |u(x)|^{p-2}u(x)\, dx = 1.$$

Finally, we point out that there is a famous so-called *nonlinear Fredholm alternative* (see [9,16,18]) which implies that the operator $J_p - \mu F_p = \mu(\lambda J_p - F_p)$ is *onto* for $\mu < \mu_1$, while it is *not onto* for $\mu = \mu_1$.

However, the coincidence and discreteness theorems given above allow us a more precise statement. The following is just a reformulation of Theorems 3 and 4.

Theorem 5 (Nonlinear Fredholm alternative). *Suppose that $J : X \to Y$ is an odd τ-homogeneous homeomorphism with $[J]_a > 0$, and $F : X \to Y$ is odd, τ-homogeneous and compact. Let $\lambda \neq 0$. Then the following four assertions are equivalent.*

(a) *The eigenvalue problem* (16) *has only the trivial solution $u = 0$.*

(b) *The operator $\lambda J - F$ is stably solvable, $[\lambda J - F]_a > 0$, and $[\lambda J - F]_q > 0$.*

(c) *The operator $\lambda J - F$ is epi on each $\Omega \in \mathfrak{O}(X)$, $[\lambda J - F]_a > 0$, and $[\lambda J - F]_b >$* 0.

(d) *The operator $\lambda J - F$ is strictly epi on some $\Omega \in \mathfrak{O}(X)$, and*

$$\inf\{||\lambda J(u) - F(u)|| : u \in \partial\Omega\} > 0.$$

We claim that the operators F_p and J_p satisfy the hypotheses of Theorem 5 in the spaces $X = W_0^{1,p}(G)$ and $Y = X^* = W^{-1,p'}(G)$. In fact, since $J_p : X \to Y$ is continuous, strictly monotone, coercive (it is here that we use the restriction

$p \geq 2!$), odd, and $(p-1)$-homogeneous, it is an *isomorphism*, by Minty's celebrated theorem [15]. Moreover, the coercivity also implies that $[J_p]_a > 0$. Finally, the operator $F_p : X \to Y$ is continuous, compact (by Krasnosel'skij's theorem and the compactness of the imbedding $X \subset L_p(G)$), odd, and also $(p-1)$-homogeneous. So Theorem 5 implies that, whenever μ is not a classical eigenvalue of (18), then the operator $J_p - \mu F_p$ is not only onto, but even stably solvable and epi. This makes it possible to obtain existence, uniqueness, and stability results for nonlinear perturbations of (18).

References

[1] J. Appell, E. De Pascale, A. Vignoli: *A comparison of different spectra for nonlinear operators*, Nonlin. Anal. TMA **40, 1** (2000), 73-90

[2] J. Appell, M. Dörfner: *Some spectral theory for nonlinear operators*, Nonlin. Anal. TMA **28, 12** (1997), 1955-1976

[3] J. Appell, E. Giorgieri, M. Väth: *Nonlinear spectral theory for homogeneous operators*, Nonlin. Funct. Anal. Appl. (to appear)

[4] P. A. Binding, P. Drábek, Y. X. Huang: *On the Fredholm alternative for the p-Laplacian*, Proc. Amer. Math. Soc **125, 12** (1997), 3555-3559

[5] M. del Pino, P. Drábek, R. Manásievich: *The Fredholm alternative at the first eigenvalue for the one dimensional p-Laplacian*, J. Differ. Equ. **151, 2** (1999), 386-419

[6] P. Drábek: *Fredholm alternative for the p-Laplacian: yes or no?*, in: Function Spaces, Diff. Oper. and Nonlin. Anal. [Ed.: V. Mustonen], Math. Inst. Acad. Sci., Prague 2000, pp. 57-64

[7] P. Drábek, G. Holubová: *Fredholm alternative for the p-Laplacian in higher dimensions*, J. Math. Anal. Appl. **263** (2001), 182-194

[8] W. Feng: *A new spectral theory for nonlinear operators and its applications*, Abstr. Appl. Anal. **2** (1997), 163-183

[9] S. Fučik: *Fredholm alternative for nonlinear operators in Banach spaces and its applications to differential and integral equations*, Comm. Math. Univ. Carol. **11, 2** (1970), 271-284

[10] M. Furi: *Stably solvable maps are unstable under small perturbations*, Zeitschr. Anal. Anw. **21, 1** (2002), 203-208

[11] M. Furi, M. Martelli, A. Vignoli: *Stably solvable operators in Banach spaces*, Atti Accad. Naz. Lincei Rend. Cl. Sci. Fis. Mat. Nat. **60** (1976), 21-26

[12] M. Furi, M. Martelli, A. Vignoli: *Contributions to the spectral theory for nonlinear operators in Banach spaces*, Annali Mat. Pura Appl. **118** (1978), 229-294

[13] M. Furi, M. Martelli, A. Vignoli: *On the solvability of nonlinear operator equations in normed spaces*, Annali Mat. Pura Appl. **128** (1980), 321-343

[14] R. I. Kachurovskij: *Regular points, spectrum and eigenfunctions of nonlinear operators* [in Russian], Dokl. Akad. Nauk SSSR **188** (1969), 274-277; Engl. transl.: Soviet Math. Dokl. **10** (1969), 1101-1105

[15] G. Minty: *Monotone nonlinear operators in Hilbert space*, Duke Math. J. **29** (1962), 341-346

[16] J. Nečas: *Sur l'alternative de Fredholm pour les opérateurs non linéaires avec applications aux problèmes aux limites*, Annali Scuola Norm. Sup. Pisa **23** (1969), 331-345

[17] J. W. Neuberger: *Existence of a spectrum for nonlinear transformations*, Pacific J. Math. **31** (1969), 157-159

[18] S. I. Pokhozhaev: *Solvability of nonlinear equations with odd operators* [in Russian], Funkts. Anal. Prilozh. **1, 3** (1967), 66-73; Engl. transl.: Funct. Anal. Appl. **1, 3** (1967), 227-233

[19] A. Rhodius: *Über numerische Wertebereiche und Spektralwertabschätzungen*, Acta Sci. Math. **47** (1984), 465-470

[20] P. SANTUCCI, M. VÄTH: *On the definition of eigenvalues for nonlinear operators*, Nonlin. Anal. TMA **40, 2** (2000), 565-576

[21] P. SANTUCCI, M. VÄTH: *Grasping the phantom: a new approach to nonlinear spectral theory*, Annali Mat. Pura Appl. (to appear)

[22] M. VÄTH: *The Furi-Martelli-Vignoli spectrum vs. the phantom*, Nonlin. Anal. TMA **47, 6** (2001), 2237-2248

Critical equations for the polyharmonic operator

Thomas Bartsch

Mathematisches Institut, Universität Giessen, Arndtstr. 2,
35392 Giessen, Germany

1. Introduction

We consider the problem

$$\begin{cases} (-\Delta)^m u = & |u|^{2^*-2}u \qquad \text{in } \Omega; \\ \qquad u \in & \mathcal{D}_0^{m,2}(\Omega). \end{cases} \tag{1.1}$$

Throughout this paper we assume $m \geq 1$, $N > 2m$, $2^* = 2N/(N-2m)$ and either $\Omega = \mathbb{R}^N$ or Ω is a smooth bounded domain in $\mathbb{R}^N$. The space $\mathcal{D}_0^{m,2}(\Omega)$ is the closure of $\mathcal{D}(\Omega)$ with respect to the norm

$$\|u\| = \left(\int_\Omega (-\Delta)^m u \cdot u\right)^{1/2} = \begin{cases} |\Delta^{m/2}u|_2 & \text{if } m \text{ is even;} \\ |\nabla\Delta^{(m-1)/2}u|_2 & \text{if } m \text{ is odd.} \end{cases}$$

Here $|\cdot|_p$ denotes the L^p-norm. Observe that $\mathcal{D}_0^{m,2}(\mathbb{R}^N) = \mathcal{D}^{m,2}(\mathbb{R}^N)$, and $\mathcal{D}_0^{m,2}(\Omega) = H_0^m(\Omega)$ if Ω is a smooth bounded domain of $\mathbb{R}^N$.

In the case $m = 1$ equation (1.1) has been studied by many people. We recall a few results which are relevant for our work on the polyharmonic case $m \geq 2$. Aubin **[1]** and Talenti **[25]** showed that the radial function

$$U(x) = \frac{\big(N(N-2)\big)^{(N-2)/4}}{(1+|x|^2)^{(N-2)/2}}$$

solves (1.1) for $m = 1$ on $\mathbb{R}^N$. Moreover, this function achieves the best Sobolev constant

$$S := \inf_{0\neq u\in\mathcal{D}^{1,2}(\mathbb{R}^N)} \frac{\|u\|}{|u|_{2^*}} = \frac{\|U\|}{|U|_{2^*}}$$

of the embedding $\mathcal{D}^{1,2}(\mathbb{R}^N) \hookrightarrow L^{2^*}(\mathbb{R}^N)$. For $x_0 \in \mathbb{R}^N$ and $\varepsilon > 0$ we set

$$U_{x_0,\varepsilon}(x) := \varepsilon^{(2-N)/2}U((x-x_0)/\varepsilon).$$

It is well known (see [17], for instance) that the $(N+2)$-dimensional manifold

$$\mathcal{M} := \{c \cdot U_{x_0,\varepsilon} : c \in \mathbb{R} \setminus \{0\}, x_0 \in \mathbb{R}^N, \varepsilon > 0\} \subset \mathcal{D}^{1,2}(\mathbb{R}^N) \setminus \{0\}$$

consists precisely of all functions achieving the best constant for the Sobolev embedding $\mathcal{D}^{1,2}(\mathbb{R}^N) \hookrightarrow L^{2^*}(\mathbb{R}^N)$. Based on this Bianchi and Egnell [7] proved the following Sobolev inequality with remainder term:

$$\|u\|^2_{\mathcal{D}^{1,2}} - S^2|u|^2_{2^*} \geq \alpha d(u, \mathcal{M})^2 \qquad \text{for } u \in \mathcal{D}^{1,2}(\mathbb{R}^N) \tag{1.2}$$

for some $\alpha > 0$ fixed. Here $d(u, \mathcal{M})$ denotes the distance of u from $\mathcal{M}$ in $\mathcal{D}^{1,2}(\mathbb{R}^N)$. This solved a problem proposed by Brézis and Lieb in [9].

The existence of an unbounded sequence of sign changing solutions of (1.1) in the case $m = 1$ is due to Weiyue Ding [11].

Concerning (1.1) on a smoothly bounded domain $\Omega \subset \mathbb{R}^N$ a famous result of Bahri and Coron [2] states the existence of at least one positive solution in the case $m = 1$ provided the singular homology group $H_d(\Omega; \mathbb{Z}/2)$ is not trivial for some $d > 0$.

Although the polyharmonic, in particular the biharmonic operator has found considerable interest in recent years very few of the above results have been extended to the case $m \geq 2$. A major difficulty is that $(-\Delta)^m$ does not satisfy the maximum principle in the case of Dirichlet boundary conditions if $m \geq 2$ (see [14]). Moreover, computations tend to be much more difficult, already in the case $m = 2$, than in the case $m = 1$. We refer the reader to the papers [15, 16] and the references therein for some recent results on the maximum principle and on nonlinear equations involving the polyharmonic operator.

In this paper we shall present and discuss the following results:

- a Sobolev inequality with remainder term for $m \geq 2$ generalizing the one of Bianchi and Egnell;
- the existence of an unbounded sequence of sign changing solutions of (1.1);
- a partial extension of the result of Bahri and Coron to the case $m = 2$.

These results are recent joint work with Matthias Schneider, Tobias Weth and Michel Willem. Details will appear in the papers [3, 4].

2. Statement of Results

We fix $m \geq 2$. The existence of the soliton solution has been proved by Swanson [24]. Setting

$$U(x) = \frac{P_{m,N}^{(N-2m)/4m}}{(1+|x|^2)^{(N-2m)/2}} \quad \text{with } P_{m,N} = \prod_{h=-m}^{m-1} (N+2h) \tag{2.1}$$

Swanson showed that $U \in E := \mathcal{D}^{m,2}(\mathbb{R}^N)$ solves (1.1) and achieves the best Sobolev constant

$$S := \inf_{0 \neq u \in E} \frac{\|u\|}{|u|_{2^*}} = \frac{\|U\|}{|U|_{2^*}}$$

of the embedding $\mathcal{D}^{m,2}(\mathbb{R}^N) \hookrightarrow L^{2^*}(\mathbb{R}^N)$. As in the introduction we set

$$U_{x_0,\varepsilon}(x) := \varepsilon^{(2m-N)/2}U((x-x_0)/\varepsilon).$$

and

$$\mathcal{M} := \{c \cdot U_{x_0,\varepsilon} : c \in \mathbb{R} \setminus \{0\}, x_0 \in \mathbb{R}^N, \varepsilon > 0\} \subset E \setminus \{0\}.$$

The results of Swanson **[24]** and Lions **[17]** imply that the $(N+2)$-dimensional manifold $\mathcal{M}$ consists precisely of all functions achieving the best constant for the Sobolev embedding $\mathcal{D}^{m,2}(\mathbb{R}^N) \hookrightarrow L^{2^*}(\mathbb{R}^N)$.

Our first result is a Sobolev inequality with remainder term.

THEOREM 2.1. *There exists a constant $\alpha > 0$ such that*

$$||u||^2 - S^2|u|_{2^*}^2 \geq \alpha d(u, \mathcal{M})^2$$

for any $u \in E = \mathcal{D}^{m,2}(\mathbb{R}^N)$. Here

$$d(u, \mathcal{M}) = \inf\{||u - v|| : v \in \mathcal{M}\}$$

is the distance of u from $\mathcal{M}$ in E.

For $m = 2$ the result is stated in **[18]**, however, the proof contains a gap (see the end of the proof of Theorem 2.2 in **[18]** on p. 81).

The second theorem deals with sign changing solutions of (1.1).

THEOREM 2.2. *There exists an unbounded sequence $(u_k)_k$ of sign changing solutions $u_k \in \mathcal{D}^{m,2}(\mathbb{R}^N)$ of* (1.1).

As a consequence of a regularity result of Luckhaus **[19]** weak solutions of (1.1) lie in $\mathcal{C}^{2m}(\mathbb{R}^N)$ hence they are classical solutions.

Finally we treat (1.1) on a smooth bounded domain $\Omega \subset \mathbb{R}^N$. It is easy to see that S is also the best Sobolev constant for the embedding $\mathcal{D}_0^{m,2}(\Omega) \hookrightarrow L^{2^*}(\Omega)$ and is not achieved. In the case $m = 1$ the Pohozaev identity shows that there is no nontrivial solution if Ω is starshaped. We do not know whether this extends to the case $m \geq 2$. The results of Pucci and Serrin **[22, 23]** imply nonexistence results for closely related equations on starshaped domains but not for (1.1). In the biharmonic case $m = 2$ it is proved in **[20, 21]** that (1.1) has no positive solution in a starshaped domain. In any case, the above results suggest that the existence of solutions of (1.1) depends on the topology or the geometry of the domain as in the case $m = 1$. The next theorem seems to be the first result in this direction for $m \geq 2$ with Dirichlet boundary conditions.

THEOREM 2.3. *There exists $q > 1$ such that* (1.1) *with $m = 2$ has a nontrivial solution on the smooth bounded domain $\Omega \subset \mathbb{R}^N$ provided there exist $\rho > 0$ and $a \in \mathbb{R}^N$ with*

$$\{x \in \mathbb{R}^N : \rho \leq |x - a| \leq \rho q\} \subset \Omega$$

and

$$\{x \in \mathbb{R}^N : |x - a| < \rho\} \not\subset \Omega.$$

The theorem should be true for arbitrary $m \geq 2$. We could prove this provided the following nonexistence result would be true:

If $u \in \mathcal{D}_0^{m,2}(\mathbb{R}_+^N)$ solves $(-\Delta)^m u = |u|^{2^*-2}u$ in $\mathbb{R}_+^N$ then $u = 0$.

Here $\mathbb{R}_+^N = \{x \in \mathbb{R}^N : x_N > 0\}$ denotes the upper half space.

In the case $m = 1$ Theorem 2.3 is due to Coron **[10]** preceding the more general result of Bahri and Coron **[2]** mentioned in the introduction. It seems to be very difficult to prove the existence of a nontrivial solution of (1.1) in the case $m \geq 2$ under the condition $H_d(\Omega; \mathbb{Z}/2) \neq 0$ for some $d > 0$. As a consequence of the lack of a maximum principle for $(-\Delta)^m$ on $H_0^m(\Omega)$ we were not able to prove the existence of a positive solution in the situation of Theorem 2.3. Moreover, the method of Bahri and Coron requires a detailed knowledge of possible "critical points at infinity". These are closely related to solutions of (1.1) on $\mathbb{R}^N$. Now, without the maximum principle one has to take all solutions of (1.1) on $\mathbb{R}^N$ into account, not only the positive ones. In view of our Theorem 2.2 from above this seems to be out of reach. A Bahri-Coron type result has been recently obtained in **[13]** for the critical biharmonic equation with Navier boundary conditions:

$$\begin{cases} \Delta^2 u = |u|^{8/(N-4)}u & \text{in } \Omega \\ u = \Delta u = 0 & \text{on } \partial\Omega. \end{cases}$$

With these boundary conditions the maximum principle holds and the argument of Bahri and Coron can be extended.

3. Sketch of Proofs

The main technical ingredient in the proof of Theorem 2.1 is the following result which is of interest in itself. For U as in (2.1) let μ_k denote the eigenvalues of the eigenvalue problem

$$(-\Delta)^m v = \mu U^{2^*-2} v, \qquad v \in E = \mathcal{D}^{m,2}(\mathbb{R}^N).$$

counted with multiplicity in increasing order.

THEOREM 3.1. $\mu_1 = 1$, $\mu_2 = \mu_3 = \ldots = \mu_{N+2} = 2^* - 1 < \mu_{N+3}$. *A corresponding choice of linearly independent eigenvectors $v_1, \ldots, v_{N+2}$ is given by*

$$v_1 = U, \qquad v_{i+1} = \frac{\partial U}{\partial y_i},\ i = 1, \ldots, N, \quad v_{N+2} = x \cdot \nabla U + \frac{N-2m}{2} U \tag{3.1}$$

For the proof of this theorem we refer the reader to **[4]**. The difficult part consists of showing that the eigenspace V_2 associated to the second eigenvalue is at most $(N+1)$-dimensional. For this we use polar coordinates and expand the elements of V_2 using the spherical harmonics $Y_l^j : S^{N-1} \to \mathbb{R}$, $l \in \mathbb{N}_0$, $j = 1, \ldots, c_l = \binom{l+N-1}{N-1} - \binom{l+N-3}{N-1}$, with radial functions $R_l^j : [0, \infty) \to \mathbb{R}$ as Fourier coefficients. From (1.1) we deduce a singular ordinary differential equation of order $2m$ for R_l^j. We then show that $R_l^j = 0$ for $l \geq 2$ and that R_l^j is uniquely determined by $R_l^j(0)$. This implies that $\dim(V_2) \leq c_0 + c_1 = N + 1$.

A simple consequence of Theorem 3.1 concerns the energy functional

$$\Phi_0 : E \to \mathbb{R}, \quad \Phi_0(u) = \frac{1}{2}\|u\|^2 - \frac{1}{2^*}|u|_{2^*}^{2^*},$$

whose critical points are the solutions of (1.1). By the results of Lions and Swanson mentioned in the introduction, the $(N+1)$-dimensional manifold

$$\mathcal{Z} = \{\pm U_{x_0,\varepsilon} : x_0 \in \mathbb{R}^N,\ \varepsilon > 0\} \subset \mathcal{M}$$

is the manifold of least energy solutions of (1.1) with least energy

$$S_0 := \left(\frac{1}{2} - \frac{1}{2^*}\right) S^{N/m}.$$

COROLLARY 3.2. *$\mathcal{Z}$ is a nondegenerate critical manifold of Φ_0 and S_0 is an isolated critical value of Φ_0.*

PROOF. That $\mathcal{Z}$ is a nondegenerate critical manifold of Φ_0 follows immediately from Theorem 3.1 because

$$D^2\Phi_0(U)(v,w) = \langle v,w\rangle - (2^*-1)\int_{\mathbb{R}^N} |U|^{2^*-2}vw$$

and

$$T_U\mathcal{Z} = \operatorname{span}\{v_2,\dots,v_{N+2}\}.$$

That S_0 is an isolated critical value follows now from $\mathcal{Z}$ being isolated, Theorem I.1 in **[17]**, and the invariance of Φ_0 under translations and dilations. □

In order to prove Theorem 2.1 we introduce some notation. Consider the multiplicative groups $\dot{\mathbb{R}} = \mathbb{R} \setminus \{0\}$ and $\mathbb{R}^+ = (0,\infty)$, and the additive group $\mathbb{R}^N$. Let $G = \dot{\mathbb{R}} \times \mathbb{R}^N \times \mathbb{R}^+$ be the (semidirect) product of these groups with the group structure defined by

$$(c,x,\varepsilon)\cdot(c',x',\varepsilon') := (cc', x+\varepsilon x', \varepsilon\varepsilon').$$

A simple computation shows that G acts on $E = \mathcal{D}^{m,2}(\mathbb{R}^N)$ via

$$(c,x,\varepsilon) * u = cu_{x,\varepsilon} : \mathbb{R}^N \to \mathbb{R},\ y \mapsto c\varepsilon^{(2m-N)/2}u\big((y-x)/\varepsilon\big).$$

By direct computation one sees that the norms $\|\cdot\|$ and $|\cdot|_{2^*}$ are invariant under the action of the subgroup $\{1\}\times\mathbb{R}^N\times\mathbb{R}^+$ of translations and dilations. We identify $c \in \dot{\mathbb{R}}$, $x \in \mathbb{R}^N$ and $\varepsilon \in \mathbb{R}^+$ with $(c,0,1) \in G$, $(1,x,1) \in G$ and $(1,0,\varepsilon) \in G$, respectively. The manifold of functions achieving the best Sobolev constant for the embedding $E \hookrightarrow L^{2^*}(\mathbb{R}^N)$ can now be written as

$$\mathcal{M} = G * U = \{g * U : g \in G\}.$$

Proof of Theorem 2.1. Suppose Theorem 2.1 is false and consider the G-invariant function

$$Q : E \setminus \mathcal{M} \to \mathbb{R}, \quad Q(u) := \frac{\|u\|^2 - S^2|u|_{2^*}^2}{d(u,\mathcal{M})^2}.$$

Then there exists a sequence $(u_n)_n$ in $E \setminus \mathcal{M}$ with $\lim_{n\to\infty} Q(u_n) = 0$. This implies

$$\|u_n\|^2 - S^2 |u_n|_{2^*}^2 \to 0$$

By homogeneity we may assume $\|u_n\| = 1$. A concentration-compactness argument of Lions ([**17**], Corollary I.2) yields a sequence of group elements $g_n = (1, x_n, \varepsilon_n) \in G$ such that the sequence $(g_n * u_n)_n$ is relatively compact in E. After passing to a subsequence we have $g_n * u_n \to u \in E$ and $\|u\|^2 = S^2 |u|_{2^*}^2$. Then $u \in \mathcal{M}$ and therefore

$$d(u_n, \mathcal{M}) = d(g_n * u_n, \mathcal{M}) \to 0. \tag{3.2}$$

We choose group elements which we again denote by $g_n = (c_n, x_n, \varepsilon_n) \in G$ such that

$$d(u_n, \mathcal{M}) = \|u_n - g_n * U\| = c_n \cdot \|g_n^{-1} * u_n - U\| = c_n \cdot d(g_n^{-1} * u_n, \mathcal{M})$$

Clearly c_n is bounded away from 0 because $\|u_n\| = 1$. Therefore (3.2) implies $d(g_n^{-1} * u_n, \mathcal{M}) = \frac{1}{c_n} d(u_n, \mathcal{M}) \to 0$. Setting $w_n := g_n^{-1} * u_n$ we have $Q(w_n) \to 0$ and $d(w_n, \mathcal{M}) = \|w_n - U\| \to 0$. The last fact implies that $w_n - U$ is orthogonal to the tangent space

$$T_U \mathcal{M} = \operatorname{span}\{U, v_2, \dots, v_{N+2}\}$$

(where the v_i's are defined in Theorem 3.1) and therefore

$$\|w_n\|^2 = \|U\|^2 + \|w_n - U\|^2.$$

Moreover, Theorem 3.1 yields

$$\|w_n - U\|^2 \geq \mu_{N+3} \int_{\mathbb{R}^N} U^{2^*-2} (w_n - U)^2.$$

Next observe that

$$\begin{aligned} |w_n|_{2^*}^{2^*} &= \int_{\mathbb{R}^N} U^{2^*} + 2^* \int_{\mathbb{R}^N} U^{2^*-1}(w_n - U) \\ &\quad + 2^*(2^*-1) \int_{\mathbb{R}^N} U^{2^*-2}(w_n - U)^2 + o(\|w_n - U\|^2). \end{aligned} \tag{3.3}$$

and

$$\int_{\mathbb{R}^N} U^{2^*-1}(w_n - U) = \int_{\mathbb{R}^N} (-\Delta)^m U (w_n - U) = 0.$$

Therefore (3.3) reduces to

$$\begin{aligned} |w_n|_{2^*}^{2^*} &= |U|_{2^*}^{2^*} + 2^*(2^*-1) \int_{\mathbb{R}^N} U^{2^*-2}(w_n - U)^2 + o(\|w_n - U\|^2) \\ &\leq |U|_{2^*}^{2^*} + \frac{2^*(2^*-1)}{\mu_{N+3}} \|w_n - U\|^2 + o(\|w_n - U\|^2) \end{aligned}$$

and thus

$$|w_n|_{2^*}^2 \leq |U|_{2^*}^2 \left(1 + \frac{2^*(2^*-1)}{|U|_{2^*}^{2^*} \mu_{N+3}} \|w_n - U\|^2 + o(\|w_n - U\|^2)\right)^{2/2^*}.$$

Using $|U|_{2^*}^{2^*-2} = \frac{\|U\|^2}{|U|_{2^*}^2} = S^2$ we obtain from these estimates

$$\begin{aligned}\|w_n\|^2 - S^2|w_n|_{2^*}^2 &\geq \|w_n - U\|^2\left(1 - S^2|U|_{2^*}^{2-2^*}\frac{2^*-1}{\mu_{N+3}}\right) + o(\|w_n - U\|^2)\\ &= \|w_n - U\|^2\left(1 - \frac{2^*-1}{\mu_{N+3}}\right) + o(\|w_n - U\|^2).\end{aligned}$$

This implies $Q(w_n) \geq 1 - \frac{2^*-1}{\mu_{N+3}} + o(1)$, in contradiction to Theorem 3.1 and $Q(w_n) \to 0$. □

Proof of Theorem 2.2. Using the stereographic projection one first shows that (1.1) can be reduced to the problem

$$\begin{cases} D_m u = |u|^{2^*-2}u & \text{on } S^N, \\ u \in H^m(S^N). \end{cases} \tag{3.4}$$

Here $S^N \subset \mathbb{R}^{N+1}$ is the unit sphere with the induced metric g, and D_m is the elliptic differential operator of order $2m$ given by

$$D_m = \prod_{k=1}^{m}\left(-\Delta_g + \frac{1}{4}(N-2k)(N+2k-2)\right),$$

with Δ_g denoting the Laplace-Beltrami-Operator on S^N. The operator D_2 is the Paneitz operator on S^N introduced by Branson [**8**] for smooth compact Riemannian manifolds (see also [**6, 12**] for further information and important estimates).

We obtain the solutions of (3.4) as critical points of the functional

$$J : H^m(S^N) \to \mathbb{R}, \quad J(u) = \frac{1}{2}\int_{S^N} D^m u \cdot u - \frac{1}{2^*}\int_{S^N} |u|^{2^*}.$$

In order to gain compactness and guarantee that the solutions change sign we use a symmetry argument similar to the one from [**5**]. We write the elements of $S^N \subset \mathbb{R}^{N+1}$ as $x = (x_1, x_2, x_3)$ with $x_1, x_2 \in \mathbb{R}^2$, $x_3 \in \mathbb{R}^{N-3}$. Now we define the Hilbert space $H \subset H^m(S^N)$ to consist of all $u \in H^m(S^N)$ such that $u(x) = u(x_1, x_2, x_3)$ depends only on $|x_1|, |x_2|, |x_3|$, and satisfies in addition $u(x_2, x_1, x_3) = -u(x_1, x_2, x_3)$. Thus any $u \in H \setminus \{0\}$ automatically changes sign. One can show by standard arguments that H is compactly embedded in $L^{2^*}(\mathbb{R}^N)$. This implies that $J|_H$ satisfies the Palais-Smale condition. The fountain theorem (see [**5, 26**]) applied to $J|_H$ now yields an unbounded sequence of critical points of $J|_H$. By the principle of symmetric criticality these are also critical points of J on $H^m(S^N)$, hence they are sign changing solutions of (3.4). □

Proof of Theorem 2.3. We consider the functional

$$\Psi : \mathcal{S} := \{u \in \mathcal{D}^{m,2}(\Omega) : |u|_{2^*} = 1\} \to \mathbb{R}, \quad \Psi_0(u) := \|u\|,$$

whose critical points are solutions of (1.1) in Ω. If $m = 2$ then Ψ satisfies the Palais-Smale condition at levels $c \in (S, 2^{4/N} S)$. This has been proved in [**15**, Lemma 6] for Navier boundary conditions. The proof for Dirichlet boundary conditions proceeds similarly.

For $R > 1$ we consider the annulus

$$A_R = \{x \in \mathbb{R}^N : 1/4R \leq |x| \leq 4R\}$$

and assume $A_R \subset \Omega$. Let

$$\beta : \mathcal{S} \to \mathbb{R}^N, \quad \beta(u) := \begin{cases} S^{-2} \displaystyle\int_\Omega x|\Delta^{m/2} u|^2 dx & \text{if } m \text{ is even,} \\ S^{-2} \displaystyle\int_\Omega x|\nabla\Delta^{(m-1)/2} u|^2 dx & \text{if } m \text{ is odd,} \end{cases}$$

be the "barycenter map". For any $\delta > 0$ there exists $\gamma > 0$ small such that β satisfies $\beta\left(\Psi^{S+\gamma}\right) \subset U_\delta(\Omega)$. We choose δ small so that $\overline{\Omega}$ is a deformation retract of $U_\delta(\Omega)$ and fix γ accordingly. Using the soliton solution and cut-off functions we can construct a map

$$H_R : [0,1) \times S^{N-1} \to \Psi^{c(R)} \subset \mathcal{S}$$

such that $\beta \circ H_R(0, x) = x$ and $\beta \circ H_R(t, x) \to x_0 \in S^{N-1}$ as $t \to 1$, for some $x_0 \in S^{N-1}$. If R is large enough we can achieve $c(R) < 2^{4/N} S$ so that we are in a range where the Palais-Smale condition holds. Suppose Ψ has no critical value in $(S, c(R))$. Then we can deform $\Psi^{c(R)}$ to $\Psi^{S+\gamma}$ with γ from above. Consequently there exists a retraction map $r : \Psi^{c(R)} \to \Psi^{S+\gamma}$. The composition $\beta \circ r \circ H_R$ induces a homotopy $[0, 1] \times S^{N-1} \to U_\delta(\Omega)$ deforming the inclusion $S^{N-1} \hookrightarrow U_\delta(\Omega)$ to a constant map. This is a contradiction to the hypothesis on the topology of Ω in Theorem 2.3. □

References

[1] Aubin, T.: Problèmes isopérimétriques et espaces de Sobolev. J. Diff. Geom. **11**, 573–598 (1976)

[2] Bahri, A. and Coron, J.M.: On a nonlinear elliptic equation involving the critical Sobolev exponent: the effect of the topology of the domain. Comm. Pure Appl. Math. **41**, 253–294 (1988)

[3] Bartsch, T., Schneider, M., and Weth, T.: Multiple solutions to a critical polyharmonic equation. Preprint.

[4] Bartsch, T., Weth, T., and Willem, M.: A Sobolev inequality with remainder term and critical equations on domains with topology for the polyharmonic operator. Calc. Var. Part. Diff. Equ. (to appear)

[5] Bartsch, T., and Willem, M.: Infinitely many nonradial solutions of a Euclidean scalar field equation. J. Funct. Anal. **117**, 447–460 (1993)

[6] Beckner, W.: Sharp Sobolev inequalities on the sphere and the Moser-Trudinger inequality. Ann. of Math. **138**, 213–242 (1993)

[7] Bianchi, G. and Egnell, H.: A note on the Sobolev inequality. J. Funct. Anal. **100**, 18–24 (1991)

[8] Branson, T.P.: Group representations arising from Lorentz conformal geometry. J. Funct. Anal. **1987**, 199-291 (1993)

[9] Brézis, H. and Lieb, E.: Sobolev inequalities with remainder terms. J. Funct. Anal. **62**, 73–86 (1985)
[10] Coron, J.M.: Topologie en cas limite des injection de Sobolev. C. R. Acad. Sci. Paris Sér. I Math. **299**, 209–212 (1984)
[11] Ding, W.: On a conformally invariant elliptic equation. Commun. Math. Phys. **107**, 331–335 (1986)
[12] Djadli, Z., Hebey, E., and Ledoux,M.: Paneitz-type operators and applications. Duke Math. J. **104**, 129–169 (2000)
[13] Ebobisse, F. and Ould Ahmedou, M.: On a nonlinear fourth order elliptic equation involving the critical Sobolev exponent. Preprint.
[14] Garabedian, P.R.: A partial differential equation arising in conformal mapping. Pacific J. Math. **1**, 485–524 (1951)
[15] Gazzola, F., Grunau, H.–C., and Squassina, M.: Existence and nonexistence results for critical growth biharmonic elliptic equations. Preprint.
[16] Grunau, H. and Sweers, G.: The maximum principle and positive principal eigenfunctions for polyharmonic equations. Lect. Notes Pure Appl. Math. **194**, 163–182 (1998)
[17] Lions, P.: The concentration-compactness principle in the calculus of variations: the limiting case, part I. Riv. Mat. Iberoamericana **1**, 145–201 (1985)
[18] Lu, G. and Wei, J.: On a Sobolev inequality with remainder terms. Proc. Amer. Math. Soc. **128**, 75–84 (1999)
[19] Luckhaus, S.: Existence and regularity of weak solutions to the Dirichlet problem for semilinear elliptic systems of higher order. J. Reine Angew. Math. **306**, 192–207 (1979)
[20] Mitidieri, E.: A Rellich type identity and applications. Comm. Part. Diff. Eq. **18**, 125–151 (1993)
[21] Oswald, P.: On a priori estimates for positive solutions of a semilinear biharmonic equation in a ball. Comment. Math. Univ. Carolinae **26**, 565-577 (1985)
[22] Pucci, P. and Serrin, J.: A general variational identity. Indiana Univ. Math. J. **35**, 681–703 (1986)
[23] Pucci, P. and Serrin, J.: Critical exponents and critical dimensions for polyharmonic operators. J. Math. Pures Appl. **69**, 55–83 (1990)
[24] Swanson, C.A.: The best Sobolev constant. Applicable Anal. **47**, 227–239 (1992)
[25] Talenti, G.: Best constants in Sobolev inequality. Annali di Mat. **110**, 353–372 (1976)
[26] Willem, M.: Minimax Theorems. Birkhäuser, Boston 1996

Standing wave solutions with a critical frequency for nonlinear Schrödinger equations

Jaeyoung Byeon[1] and Zhi-Qiang Wang[2]

[1] *Department of Mathematics, POSTECH Pohang, Kyungbuk 790-784, Republic of Korea*

[2] *Department of Mathematics and Statistics, Utah State University, Logan, UT 84322*

ABSTRACT. We present some recent works ([BW1,BW2]) of the authors on standing waves for nonlinear Schrödinger equations with a potential V

$$i\hbar\frac{\partial\psi}{\partial t}+\frac{\hbar^2}{2}\Delta\psi-V(x)\psi+f(\psi)=0 \quad \text{in } \mathbb{R}^N.$$

A solution of the form $\psi(x,t)=\exp(-iEt/\hbar)v(x)$ is called a standing wave. There have been numerous works on the standing waves with a frequency $E<\inf V$. Here some existence results and new features of standing waves with a critical frequency $E=\inf V$ will be illustrated.

The evolution of a free non-relativistic quantum particle is described by linear Schrödinger equations, which is one of the main objects in quantum mechanics. On the other hand, for a group of identical particles interacting each other in ultracold states, in particular, Bose-Einstein condensates, their evolution is described, via Hartree approximation, to an excellent degree of accuracy by nonlinear Schrödinger equations(refer [Me]). The equation arises in many fields of physics, in particular, when we describe the propagation of light in some nonlinear optical materials; the nonlinear Schrödinger equations in nonlinear optics are reduced from Maxwell's equations(refer [Mi]). The nonlinear Schrödinger equation is typically of the form

$$i\hbar\frac{\partial\psi}{\partial t}+\frac{\hbar^2}{2}\Delta\psi-V(x)\psi+f(\psi)=0, \tag{1}$$

where $\hbar$ denotes the Plank constant, i is the imaginary unit. We are concerned with the existence of standing waves of the nonlinear Schrödinger equation (1)

e-mail: `jbyeon@postech.ac.kr` `wang@math.usu.edu`.

for small $\hbar$. For small $\hbar > 0$, these standing wave solutions are refereed as semi-classical states. Here a solution of the form $\psi(x,t) = \exp(-iEt/\hbar)v(x)$ is called a standing wave. Suppose that $f(\exp(i\theta)t) = f(t)$ for $\theta, t \in \mathbb{R}$. Then, a function $\psi(x,t) \equiv \exp(-iEt/\hbar)\ v(x)$ is a standing wave solution of (1) if and only if the function v satisfies

$$\frac{\hbar^2}{2}\Delta v - (V(x) - E)v + f(v) = 0, \quad x \in \mathbb{R}^N. \tag{2}$$

In this paper, we are concerned with problem (2) for the case

$$\inf_{x\in\mathbb{R}^N} V(x) = E.$$

In this case, v is called a standing wave solution with a critical frequency E. We shall survey the recent works [BW1] and [BW2].

To motivate our discussion, let us first review some of the related results in the literature for the case $\inf_{x\in\mathbb{R}^N} V(x) > E$. Under condition $\inf_{x\in\mathbb{R}^N} V(x) > E$, problem (2) has a ground state solution (possibly stable solution) for $\hbar > 0$ small (refer [R]) when $\inf_{x\in\mathbb{R}^N} V(x) < \liminf_{|x|\to\infty} V(x)$ and the f satisfies some conditions; furthermore, for $a > 0$, the following well-understood problem in $\mathbb{R}^N$ plays a crucial role for constructing solutions of equation (2) and is considered the limiting equation for (2) as $\hbar \to 0$,

$$\Delta u - au + f(u) = 0. \tag{3}$$

This problem has a unique ground state solution $w > 0$.

On the other hand, for a typical nonlinearity $f(t) = |t|^{p-1}t$ and

$$\inf_{x\in\mathbb{R}^N} V(x) < E$$

we can show easily that there are no ground state solutions of problem (2) if $\hbar > 0$ is sufficiently small; moreover, there is no nice limiting problem as in the case $E < \inf_{x\in\mathbb{R}^N} V(x)$. In this sense, we say that $E = \inf_{x\in\mathbb{R}^N} V(x)$ is a *critical frequency* (or energy) for nonlinear Schrödinger equation (1) or problem (2).

There have been enormous investigations on problem (2) under condition

$$\inf_{x\in\mathbb{R}^N} V(x) > E.$$

Floer and Weinstein proved in [FW] that for sufficiently small $\hbar > 0$, there exists a solution $u_\hbar$ of (2) with $\liminf_{\hbar\to 0} \max_{x\in\mathbb{R}^N} |u_\hbar(x)| > 0$ which is concentrated around a non-degenerate critical point of V when $N = 1$, $E < \inf_{\mathbb{R}^N} V(x)$, $f(t) = |t|^2 t$ and V is a bounded function. They used a Lyapunov-Schmidt reduction method to obtain the result. Further investigations and developments have been carried out by Oh[O1-3], Wang[W], Rabinowitz[R], del Pino and Felmer[DF1-3], Ambrosetti, Badiale and Cingolani[ABC], Gui[G], Li[L], Dancer and Yan [DY], Kang and Wei[KW] and many others. (Refer also some works in [B] and [Wa] for a radial potential V.)

In all the above mentioned works, they use the ground state solution w of the limiting equation stated above as a building block to construct single bump or multi-bump solutions for (2) with each bump looking like a translated ground state solution w. This ground state solution w enjoys many nice properties which have been used in an essential way in the above works. Especially, the Lyapunov-Schmidt reduction method relies upon the nondegeneracy property of w. The limiting profiles of the perturbed equations are essentially determined by the ground state solution w of the limiting equation.

In contrast, for the case of $E = \inf_{x\in\mathbb{R}^N} V(x)$ the situation changes drastically and depends upon the local behavior of V near its global minimum where $V = E$. Moreover, the limiting equations have many different forms, of which some are defined on R^N with homogeneous potentials and some are defined on bounded domains which could be of arbitrary shapes. These new features with the limiting equations provide new phenomena for the limiting profiles of the solutions of (2). For example, for all solutions obtained so far under condition $\inf_{R^N} V(x) > E$, the maximum values of the solutions are bounded away from zero (in fact, asymptotically they are proportional to the maximum value of w). On the other hand, in the case of $E = \min_{x\in\mathbb{R}^N} V(x)$, there is a solution of (2) whose maximum value goes to 0 as $\hbar \to 0$. In fact, if $f(u) = |u|^{p-1}u, p \in (1,(N+2)/(N-2))$ and A is an isolated component of $\{x \in \mathbb{R}^N | V(x) = E\}$, there is a solution $v_\hbar$ of (2) such that $\lim_{\hbar\to 0} \|v_\hbar\|_{L^\infty(\mathbb{R}^N)} = 0$, $\liminf_{\hbar\to 0} \hbar^{-2/(p-1)}\|v_\hbar\|_{L^\infty(\mathbb{R}^N)} > 0$ and $v_\hbar$ is exponentially small on $\mathbb{R}^N \setminus U$ as $\hbar \to 0$, where $\overline{A} \subset U$. This is in striking contrast with the case $\min_{x\in\mathbb{R}^N} V(x) > E$.

Existence of standing waves with a critical frequency

We rewrite equation (2) in the following form:

$$\begin{array}{ll} \varepsilon^2\Delta v - V(x)v + v^p = 0,\ v > 0 & \text{in} \quad \mathbb{R}^N \\ \lim_{|x|\to 0} v(x) = 0. & \end{array} \tag{4}$$

Here we assume that $1 < p < \infty$ for $N \le 2$ and $1 < p < (N+2)/(N-2)$ for $N \ge 3$. We assume that the potential V satisfies the following conditions:

(V1) V is a continuous nonnegative function on $\mathbb{R}^N$;

(V2) $\liminf_{|x|\to\infty} V(x) > 0$;

(V3) the zero set of $V, \mathcal{Z} \equiv \{x \in \mathbb{R}^N | V(x) = 0\}$ is non-empty.

Let A be an isolated component of $\mathcal{Z}$. We define $A^\delta = \{x \in \mathbb{R}^N | \mathrm{dist}(x, A) \le \delta\}$, and $A^\delta_\varepsilon = \{x \in \mathbb{R}^N | \varepsilon x \in A^\delta\}$. We assume that

(A) for some $\delta > 0$, $A^{8\delta} \cap (\mathcal{Z} \setminus A) = \emptyset$.

We define an energy functional

$$\Gamma_\varepsilon(u) \equiv \frac{1}{2}\int_{\mathbb{R}^N} \varepsilon^2|\nabla u|^2 + Vu^2 dx - \frac{1}{p+1}\int_{\mathbb{R}^N} |u|^{p+1}dx.$$

Positive critical points of $\Gamma_\varepsilon(u)$ are solutions of equations (4).

Then, we obtain the following result on the existence of standing waves.

Theorem 1 ([BW1]) *Suppose that (V1-3) and (A) hold. Then for sufficiently small $\varepsilon > 0$, there exists a solution v_ε of (4) such that*

(i) $\lim_{\varepsilon\to 0}\varepsilon^{-N}\Gamma_\varepsilon(v_\varepsilon)=0$;
(ii) $\lim_{\varepsilon\to 0}\|v_\varepsilon\|_{L^\infty(\mathbb{R}^N)}=0,\ \liminf_{\varepsilon\to 0}\varepsilon^{-2/(p-1)}\|v_\varepsilon\|_{L^\infty(\mathbb{R}^N)}>0$;
(iii) for each $\delta'>0$, *there exist constants* $C,c>0$ *such that*

$$v_\varepsilon(x)\le C\exp(-\frac{c}{\varepsilon}dist(x,A^{\delta'})).$$

An existence of a least energy solution for (4) is rather easy to prove. The least energy solution satisfies the properties (i),(ii) and (iii) with $\mathcal{Z}$ instead of A.

The idea of proving Theorem 1 is to use local minimization arguments together with elliptic estimates. For the homogeneous nonlinearity v^p we can formulate the problem as one of local minimization with a constraint. We refer the details to [BW1].

Asymptotic profiles for standing waves with a critical frequency

The asymptotic behavior of localized solutions given in Theorem 1 depends in a very delicate way on some local properties of V around A. We distinguish three cases here: (i) The flat case, where the interior of A, $int(A)$, is nonempty. (ii) The finite case, where A is a single point and V behaves like a finite order polynomial near A. (iii) The infinite case, where A is a single point and V is exponentially flat near A. Though these three cases do not cover all possible local behaviors of V around A, they are most typical models. Finer analysis is needed for more complicate cases.

The flat case. We consider a case in which the set of interior points of $A, int(A)$, is not empty, and $A=\overline{int(A)}$. Let $int(A)=\cup_{i\in J}A_i$, where $\{A_i\}_{i\in J}$ are connected components of $int(A)$. For each $i\in J$, we consider the following problem

$$\begin{aligned}\Delta u+u^p &= 0 \quad \text{in } A_i\\ u &> 0 \quad \text{in } A_i\\ u &= 0 \quad \text{on } \partial A_i,\end{aligned} \tag{5-i}$$

Each problem (5-i) has a least energy solution U_i. Let v_ε be a solution given in Theorem 1.

Theorem 2 ([BW1]) *The following holds:*

$$\lim_{\varepsilon\to 0}\varepsilon^{-2(p+1)/(p-1)}\Gamma_\varepsilon(v_\varepsilon)=\frac{1}{2}\int_{A_j}|\nabla U_j|^2dx-\frac{1}{p+1}\int_{A_j}(U_j)^{p+1}dx.$$

Moreover, a normalized function $\varepsilon^{2/(1-p)}v_\varepsilon$ *converges pointwise to some least energy solution* U_j *of (5-j) on* A_j *and to* 0 *on* $\mathbb{R}^N\setminus A_j$. *For each* $\delta>0$, *the convergence is uniform on* $\{x\in\mathbb{R}^N\,|\,dist(x,\partial A)\ge\delta\}$.

The finite case. In this section, we will illustrate asymptotic behavior of v_ε for certain potential functions which are *asymptotically homogeneous* at zero.

Definition 3 For $m \in (0,\infty)$, a continuous function $P : \mathbb{R}^N \setminus \{0\} \to (0,\infty)$ is called a $m-$homogeneous positive function if $P(x) > 0$ for $x \neq 0$ and $P(tx) = t^m P(x)$ for $t \in [0,\infty), x \in \mathbb{R}^N$.

This is a generalization of a homogeneous polynomial P_{2m} of order $2m$ satisfying $P_{2m}(x) > 0$ for $x \neq 0$. Let P be a $m-$homogeneous positive function. We define a norm $\|u\|_P \equiv \left(\int_{\mathbb{R}^N} |\nabla u|^2 + Pu^2 dx \right)^{1/2}$, and a space H the completion of $C_0^\infty(\mathbb{R}^N)$ with respect to the norm $\|\cdot\|_P$. It is easy to see the following.

Proposition 4 *There exists a minimizer U_P of the following minimization problem*

$$I_P \equiv \inf \left\{ \int_{\mathbb{R}^N} |\nabla u|^2 + Pu^2 dx \middle| \int_{\mathbb{R}^N} u^{p+1} dx = 1, u \in H \right\}.$$

We see that a scaled function $w \equiv (I_P)^{\frac{1}{p-1}} w_0$ satisfies the following equation

$$\Delta w - Pw + w^p = 0 \qquad \text{in } \mathbb{R}^N. \tag{6}$$

Now we give an asymptotic profile and an energy estimate of v_ε.

Theorem 5 ([BW1]) *Let x_0 be an isolated zero point of V. Suppose that $V(x + x_0) = P(x) + Q(x)$, where P is an $m-$homogeneous positive function and*

$$\lim_{|x|\to 0} |x|^{-m} |Q(x)| = 0.$$

Let v_ε be a solution of (4) around x_0 given in Theorem 1. Then, there exists a least energy solution w of (6) such that a re-scaled function $\varepsilon^{-\frac{2}{p-1}\frac{m}{m+2}} v_\varepsilon(\varepsilon^{\frac{2}{m+2}} x)$ converges to w uniformly on $\mathbb{R}^N$. Moreover, the following energy estimate holds

$$\lim_{\varepsilon\to 0} \varepsilon^{-\frac{2m}{m+2}\frac{p+1}{p-1} - \frac{2N}{m+2}} \Gamma_\varepsilon(v_\varepsilon) = I_P^{\frac{p+1}{p-1}} (1/2 - 1/(p+1)).$$

The infinite case. We will illustrate the asymptotic behavior of a localized solution around x_0 when $V(x)$ is typically of form $\exp(-1/|x - x_0|^m)$. In fact, we shall consider more general situation where the level sets of V can be non-convex sets, but, strict star-shaped. More precisely, let Ω be a bounded domain in $\mathbb{R}^N$. We assume that there exists a continuous map $r : \mathbb{R}^N \setminus \{0\} \to (0,\infty)$ satisfying

$$\begin{cases} x/t \in \mathbb{R}^N \setminus \overline{\Omega}, & t \in (0, r(x)), \\ x/t \in \partial\Omega, & t = r(x), \\ x/t \in \Omega, & t \in (r(x), \infty). \end{cases}$$

It is easy to see that such an Ω should be a strictly star-shaped domain. Then, for any $x \in \mathbb{R}^N \setminus \{0\}$, we can find a unique pair $(r(x), s(x)) \in (0,\infty) \times \partial\Omega$ with $x = r(x)s(x)$.

Definition 6 A continuous function $b : \mathbb{R}^N \to [0,\infty)$ is called an Ω quasi homogeneous function if

(i) $b(r(x)s(x))$ depends only on $r(x)$;
(ii) $b(r)$ is strictly increasing with respect to $r \in [0,\infty)$; and
(iii)

$$\lim_{r\to 0}\frac{b(cr)}{b(r)}\quad\begin{cases} <1 & \text{for } c<1,\\ >1 & \text{for } c>1.\end{cases}$$

Moreover, a continuous function $a : \mathbb{R}^N \setminus \{0\} \to (0,\infty)$ is called an asymptotically (Ω, b) quasi-homogeneous function if there is an Ω quasi-homogeneous function b satisfying $\lim_{|x|\to 0}\frac{a(x)}{b(x)} = 1$.

We consider a following limiting problem

$$\begin{aligned} \Delta w + w^p &= 0 && \text{in } \Omega\\ w &> 0 && \text{in } \Omega\\ w &= 0 && \text{on } \partial\Omega. \end{aligned} \tag{7}$$

Theorem 7 ([BW1]) *Let x_0 be an isolated zero point of V. Suppose that $V(x+x_0) = \exp(-\frac{1}{a(x)})$ for $|x|<1$, where a is an asymptotically (Ω,b) quasi-homogeneous function. Define*

$$g(\varepsilon) = \frac{1}{b^{-1}(\frac{-1}{\log\varepsilon^2})}.$$

Let v_ε be a solution of (4) around x_0 given in Theorem 1. Then, there exists a least energy solution w of (7) such that for $d>0$, a re-scaled function $(\varepsilon g(\varepsilon))^{-\frac{2}{p-1}}v_\varepsilon(\frac{x}{g(\varepsilon)})$ converges to $\overline{w}$ uniformly on $\{x\in\mathbb{R}^N \,|\, dist(x,\partial\Omega)\ge d\}$ as $\varepsilon\to 0$, where

$$\overline{w}(x) = \begin{cases} w(x) & \text{for } x\in\Omega,\\ 0 & \text{for } x\notin\Omega\end{cases}$$

Moreover, the following energy estimate holds

$$\lim_{\varepsilon\to 0}(\varepsilon g(\varepsilon))^{-\frac{2(p+1)}{(p-1)}}g(\varepsilon)^{-N}\Gamma_\varepsilon(v_\varepsilon) = \int_{\mathbb{R}^N}\frac{1}{2}|\nabla w|^2 - \frac{1}{p+1}w^{p+1}dx.$$

Here we observe a new phenomenon that is a highly contrast to that for the case $\min_{x\in\mathbb{R}^N}V(x) > E$ when V is exponentially flat near an isolated zero. For example, we consider $V(x) = \exp(-\frac{1}{a(x)})$ for $|x|<1$, where a is an asymptotically (Ω,b) quasi-homogeneous function. Assume that $V(x_1,x_2,\cdots,x_n) = V(-x_1,x_2,\cdots,x_n)$, and that

$$\Omega \equiv B((-1+\lambda,0,\cdots,0),1)\cup B((1-\lambda,0,\cdots,0),1).$$

For sufficiently small $\lambda>0$, a least energy solution w of (6) in the class of even functions $\{v(x_1,x_2\cdots,x_n) = v(-x_1,x_2,\cdots,x_n)\}$ has exactly two maximum points. Then, from Theorem 7, we see that a least energy solution of (3) in the class of even functions has at least two local maximum points $h_\varepsilon^1, h_\varepsilon^2$ satisfying $\lim_{\varepsilon\to 0}|h_\varepsilon^1 - h_\varepsilon^2| =$

0 It is well known that such kind of phenomenon does not occur around a local minimum point if $\min_{x\in\mathbb{R}^N} V(x) > E$.

Existence of standing waves with a critical frequency for general nonlinearity

The proof of Theorem 1 (see [BW1]) depends on the homogeneous property of a nonlinear function $f(u) = |u|^{p-1}u$. On the other hand, when $\inf V > E$, via a variational approach, the existence of standing waves have been obtained for quite a general f (see [R],[DF1][G] and [JT] and references therein). In the case that $\inf V > E$, a restriction on the growth of f is inevitable. On the other hand, in the case that $\inf V = E$, we obtained the existence of standing waves with a critical frequency concentrating around global minimum points of V without any restriction on the growth of f at ∞. Instead od the growth condition for u large, we need a condition on the decay rate of f at 0, and we construct localized solutions concentrating around local, not necessarily global, minimum points of V.

More precisely, we assume that there are a positive number $m > 0$, a compact set A and an open bounded set $U \supset A$ satisfying

$$m \equiv V(x) < V(y), x \in A, y \in \partial U,$$

and that a continuous function $f : \mathbb{R} \to \mathbb{R}^+$ satisfies the following conditions depending upon $m, \mathcal{Z}$ and A. Recall that $\mathcal{Z} = \{x \mid V(x) = 0\}$. For any set $B \subset \mathbb{R}^N$ and $\delta > 0$, we define $B^\delta \equiv \{x \in \mathbb{R}^N | \mathrm{dist(x, B)} \leq \delta\}$.

(f1) $\lim_{t\to 0} f(t)/t = 0$ if $\mathcal{Z}\cap(\mathbb{R}^N\setminus A) = \emptyset$, and $\limsup_{t\to 0} f(t)/t^{1+\mu} < \infty$ for some $\mu > 0$ if $\mathcal{Z} \cap (\mathbb{R}^N \setminus A) \neq \emptyset$.

(f2) There exists some $p \in (1, \infty)$ for $N = 2$ and $p \in (1, (N+2)/(N-2))$ for $N \geq 3$ such that $\limsup_{t\to\infty} f(t)/t^p < \infty$ if $m > 0$, and $\liminf_{t\to 0} \frac{1}{t^{p+1}} \int_0^t f(s)ds > 0$ if $m = 0$.

(f3) $f(t) = 0$ for $t \leq 0$, and there exists $q > 2$ such that $0 < qF(t) \equiv q\int_0^t f(s)ds \leq f(t)t$ for $t > 0$.

Theorem 8 ([BW2]) *Let $N \geq 2$. Suppose that (V1-3) and (f1-3) hold. Then for sufficiently small $\varepsilon > 0$, there exists a positive solution v_ε of (4) such that for any $\delta > 0$, there exist $c, C > 0$ satisfying*

(i) $v_\varepsilon(x) \leq C\exp(-c\,\mathrm{dist(x, A^\delta)}/\varepsilon)$;

(ii) if $m > 0$, there exists a unique maximum point x_ε of v_ε such that $\lim_{\varepsilon\to 0} dist(x_\varepsilon, A) = 0$, and that $w_\varepsilon(x) \equiv v_\varepsilon(\varepsilon(x - x_\varepsilon))$ converges(up to a subsequence) uniformly to a positive, least energy solution of (3) with $a = m$;

(iii) if $m = 0$, it holds that $\lim_{\varepsilon\to 0} \|v_\varepsilon\|_{L^\infty} = 0$ and $\lim_{\varepsilon\to 0} \varepsilon^{-\frac{2}{\mu}} \|v_\varepsilon\|_{L^\infty} > 0$, where $\mu > 0$ is in (f1).

The proof is based on a new variational setting, using both the idea of penalization and the idea of localization. While the idea of localization has been used in proving Theorem 1 for a minimization problem with constraints, we develop further the idea for the free functional, since in general a minimization argument does not work for general nonlinearity f. We refer the details to [BW2].

References

[ABC] A. Ambrosetti, M. Badiale and S. Cingolani, Semiclassical states of nonlinear Schrödinger equations, Arch. Rational Mech. Anal. 140, (1997), 285–300.

[B] J. Byeon, Standing waves for nonlinear Schrödinger equations with a radial potential, Nonlinear Analysis,TMA, 50 (2002),1135–1151.

[BW1] J. Byeon and Z.-Q. Wang, Standing waves with a critical frequency for nonlinear Schrödinger equations, To appear in Arch. Rational Mech. Anal.

[BW2] J. Byeon and Z.-Q. Wang, Standing waves with a critical frequency for nonlinear Schrödinger equations, II., To appear in Calculus of Variations and PDE.

[DF1] M. del Pino and P.L. Felmer, Local mountain passes for semilinear elliptic problems in unbounded domains, Calculus of Variations and PDE, 4(1996), 121–137.

[DF2] M. del Pino and P.L. Felmer, Semi-classical states for nonlinear Schrödinger equations, J. Functional Analysis, 149(1997), 245–265,

[DF3] M. del Pino and P.L. Felmer, Multi-peak bound states for nonlinear Schrödinger equations, Ann. Inst. Henri Poincaré, 15 (1998), 127–149,

[DY] E.N. Dancer and S. Yan, On the existence of multipeak solutions for nonlinear field equations on $\mathbf{R}^N$, Discrete Contin. Dynam. Systems, 6 (2000), 39–50,

[FW] A. Floer and A. Weinstein, Nonspreading wave packets for the cubic Schrödinger equations with a bounded potential, J. Functional Analysis, 69 (1986), 397–408.

[G] C. Gui, Existence of multi-bump solutions for nonlinear Schrödinger equations via variational method, Comm. Partial Differential Equations, 21 (1996), 787–820.

[JT] L. Jeanjean and K. Tanaka, Singularly perturbed elliptic problems with superlinear or asymptotically linear nonlinearities, preprint.

[KW] X. Kang and J. Wei, On interacting bumps of semi-classical states of nonlinear Schrödinger equations, Adv. Differential Equations, 5 (2000), 899–928.

[L2] Y.Y. Li, On a singularly perturbed elliptic equation, Adv. Differential Equations, 2 (1997), 955–980.

[Me] P. Meystre, Atom Optics, Springer, (2001).

[Mi] D.L. Mills, Nonlinear optics, Springer, (1998).

[O1] Y.G. Oh, Existence of semiclassical bound states of nonlinear Schrödinger equations with potentials of the class $(V)_a$, Comm. Partial Differential Equations, 13 (1988), 1499–1519.

[O2] Y.G. Oh, Correction to: Existence of semiclassical bound states of nonlinear Schrödinger equations with potentials of the class $(V)_a$, Comm. Partial Differential Equations, 14 (1989),833–834.

[O3] Y.G. Oh, On positive multi-lump bound states of nonlinear Schrödinger equations under multiple well potential, Comm. Math. Phys., 131 (1990), 223–253.

[R] P.H. Rabinowitz, On a class of nonlinear Schrödinger equations, Z. Angew. Math. Phys., 43 (1992), 270–291.

[W] X. Wang, On concentration of positive bound states of nonlinear Schröinger equations, Comm. Math. Phys., 153 (1993), 229–244.

[Wa] Z.-Q. Wang, Existence and symmetry of multi-bump solutions for nonlinear Schrödinger equations, J. Differential Equations, 159 (1999), 102–137.

Infinitely many positive energy solutions for semilinear elliptic equations with concave and convex nonlinearity

Daoming Cao and Pigong Han

Institute of Applied Mathematics, AMSS, Chinese Academy of Sciences, Beijing 100080, P.R.of China

ABSTRACT. In this paper, we study Dirichlet problem for a class of semilinear elliptic equations $-\Delta u+\lambda u=|u|^{2^*-2}u+\mu|u|^{q-2}u$ in Ω, $u\in H_0^1(\Omega)$, where $\Omega\subset R^N$ is a bounded domain, $N\geq 3$, $2^*=\frac{2N}{N-2}$, $1<q<2$, $\mu>0$, $\lambda\in(-\lambda_1(\Omega),0]$, $\lambda_1(\Omega)$ is the first eigenvalue of the Laplace operator $-\Delta$ with Dirichlet boundary conditions. By using variational method, we prove the existence of infinitely many solutions with positive energy for small $\mu>0$, when Ω exhibits suitable symmetry.

1. Introduction ans main results

Let Ω is a bounded domain in R^N, $N\geq 3$, $2^*=\frac{2N}{N-2}$ is the critical exponent for the Sobolev embedding $H_0^1(\Omega)\hookrightarrow L^{2^*}(\Omega)$.

In this paper, we consider the following semilinear elliptic problem

$$\begin{cases} -\Delta u+\lambda u=|u|^{2^*-2}u+\mu|u|^{q-2}u, & x\in\Omega,\\ u\in H_0^1(\Omega), \end{cases} \tag{1.1}$$

where $\lambda\in(-\lambda_1(\Omega),0]$, $\lambda_1(\Omega)$ is the first eigenvalue of the Laplace operator $-\Delta$ with zero Dirichlet boundary condition, $1<q<2$, $\mu>0$.

$u\in H_0^1(\Omega)$ is said to be a solution of (1.1) if u satisfies

$$\int_\Omega(\nabla u\cdot\nabla v+\lambda uv-|u|^{2^*-2}uv-\mu|u|^{q-2}uv)dx=0,\quad \forall v\in H_0^1(\Omega).$$

Key words and phrases. existence of solutions, semilinear elliptic equation, critical point.

Supported by Special Funds For Major States Basic Research Projects of China (G1999075107) and Knowledge Innovation Funds of CAS in China.

Hence the nontrivial solutions of (1.1) are equivalent to the nonzero critical points of the energy functional

$$J(u) = \frac{1}{2}\int_\Omega (|\nabla u|^2 + \lambda u^2)dx - \frac{1}{2^*}\int_\Omega |u|^{2^*}dx - \frac{\mu}{q}\int_\Omega |u|^q dx, \ \forall u \in H_0^1(\Omega). \quad (1.2)$$

Note that 2^* is critical Sobolev exponent, the embedding $H_0^1(\Omega) \hookrightarrow L^{2^*}(\Omega)$ is not compact, which leads to that J does not satisfy the Palais-Smale (P.S. for short) condition. This results in difficulty in obtaining existence of solutions, especially for existence of infinitely many solutions with positive energy.

In the last twenty years, Dirichlet problems for semilinear equations have been studied extensively. By variant methods, for example, upper-lower solution method, variational method, and degree theory etc, people have obtained many important and interesting results.

The Dirichlet problem with concave and convex nonlinearity

$$\begin{cases} -\Delta u = |u|^{p-2}u + \mu|u|^{q-2}u, & x \in \Omega, \\ u \in H_0^1(\Omega), \end{cases} \quad (1.3)$$

has been treated by A.Ambrosetti, H.Brezis and G.Cerami [1]. Let

$$I(u) = \frac{1}{2}\int_\Omega |\nabla u|^2 dx - \frac{1}{2^*}\int_\Omega |u|^{2^*}dx - \frac{\mu}{q}\int_\Omega |u|^q dx, \ \forall u \in H_0^1(\Omega),$$

the energy functional corresponding to (1.3). The main results in [1] are the following (where Ω is a general bounded domain).

1. For all $1 < q < 2 < p$, there exists $\Lambda > 0$ such that , for all $\mu \in (0, \Lambda)$, problem (1.3) has a minimal solution u_μ satisfying $I(u_\mu) < 0$.
2. Let $1 < q < 2 < p \leq 2^*$, then, for all $\mu \in (0, \Lambda)$, problem (1.3) has a second solution $v_\mu > u_\mu$.
3. Let $1 < q < 2 < p \leq 2^*$, then, there exists $\mu^* > 0$ such that for all $\mu \in (0, \mu^*)$, problem (1.3) has infinitely many solutions such that $I(u) < 0$.
4. Let $1 < q < 2 < p < 2^*$, then, for all $\mu \in (0, \mu^*)$, problem (1.3) has also infinitely many solutions such that $I(u) > 0$.

In the last section of [1], some open problems are proposed. The second one is that whether problem (1.3) has infinitely many solutions with positive energy, when $p = 2^*$, for $\mu > 0$ small enough. In this paper, for some special domain Ω, we give the open problem an affirmative (positive) answer (see Theorem 1.1).

Following the argument of [1][4-5] , it is not difficult to obtain infinitely many solutions for (1.1) with negative energy (that is, $J(u) < 0$) since $J(u)$ satisfies $(P.S.)_c$ condition for $c < 0$. However, due to the fact that $J(u)$ satisfies $(P.S.)_c$ condition only for $c \in (-\infty, \frac{1}{N}S^{\frac{N}{2}} - k_0\mu^{\frac{2^*}{2^*-q}})$ (see Lemma 2.1), it is not easy to obtain solutions with positive energy and even difficult to obtain existence of infinitely many solutions.

Assuming that the domain Ω satisfies certain symmetry, D. Fortunato and E. Jannelli [10] considered problem (1.1) with $\mu = 0$, and obtained infinitely many solutions.

Since the lack of compactness and that we are interested only in solutions with positive energy, fountain theorem and its dual theorem (see [4-5][14]) are not applicable to the case in this paper. We take some ideas from [1][7][10], and by variational method, obtain the existence of infinitely many different solutions u with positive energy, that is, $J(u) > 0$.

To state our results, as in [10], we need to introduce some notations.

Definition 1.1 The domain Ω is said to be a cylinder in R^N if there exists an open bounded subset ω in R^{N-1} and $a\ , b \in R$ with $a < b$ such that $\Omega = \omega \times (a,\ b)$.

Let V be a 2-dimensional subspace of R^N and $V^\perp$ be its orthogonal complement, moreover, let $\sum_V$ be the group of the rotations of V, and any $\sigma \in \sum_V$ induces a map $\lambda_\sigma :\ R^N \longrightarrow R^N$ defined by

$$\lambda_\sigma(y_1\ , y_2) = (\sigma(y_1),\ y_2), \quad \forall y_1 \in V\ , y_2 \in V^\perp.$$

With these notations, we give the following definition

Definition 1.2 We say that domain Ω in R^N has a rotational symmetry, if there exists a 2-dimensional subspace V of R^N such that $\lambda_\sigma(\Omega) = \Omega$ for any $\sigma \in \sum_V$.

Obviously the unit ball in R^N is a domain with rotational symmetry. Without loss of generality, we always assume $a = 0$, $b = \pi$ in Definition 1.1. $\forall x \in \Omega$, set $x = (x'\ , t)$, $x' \in \omega$, $t \in (0,\ \pi)$ and $\Delta = \Delta_{x'} + \frac{\partial^2}{\partial t^2}$, where $\Delta_{x'} = \sum_{i=1}^{N-1} \frac{\partial^2}{\partial x_i^2}$. We denote by $\mu_j\ , j = 1,\ 2, \cdots (\mu_1 < \mu_2 \leq \cdots)$ the eigenvalues of the operator $-\Delta_{x'}$: $H_0^1(\omega) \longrightarrow H^{-1}(\omega)$ and by $e_j(x')$ the corresponding normalized eigenfunctions.

The eigenvalues of the operator $-\Delta : H_0^1(\Omega) \longrightarrow H^{-1}(\Omega)$ and the corresponding normalized eigenfunctions are $\lambda_{jk} = \mu_j + k^2$, $j,\ k = 1,\ 2, \cdots$, and $e_{jk}(x'\ , t) = e_j(x') \sin kt$, $(x'\ , t) \in \omega \times (0,\ \pi)$, respectively.

Set $V_m = \{u \in H_0^1(\Omega) \mid u_{jk} = 0 \text{ if } \frac{k}{m} \notin \mathbb{N}\}$, $m \in \mathbb{N}$, where u_{jk} $(j,\ k \in \mathbb{N})$ denote the Fourier coefficients of u along e_{jk}. It is easy to verify that V_m is a subspace of $H_0^1(\Omega)$. We denote by $J|_{V_m}$ the restriction of J to V_m and by $J'|_{V_m}$ its *Fréchêt* derivative. Obviously, $J|_{V_m} \in C^1(V_m,\ \mathbb{R})$, where the functional J is defined in (1.2).

Let

$$\Gamma_m = \{\gamma \in C([0,1],\ V_m) |\ \gamma(0) = 0,\ J|_{V_m}(\gamma(0)) < 0\},$$

$$c_m := \inf_{\gamma \in \Gamma_m} \max_{t \in [0,1]} J|_{V_m}(\gamma(t)).$$

In addition, we also use the following observation of extension (see Appendix in [10]):

A smooth function on $[0,\ \pi]$ satisfying $h(0) = h(\pi) = 0$, has an expansion in terms of $\{\sin kmt\}_{k=1}^{\infty}$ if and only if

$$\begin{cases} h(t+\frac{2\pi}{m}) = h(t), & 0 \leq t \leq \pi - \frac{2\pi}{m}, \\ h(\frac{\pi}{m}+t) = -h(\frac{\pi}{m}-t), & 0 \leq t \leq \frac{\pi}{m}. \end{cases} \tag{1.4}$$

We will use the following variant of the mountain pass theorem of Ambrosetti and Rabinowitz [2] without the (P.S) condition. i.e.,

Let J be a C^1 functional on a Banach space X. Suppose there exists a neighbourhood U of 0 in X and a constant α such that $J(u) \geq \alpha$ for every u in the boundary of U. $J(0) < \alpha$ and $J(v) < \alpha$ for some $v \notin U$. Set $c = \inf_{\gamma \in \Gamma} \max_{t \in [0,1]} J(\gamma(t)) \geq \alpha$, where Γ denotes the class of continuous paths jointing 0 and v. Then there exists a sequence (u_k) in X such that as $k \to \infty$

$$J(u_k) \to c\ ,\ \ J'(u_k) \to 0\ \text{ in } X^*.$$

The functional $J \in C^1(X, \mathbb{R})$ is said to satisfy the $(P.S)_c$ condition if any sequence $\{u_k\} \subset X$ such that as $k \to \infty$

$$J(u_k) \to c\ ,\ \ J'(u_k) \to 0\ \text{ strongly in } X^*,$$

contains a subsequence converging in X to a critical point of J.

Our main result is the following

Theorem 1.1 *Let $N \geq 4$, $\lambda \in (-\lambda_1(\Omega),\ 0)$, $q \in (1,\ 2)$ or $N \geq 6$, $\lambda = 0$, $q \in (\frac{N^2}{(N-1)(N-2)},\ 2)$, and suppose that Ω is either a cylinder or a domain with rotational symmetry in the sense of Definition 1.2. Then, there exists $\mu^* > 0$ such that for all $0 < \mu < \mu^*$, problem (1.1) has infinitely many different solutions with positive energy.*

As in [7], the case $N = 3$ is much more delicate. Nevertheless, if Ω is a unit ball centered at the origin, we have

Theorem 1.2 *Let $N = 3$, $\lambda \in (-\lambda_1(\Omega),\ -\frac{1}{4}\lambda_1(\Omega))$, and Ω be a unit ball centered at the origin. Then, there exists $\mu^{**} > 0$ such that for all $0 < \mu < \mu^{**}$, problem (1.1) has infinitely many different solutions with positive energy.*

For simplicity, we denote by

$$\|u\| = \left(\int_\Omega |\nabla u|^2 dx\right)^{\frac{1}{2}},\quad |u|_p = \left(\int_\Omega |u|^p dx\right)^{\frac{1}{p}},\ \forall u \in H_0^1(\Omega)$$

the norms in $H_0^1(\Omega)$, $L^p(\Omega)$, $p \geq 1$, respectively. And C denote (possibly different) positive constants.

2. Proofs of Theorems 1.1 and 1.2

Let $S := \inf\{\|u\|^2/|u|_{2^*}^2 \mid u \in H_0^1(\Omega)\backslash\{0\}\}$ and $S_m := \inf\{\|u\|^2/|u|_{2^*}^2 \mid u \in V_m\backslash\{0\}\}$ denote the best constants for the imbedding

$$H_0^1(\Omega) \hookrightarrow L^{2^*}(\Omega),\ \text{ and } V_m \hookrightarrow L^{2^*}(\Omega),$$

respectively.

Then we have (see [10])

$$S_m = m^{\frac{2}{N}} S, \quad m = 1,\ 2, \cdots . \tag{2.1}$$

It is easy to verify that there exists $k_0 > 0$ independent of $\mu > 0$ such that

$$\min_{t>0}\left(\frac{1}{N}t^{2^*} - \mu\left(\frac{1}{q} - \frac{1}{2}\right)|\Omega|^{1-\frac{q}{2^*}} t^q\right) = -k_0 \mu^{\frac{2^*}{2^*-q}}. \tag{2.2}$$

Lemma 2.1 *For the k_0 defined in (2.2) and c satisfying*

$$-\infty < c < \frac{1}{N} S_m^{\frac{N}{2}} - k_0 \mu^{\frac{2^*}{2^*-q}}, \tag{2.3}$$

the functional $J|_{V_m}$ satisfies $(P.S.)_c$ condition for all $\lambda \in (-\lambda_1(\Omega),\ 0)$.

Proof Assume $\{u_k\} \subset V_m$ is a sequence such that as $k \to \infty$

$$J|_{V_m}(u_k) \to c\ , \quad J'|_{V_m}(u_k) \to 0 \ \text{ strongly in } X^*,$$

that is,

$$\frac{1}{2}\int_\Omega (|\nabla u_k|^2 + \lambda |u_k|^2) dx - \frac{1}{2^*}\int_\Omega |u_k|^{2^*} dx - \frac{\mu}{q}\int_\Omega |u_k|^q dx = c + o(1), \tag{2.4}$$

$$\int_\Omega (\nabla u_k \cdot \nabla v + \lambda u_k v) dx - \int_\Omega |u_k|^{2^*-2} u_k v dx - \mu \int_\Omega |u_k|^{q-2} u_k v dx = o(\|v\|), \quad \forall v \in V_m. \tag{2.5}$$

Let $v = u_k$ in (2.5), and we have

$$\int_\Omega (|\nabla u_k|^2 + \lambda |u_k|^2) dx - \int_\Omega |u_k|^{2^*} dx - \mu \int_\Omega |u_k|^q dx = o(\|u_k\|). \tag{2.6}$$

From (2.4) and (2.6), we obtain

$$\left(\frac{1}{2} - \frac{1}{2^*}\right)\int_\Omega (|\nabla u_k|^2 + \lambda |u_k|^2) dx - \mu\left(\frac{1}{q} - \frac{1}{2^*}\right)\int_\Omega |u_k|^q dx \le C(1 + \|u_k\|).$$

Note that $1 < q < 2$ and $\lambda \in (-\lambda_1(\Omega),\ 0)$, it therefore follows that $\|u_k\| \le C$. By choosing a subsequence if necessary, we may assume that, as $k \longrightarrow \infty$

$$\begin{aligned}
&u_k \rightharpoonup u \ \text{ weakly in } \ V_m,\\
&u_k \rightharpoonup u \ \text{ weakly in } \ L^{2^*}(\Omega),\\
&u_k \to u \ \text{ strongly in } \ L^q(\Omega),\\
&u_k \to u \ \text{ strongly in } \ L^s(\Omega) \text{ for } 1 < s < 2^*,\\
&u_k \to u \ \text{ a.e on } \ \Omega.
\end{aligned}$$

It follows from (2.5) that u satisfies

$$\int_\Omega (\nabla u \cdot \nabla v + \lambda u v) dx - \int_\Omega |u|^{2^*-2} u v dx - \mu \int_\Omega |u|^{q-2} u v dx = 0, \quad \forall v \in V_m. \tag{2.7}$$

and

$$J|_{V_m}(u) = (\frac{1}{2} - \frac{1}{2^*})\int_\Omega |\nabla u|^2 dx - \mu(\frac{1}{q} - \frac{1}{2})\int_\Omega |u|^q dx. \tag{2.8}$$

Set $v_k = u_k - u$. Brezis-Lieb lemma [6] leads to

$$\int_\Omega |u_k|^{2^*} dx = \int_\Omega |v_k|^{2^*} dx + \int_\Omega |u|^{2^*} dx + o(1).$$

Obviously,

$$\int_\Omega |\nabla u_k|^2 dx = \int_\Omega |\nabla v_k|^2 dx + \int_\Omega |\nabla u|^2 dx + o(1).$$

Hence we obtain

$$J|_{V_m}(u) + \frac{1}{2}|\nabla v_k|_2^2 - \frac{1}{2^*}|v_k|_{2^*}^{2^*} = c + o(1). \tag{2.9}$$

Since $\langle J'|_{V_m}(u_k), u_k\rangle \to 0$ as $k \to \infty$, we obtain as $k \longrightarrow \infty$

$$||v_k||^2 - |v_k|_{2^*}^{2^*} \longrightarrow -||u||^2 - \lambda|u|_2^2 + |u|_{2^*}^{2^*} + \mu|u|_q^q = -\langle J'|_{V_m}(u), u\rangle = 0.$$

We may therefore assume that, as $k \longrightarrow \infty$

$$||v_k||^2 \to b, \quad |v_k|_{2^*}^{2^*} \to b,$$

where b is a nonnegative constant.

By Sobolev inequality, we have $||v_k||^2 \geq S_m|v_k|_{2^*}^2$ and so $b \geq S_m b^{\frac{2}{2^*}}$. If $b = 0$, the proof is complete. Assume $b \geq S_m^{\frac{N}{2}}$, from (2.8) and (2.9), we have

$$\begin{aligned} c &= J|_{V_m}(u) + (\frac{1}{2} - \frac{1}{2^*})b \\ &= (\frac{1}{2} - \frac{1}{2^*})(b + |u|_{2^*}^{2^*}) - \mu(\frac{1}{q} - \frac{1}{2})|u|_q^q \\ &\geq \frac{1}{N}S_m^{\frac{N}{2}} + \frac{1}{N}|u|_{2^*}^{2^*} - \mu(\frac{1}{q} - \frac{1}{2})|\Omega|^{1-\frac{q}{2^*}}|u|_{2^*}^q \\ &\geq \frac{1}{N}S_m^{\frac{N}{2}} - k_0\mu^{\frac{2^*}{2^*-q}}. \end{aligned}$$

We obtain a contradiction since $c < \frac{1}{N}S_m^{\frac{N}{2}} - k_0\mu^{\frac{2^*}{2^*-q}}$.

Lemma 2.2 *Let $N \geq 4$, $\lambda \in (-\lambda_1(\Omega), 0)$. There exists $\mu^* > 0$ such that for any $0 < \mu < \mu^*$, there is a nonnegative function $v \in H_0^1(\Omega)\backslash\{0\}$ satisfying*

$$\sup_{t \geq 0} J(tv) < \frac{1}{N}S^{\frac{N}{2}} - k_0\mu^{\frac{2^*}{2^*-q}}. \tag{2.10}$$

If $N \geq 6$, $\lambda = 0$, then there exists $\mu_* > 0$ such that for any $\mu \in (0, \mu_*)$ and $\frac{N^2}{(N-1)(N-2)} < q < 2$, we also have a nonnegative function $v \in H_0^1(\Omega)\backslash\{0\}$ such that (2.10) holds.

Proof Let $\psi(x) \in C_0^\infty(\Omega)$ is a cut-off function which satisfies $\psi \equiv 1$ in a neighbourhood of x_0, $x_0 \in \Omega$ and define, for $\epsilon > 0$, $u_\epsilon(x) = \frac{\psi(x)}{(\epsilon + |x - x_0|^2)^{\frac{N-2}{2}}}$.

From [7], we have for $N \geq 4$

$$|\nabla u_\epsilon|_2^2 = K_1 \epsilon^{-\frac{N-2}{2}} + O(1),$$

$$|u_\epsilon|_{2^*}^2 = K_2 \epsilon^{-\frac{N-2}{2}} + O(\epsilon),$$

$$|u_\epsilon|_2^2 = \begin{cases} K_3 |\log \epsilon| + O(1), & N = 4, \\ K_3 \epsilon^{-\frac{N-4}{2}} + O(1), & N \geq 5, \end{cases}$$

for $N \geq 5$,

$$|u_\epsilon|_q^q = \begin{cases} O(\epsilon^{\frac{N-(N-2)q}{2}}), & \frac{N}{N-2} < q < 2, \\ O(|\log \epsilon|), & q = \frac{N}{N-2}, \\ O(1), & 1 < q < \frac{N}{N-2}, \end{cases}$$

where K_1, K_2 and K_3 denote positive constants which depend only on N and $K_1/K_2 = S$.

Assume the maximum of $J(tu_\epsilon)$ is achieved at a point t_ϵ.i.e.,

$$\begin{aligned} \sup_{t \geq 0} J(tu_\epsilon) &= J(t_\epsilon u_\epsilon) \\ &= \frac{t_\epsilon^2}{2} \int_\Omega (|\nabla u_\epsilon|^2 + \lambda |u_\epsilon|^2) dx - \frac{t_\epsilon^{2^*}}{2^*} \int_\Omega |u_\epsilon|^{2^*} dx - \frac{\mu t_\epsilon^q}{q} \int_\Omega |u_\epsilon|^q dx \\ &\leq \frac{t_\epsilon^2}{2} \int_\Omega (|\nabla u_\epsilon|^2 + \lambda |u_\epsilon|^2) dx - \frac{t_\epsilon^{2^*}}{2^*} \int_\Omega |u_\epsilon|^{2^*} dx \\ &\leq \sup_{t \geq 0} \left(\frac{t^2}{2} \int_\Omega (|\nabla u_\epsilon|^2 + \lambda |u_\epsilon|^2) dx - \frac{t^{2^*}}{2^*} \int_\Omega |u_\epsilon|^{2^*} dx \right) \\ &= \frac{1}{N} \left(\frac{|\nabla u_\epsilon|_2^2 + \lambda |u_\epsilon|_2^2}{|u_\epsilon|_{2^*}^2} \right)^{\frac{N}{2}}. \end{aligned} \tag{2.11}$$

For $N \geq 5$,

$$\begin{aligned} \sup_{t \geq 0} J(tu_\epsilon) &\leq \frac{1}{N} \left(\frac{K_1 \epsilon^{-\frac{N-2}{2}} + \lambda K_3 \epsilon^{-\frac{N-4}{2}} + O(1)}{K_2 \epsilon^{-\frac{N-2}{2}} + O(\epsilon)} \right)^{\frac{N}{2}} \\ &= \frac{1}{N} S^{\frac{N}{2}} + \frac{1}{2} \lambda S^{\frac{N}{2}} K_1^{-1} K_3 \epsilon + O(\epsilon^{\frac{N-2}{2}}). \end{aligned} \tag{2.12}$$

Hence, from (2.12), we may choose $\epsilon_1 > 0$ such that

$$\sup_{t \geq 0} J(tu_\epsilon) < \frac{1}{N} S^{\frac{N}{2}} + \frac{1}{4} \lambda S^{\frac{N}{2}} K_1^{-1} K_3 \epsilon, \quad \text{for } 0 < \epsilon < \epsilon_1.$$

It remains only to verify that

$$\frac{1}{N} S^{\frac{N}{2}} + \frac{1}{4} \lambda S^{\frac{N}{2}} K_1^{-1} K_3 \epsilon < \frac{1}{N} S^{\frac{N}{2}} - k_0 \mu^{\frac{2^*}{2^*-q}}, \quad \text{for } 0 < \epsilon < \epsilon_1. \tag{2.13}$$

Obviously, there exists a small positive constant μ_1 such that for any $0 < \mu < \mu_1$, (2.13) holds.

Similarly, for $N = 4$, there exists $\epsilon_2 > 0,\ \mu_2 > 0$ such that for any $0 < \mu < \mu_2$

$$\sup_{t\geq 0} J(tu_\epsilon) \leq \frac{1}{N}S^{\frac{N}{2}} + \frac{1}{4}\lambda S^{\frac{N}{2}}K_1^{-1}K_3\epsilon^{\frac{N-2}{2}}|\log\epsilon| < \frac{1}{N}S^{\frac{N}{2}} - k_0\mu^{\frac{2^*}{2^*-q}},\quad \text{for } 0<\epsilon<\epsilon_2. \tag{2.14}$$

Set $\epsilon_0 = \min\{\epsilon_1,\ \epsilon_2\} > 0,\ \mu^* = \min\{\mu_1,\ \mu_2\} > 0$, then for any $0 < \mu < \mu^*$, we have for $N \geq 4$

$$\sup_{t\geq 0} J(tu_\epsilon) < \frac{1}{N}S^{\frac{N}{2}} - k_0\mu^{\frac{2^*}{2^*-q}},\quad \text{for } 0<\epsilon<\epsilon_0.$$

If $N \geq 6$, $\lambda = 0$. Set $v_\epsilon = \frac{u_\epsilon}{|u_\epsilon|_{2^*}}$, then $|\nabla v_\epsilon|_2^2 = S + O(\epsilon^{\frac{N-2}{2}})$. Assume the maximum of $J(tu_\epsilon)$ is achieved at a point t_ϵ.i.e.,

$$\begin{aligned}\sup_{t\geq 0} J(tv_\epsilon) &= J(t_\epsilon v_\epsilon)\\ &= \frac{t_\epsilon^2}{2}\int_\Omega |\nabla v_\epsilon|^2 dx - \frac{t_\epsilon^{2^*}}{2^*} - \frac{\mu t_\epsilon^q}{q}\int_\Omega |v_\epsilon|^q dx.\end{aligned} \tag{2.15}$$

Since the derivative of the function $t \mapsto J(tv_\epsilon)$ vanishes at $t = t_\epsilon$, we have

$$t_\epsilon|\nabla v_\epsilon|_2^2 - t_\epsilon^{2^*-1} - \mu t_\epsilon^{q-1}|v_\epsilon|_q^q = 0, \tag{2.16}$$

and therefore

$$t_\epsilon \leq |\nabla v_\epsilon|^{\frac{2}{2^*-2}}. \tag{2.17}$$

Now we claim as $\epsilon \longrightarrow 0$

$$t_\epsilon \longrightarrow S^{\frac{1}{2^*-2}}. \tag{2.18}$$

First we prove that t_ϵ is bounded below. Otherwise if $t_\epsilon \longrightarrow 0$, as $\epsilon \longrightarrow 0$.

It is easy to verify that $|v_\epsilon|_q^q = \frac{|u_\epsilon|_q^q}{|u_\epsilon|_{2^*}^q} \longrightarrow 0$, as $\epsilon \longrightarrow 0$. Therefore we have $J(t_\epsilon v_\epsilon) \longrightarrow 0$, as $\epsilon \longrightarrow 0$.

Set $t_0 = \left(\frac{NS}{N-2}\right)^{\frac{1}{2^*-2}} > 0$, then there exists $\alpha > 0$ such that $\alpha < \lim\limits_{\epsilon\longrightarrow 0} J(t_0 v_\epsilon) \leq \lim\limits_{\epsilon\longrightarrow 0} J(t_\epsilon v_\epsilon) = 0$.

We obtain a contradiction. Hence we may assume $t_\epsilon \geq \beta > 0$, and then (2.18) follows from (2.16) immediately.

On the other hand, it is easy to check that $|v_\epsilon|_q^q \geq C\epsilon^{\frac{N-(N-2)q}{2}},\quad 1 < q < 2$.
Therefore, from (2.15), we have

$$\begin{aligned}\sup_{t\geq 0} J(tv_\epsilon) &\leq \max_{t\geq 0}\left(\frac{t^2}{2}\int_\Omega |\nabla v_\epsilon|^2 dx - \frac{t^{2^*}}{2^*}\right) - \frac{\mu t_\epsilon^q}{q}\int_\Omega |v_\epsilon|^q dx\\ &\leq \frac{1}{N}S^{\frac{N}{2}} + O(\epsilon^{\frac{N-2}{2}}) - \frac{\mu\beta^q C_1}{q}\epsilon^{\frac{N-(N-2)q}{2}}.\end{aligned}$$

Now we only need to verify that

$$\frac{1}{N}S^{\frac{N}{2}} + O(\epsilon^{\frac{N-2}{2}}) - \frac{\mu\beta^q C_1}{q}\epsilon^{\frac{N-(N-2)q}{2}} < \frac{1}{N}S^{\frac{N}{2}} - k_0\mu^{\frac{2^*}{2^*-q}}. \tag{2.19}$$

After a detailed calculation, we obtain (2.19) only for $N \geq 6$, $\frac{N^2}{(N-1)(N-2)} < q < 2$ and $0 < \mu < \mu_*$, where μ_* is a small positive constant. The proof of Lemma 2.2 is complete.

Lemma 2.3 *Let $N \geq 4$, $\lambda \in (-\lambda_1(\Omega),\ 0)$, for any $\mu \in (0,\ \mu^*)$ and $1 < q < 2$ or $N \geq 6$, $\lambda = 0$, for any $\mu \in (0,\ \mu_*)$ and $\frac{N^2}{(N-1)(N-2)} < q < 2$, there exists a nonnegative function $\bar{v} \in V_m \backslash \{0\}$ satisfying*

$$\sup_{t \geq 0} J(t\bar{v}) < \frac{1}{N} S_m^{\frac{N}{2}} - k_0 \mu^{\frac{2^*}{2^*-q}}, \tag{2.19}$$

where μ^*, μ_* are given in Lemma 2.2 and S_m is defined in (2.1), $m = 1,\ 2, \cdots$.

Proof We first consider the case that Ω is a cylinder. Set $\Omega_m = \omega \times (0,\ \frac{\pi}{m})$, $x_0 \in \Omega_m$, and consider $\psi_\epsilon(x) = \frac{\psi(x)}{(\epsilon+|x-x_0|^2)^{\frac{N-2}{2}}}$, $x \in \Omega_m$. where $\psi \in C_0^\infty(\Omega_m)$ is a nonnegative function such that $\psi \equiv 1$ in some neighborhood of $x_0 \in \Omega_m$.

Since $\psi_\epsilon \in H_0^1(\Omega_m)$, we have

$$\psi_\epsilon(x',\ t) = \sum_{j,k \in \mathbb{N}} c_{jk} e_j(x') \sin kmt,\ (x',\ t) \in \Omega_m,$$

where c_{jk} are the Fourier coefficients of ψ_ϵ along $e_j(x') \sin kmt$. From (1.4), ψ_ϵ can be extended to Ω, denote this extension by $\bar{\psi}_\epsilon$. Obviously, $\bar{\psi}_\epsilon \in V_m$.

Since ψ_ϵ can be regarded also as a function in $H_0^1(\Omega)$. By Lemma 2.2, we have

$$\sup_{t \geq 0} J(t\psi_\epsilon) < \frac{1}{N} S^{\frac{N}{2}} - k_0 \mu^{\frac{2^*}{2^*-q}}, \quad \text{for } 0 < \epsilon < \epsilon_0. \tag{2.20}$$

Moreover, it is easy to check that

$$J|_{V_m}(t\bar{\psi}_\epsilon) = mJ(t\psi_\epsilon), \quad \text{for any } t \geq 0. \tag{2.21}$$

Hence from(2.20)(2.21), we obtain

$$\sup_{t \geq 0} J|_{V_m}(t\bar{\psi}_\epsilon) < \frac{1}{N} S_m^{\frac{N}{2}} - k_0 \mu^{\frac{2^*}{2^*-q}}, \quad \text{for } 0 < \epsilon < \epsilon_0.$$

Next we consider the case that Ω is an bounded domain with rotational symmetry.

Let $\{v_1,\ v_2, \cdots, v_N\}$ be the canonical base in R^N and $V = span\{v_{N-1},\ v_N\}$. Set

$$x_{N-1} = \rho \cos t, \quad x_N = \rho \sin t, \quad \rho > 0,\ t \in [0,\ 2\pi].$$

Due to the rotational symmetry of Ω, there exists an open, bounded subset of R^{N-1} such that $\Omega = \omega \times (0,\ 2\pi)$. Then a complete system of eigenfunctions of the Laplace operator $-\Delta : H_0^1(\Omega) \longrightarrow H_0^1(\Omega)$ is given by (see Appendix in [10])

$$\{e_j(x') \sin kt,\ e_j(x') \cos kt \mid (x',\ t) \in \Omega,\ j,\ k \in \mathbb{N}\}.$$

Set $V_m = \{u \in H_0^1(\Omega) \mid u_{jk} = 0 \text{ if } \frac{k}{m} \notin \mathbb{N} \text{ and } \bar{u}_{jk} = 0 \text{ for any } j,\ k \in \mathbb{N}\}$, where u_{jk}, $\bar{u}_{jk}$ are the Fourier coefficients of u along $e_j(x')\sin kt$, $e_j(x')\cos kt$, respectively. The next proof is analogous to the proof of the case that Ω is a cylinder.

Proof of Theorem 1.1 We first consider the case that Ω is a cylinder. $\forall u \in V_m$,

$$\begin{aligned} J|_{V_m}(u) &= \frac{1}{2}\int_\Omega (|\nabla u|^2 + \lambda |u|^2)dx - \frac{1}{2^*}\int_\Omega |u|^{2^*}dx - \frac{\mu}{q}\int_\Omega |u|^q dx \\ &= \frac{1}{2}\sum_{j,\,k\in\mathbb{N}} (\mu_j + k^2m^2 - \lambda)|u_{jk}|^2 - \frac{1}{2^*}|u|_{2^*}^{2^*} - \frac{\mu}{q}|u|_q^q \\ &\geq \frac{m^2}{2}\sum_{j,\,k\in\mathbb{N}} \lambda_{jk}|u_{jk}|^2 - \frac{1}{2^*}|u|_{2^*}^{2^*} - \frac{\mu^*}{q}|\Omega|^{1-\frac{q}{2^*}}|u|_{2^*}^q \\ &\geq \frac{m^2}{2}||u||^2 - \frac{1}{2^*}||u||^{2^*}S_m^{-\frac{2^*}{2}} - \frac{\mu^*}{q}|\Omega|^{1-\frac{q}{2^*}}||u||^q S_m^{-\frac{q}{2}}. \end{aligned}$$

By (2.1), S_m increases with m, hence $\forall R > 0$, there exist $m \in \mathbb{N}$, $\rho > 0$ such that

$$b := \inf_{||u||=\rho} J|_{V_m}(u) \geq R > 0 = J|_{V_m}(0). \tag{2.22}$$

In addition, as $t \longrightarrow \infty$

$$J|_{V_m}(t\bar{v}) = \frac{t^2}{2}\int_\Omega (|\nabla \bar{v}|^2 + \lambda|\bar{v}|^2)dx - \frac{t^{2^*}}{2^*}\int_\Omega |\bar{v}|^{2^*}dx - \frac{\mu t^q}{q}\int_\Omega |\bar{v}|^q dx \longrightarrow -\infty,$$

where the function $\bar{v}$ is from Lemma 2.3.

Hence, there exists $t_0 > 0$ such that $||t_0\bar{v}|| > \rho$ and $J|_{V_m}(t_0\bar{v}) < 0$. By the mountain pass theorem and Lemma 2.3, there exists a sequence $(u_k) \subset V_m$ satisfying as $k \longrightarrow \infty$

$$J|_{V_m}(u_k) \longrightarrow c_m > 0\ ,\quad J'|_{V_m}(u_k) \longrightarrow 0 \text{ strongly in } V_m^*. \tag{2.23}$$

In addition, $c_m \leq \sup_{t\in[0,1]} J(tt_0\bar{v}) \leq \sup_{t\geq 0} J(t\bar{v}) < \frac{1}{N}S_m^{\frac{N}{2}} - k_0\mu^{\frac{2^*}{2^*-q}}$, by Lemma 2.1, $\{u_k\}$ has a converging subsequence, still denoted by $\{u_k\}$, such that

$$u_k \longrightarrow u \text{ strongly in } V_m.$$

Consequently u is a critical point of $J|_{V_m}$ and c_m is a critical value of the functional $J|_{V_m}$.

Now, we prove that u solves (1.1), i.e.,

$$\langle J'(u),\ \varphi\rangle = 0, \quad \forall\varphi \in H_0^1(\Omega). \tag{2.24}$$

In addition, we write

$$\begin{aligned} \langle J'(u),\ \varphi\rangle &= \langle J'(u),\ \varphi_1\rangle + \langle J'(u),\ \varphi_2\rangle, \\ &\quad \forall\varphi \in H_0^1(\Omega),\quad \varphi = \varphi_1 + \varphi_2,\quad \varphi_1 \in V_m,\ \varphi_2 \in V_m^\perp. \end{aligned} \tag{2.25}$$

By (2.23), as $k \longrightarrow \infty$

$$\langle J'|_{V_m}(u_k),\ \varphi_1\rangle \longrightarrow \langle J'(u),\ \varphi_1\rangle = 0$$

So it is only to prove that $\langle J'(u),\ \varphi_2\rangle = 0$.

From (1.4), it is easy to verify that, if $h(t)$ has an expansion in terms of $\{\sin kmt\}_{k=1}^{\infty}$, then the same is true for $(g \circ h)(t)$, when g is any continuous odd function. By density argument, it is enough to prove

$$\langle J'(u),\ \varphi_2\rangle = 0, \quad \forall u \in C_0^{\infty}(\Omega) \cap V_m. \tag{2.26}$$

Now, let $u \in C_0^{\infty}(\Omega) \cap V_m$, then $(-\Delta u + \lambda u) \in C_0^{\infty}(\Omega) \cap V_m$, moreover, $|u|^{2^*-2}u$, $|u|^{q-2}u$ can be written as

$$|u|^{2^*-2}u = \sum_{k=1}^{\infty} \alpha_k(x') \sin kmt, \quad \text{and} \quad |u|^{q-2}u = \sum_{k=1}^{\infty} \beta_k(x') \sin kmt, \quad \text{respectively.}$$

where $\alpha_k(x')$, $\beta_k(x')$ are suitable continuous functions. And from the above argument, we obtain (2.26) immediately.

Next we consider the case that Ω is a bounded domain with rotational symmetry. We choose V_m as in Lemma 2.3, and the proof is analogous to the above. Thus the proof of Theorem 1.1 is complete due to the arbitrariness of R in (2.22).

Proof of Theorem 1.2 $\forall \epsilon > 0$, define $u_\epsilon(x) = \frac{\psi(|x|)}{(1+|x|^2)^{\frac{1}{2}}}$, where ψ is a fixed function such that $\psi(0) = 1$, $\psi'(0) = 0$, and $\psi(1) = 0$. Then the following estimates hold (see [7])

$$|\nabla u_\epsilon|_2^2 = K_1 \epsilon^{-\frac{1}{2}} + \omega \int_0^1 |\psi'(r)|^2 dr + O(\epsilon^{\frac{1}{2}}),$$
$$|u_\epsilon|_6^2 = K_2 \epsilon^{-\frac{1}{2}} + O(\epsilon^{\frac{1}{2}}),$$
$$|u_\epsilon|_2^2 = \omega \int_0^1 \psi^2(r) dr + O(\epsilon^{\frac{1}{2}}),$$

where K_1 and K_2 are positive constants such that $\frac{K_1}{K_2} = S$ and ω is the area of the unit sphere S^2.

Following the proof of Lemma 2.2, there exist $\epsilon_3 > 0$, $\mu^{**} > 0$ such that for all $0 < \mu < \mu^{**}$

$$\begin{aligned} \sup_{t \ge 0} J(tu_\epsilon) &\le S^{\frac{3}{2}} + \frac{3\omega S^{\frac{1}{2}}}{4K_2}(\lambda + \frac{1}{4}\lambda_1(\Omega))\epsilon^{\frac{1}{2}} + O(\epsilon) \\ &< S^{\frac{3}{2}} + \frac{3\omega S^{\frac{1}{2}}}{8K_2}(\lambda + \frac{1}{4}\lambda_1(\Omega))\epsilon^{\frac{1}{2}} < S^{\frac{3}{2}} - k_0 \mu^{\frac{6}{6-q}}, \quad \text{for } 0 < \epsilon < \epsilon_3. \end{aligned}$$

Theorem 1.2 can be proved by analogous arguments to the proof of Theorem 1.1.

References

[1] A.Ambrosetti, H.Brezis and G.Cerami, Combined effects of concave and convex nonlinearities in some elliptic problems, *J. Funct. Anal.*, 122(1994),519-543.

[2] A.Ambrosetti and P.Rabinowitz, Dual variational methods in critical point theory and applications, *J. Funct. Anal.*, 14(1973),349-381.

[3] A.Ambrosetti and M.Struwe, A note on the problem $-\Delta u = \lambda u + |u|^{2^*-2}u$, *Manuscripta Math.*, 54(1986),373-379.

[4] T.Bartsch and M.Willem , Infinitely many nonradial solutions of a Euclidean scalar field equation, *J. Funct. Anal.*, 117(1993),447-460.

[5] T.Bartsch and M.Willem , On an elliptic equation with concave and convex nonlinearity, *Proc. Amer. Math. Soc.*, 123(1995),3555-3561.

[6] H.Brezis and E.Lieb, Relation between pointwise convergence of functions and convergence of functionals, *Proc. Amer. Math. Soc.*, 88(1983),486-490.

[7] H.Brezis and L.Nirenberg, Positive solutions of nonlinear elliptic equations involving critical Sobolev exponent,*Comm. Pure Appl. Math.*, 36(1983),437-478.

[8] G.Cermi, D.Fortunato and M.Struwe, Bifurcation and multiplicity results for nonlinear elliptic problems involving critical Sobolev exponents, *Ann. Inst. H. Poincaré Anal. Nonlinéaire.*, 1(1984),341-350.

[9] G.Cerami, S.Solimini and M.Struwe, Some existence results for superlinear elliptic boundary value problems involving critical exponents, *J. Funct. Anal.*, 69(1986),289-306.

[10] D.Fortunato and E.Jannelli, Infinitely many solutions for some nonlinear elliptic problems in symmetrical domains, *Proc. Roy. Soc. Edinburg.*, 105A(1987),205-213.

[11] J.Garcia Azorero and I.Peral Alonso, Multiplicity of solutions for elliptic problems with critical exponent or with a nonsymmetric term, *Trans. Amer. Math. Soc.*, 323(1991), 877-895.

[12] G.Tarantello, Nodal solutions of semilinear elliptic equations with critical exponent, *Diff. Inte. Equats.*, 5(1992),25-42.

[13] G.Tarantello, On nonhomogeneous elliptic equations involving critical Sobolev exponent, *Ann. Inst. H. Poincaré Anal. Nonlinéaire.*, 9(1992),281-304.

[14] M.Willem, Minimax Theorems,Birkhäuser,Boston,1996.

Heat method in nonlinear elliptic equations

Kung-Ching Chang

LMAM, School of Mathematical Sciences, Peking University, Beijing 100871, China

ABSTRACT. The heat flow is applied to replace the gradient flow for semilinear elliptic variational problems in the critical point theory. The method is also extended to study the Plateau problem for minimal surfaces.

It is well known that there is a natural connection between the heat equations and the elliptic equations. Let Ω be a bounded domain in R^n with smooth boundary, $\Omega^T = [0,T] \times \Omega$, v and u be functions defined on Ω and Ω^T respectively. Given continuous functions f and u_0 on Ω, assume that u and v satisfy the following equations:

$$\begin{aligned} \partial_t u = \Delta u + f \quad &\text{in} \quad \Omega^T, \\ u = 0 \quad &\text{on} \quad (0,T) \times \partial\Omega, \\ u(0,\cdot) = \phi \quad &\text{in} \quad \Omega. \end{aligned} \tag{1}$$

and

$$\begin{aligned} -\Delta v = f, \quad &\text{in} \quad \Omega, \\ v = 0 \quad &\text{on} \quad \partial\Omega, \end{aligned} \tag{2}$$

respectively. One has

$$u(t,\cdot) \to v, \tag{3}$$

as $t \to +\infty$.

Physically, this means that the heat diffusion tends to its equilibrium.

If f depends on u, i.e., one considers the diffusion with a reaction term, then it is more complicate. However, if f is sublinear, i.e., $f(u) \leq Cu + M$, where $C \leq \lambda_1$,

Supported by NSFC, MCME.

the first eigenvalue of the Laplacian and M is a constant, then the same conclusion holds. In fact, let us define the following functional on the Sobolev space $H_0^1(\Omega)$:

$$E(v) = \int_\Omega (\frac{1}{2}|\nabla v|^2 - F(v)),$$

where F is a primitive of f. In this case, v is a minimizer of E. And E is a strict Lyapunov function of the heat equation. How are the other cases?

At the first glance, one may hesitate to give an answer, because there are two questions: (1) Does the solution u globally exist? (2) Even if u exists globally, does it have a limit? It is well known that the solution u may blow up at finite time, e.g., $f(u) = u^\alpha$ for $\alpha > 1$ (Fujita), and that it is easy to construct f (even linear) such that the solution u is of exponentially growth.

However, E is again a strict Lyapunov functional:

$$E(\phi) - E(u(t,\cdot)) = \int_0^t \int_\Omega |\partial_s u(s,x)|^2 dx ds. \tag{4}$$

The purpose of this paper is to construction the deformation of the level sets for the Lyapunov functional E by the use of the solutions u of the heat equation.

Before going on, let us briefly review the topological- variational method (Minimax Principle and Morse Theory), in which the deformation plays an important role. Namely, let K be the critical set of E, and let the level set $E_a = \{v \in H_0^1(\Omega) \mid E(v) \le a\}$ for $a \in R^1$.

The following two Deformation Principles are crucial: for any two real numbers $a < d$,

Deformation 1. If $K \cap E^{-1}[a,d] = \emptyset$, then E_a is a strong deformation retract of E_d.

Deformation 2. If $K \cap E^{-1}(a,d] = \emptyset$, and if $K_a = K \cap E^{-1}(a)$ is isolated; then E_a is a strong deformation retract of E_d.

In fact, Deformation 2 implies Deformation 1, but for pedagogical purpose, we state them separately. Thus, if one knows that E_a and E_d have different topologies, then there must be a critical value in $[a,d]$. Deformation 1 is the base for Minimax Principle. And Deformation 2 is used in the Morse theory for isolated critical points, see for instance [Ch1].

In the literature of the study, the deformation is constructed by the negative gradient flow of E in $H_0^1(\Omega)$:

$$\partial_t w(t,\cdot) = -w(t,\cdot) + (-\Delta)^{-1} f(w(t,\cdot)).$$

This is a differential integral equation. The global existence of this flow w is guaranteed by the Palais Smale Condition. A natural question is why not the heat flow? Although there are two problems mentioned above, in fact, in order to get the First Deformation Principle, only the following two statements are needed. Let us denote $O(\phi)$ the orbit $\{u(t,\cdot)|0 \le t < T(\phi)\}$, where $T(\phi)$ is the maximal existence interval of the solution of (1); and $\omega(\phi)$ the limit set of $O(\phi)$ if $T(\phi) = \infty$.

Statement 1. For any orbit $O(\phi)$, if it blows up at finite time T, then $\lim_{t\to T-0} E(u(t,\cdot)) = -\infty$.

Statement 2. If the orbit $O_(\phi)$ exists globally (i.e., $T(\phi) = +\infty$) and is in $E^{-1}[a,d]$, then $\omega(\phi) \neq \emptyset$ and contains a critical point of E.

Indeed, if Statements 1 and 2 are true, then, provided by the H^1 continuous dependence of the heat flow to initial data ϕ (for subcritical nonlinearity f) and the transversality to the level set, Deformation 1 holds. The deformation can be constructed by the heat flow (similar to the gradient flow, see [Ch1]). As to prove Deformation 2, we need further (see also [Ch1]):

Statement 3. If the orbit O_ϕ exists globally, and is in $E^{-1}[a,d]$, then the limit $\lim_{t\to+\infty} u(t)$ exists and is at the level set $E^{-1}(a)$.

Now, we turn to study the three Statements for heat equation (1). Obviously, for any f, if it is at most of linear growth in u, then the solution of (1) exists globally. In this case, Statement 1 is trivial. Thus, for Statement 1, we are only interested in the superlinear case. Assume

$$|f(u)| \leq A(1+|u|^p), \quad \text{where} \quad 1 < p < \frac{N+2}{N-2}. \tag{5}$$

as $dim(\Omega) = N > 2$, otherwise, no limitation of p, and A is a constant.

Moreover, we assume

$$(p+1)F(u) \leq uf(u) \quad \text{for} \quad |u| \geq M, \tag{6}$$

where M is a constant. The following result should be known by experts, (see e.g., [G], [DD]) but, to my knowledge, has not been clearly stated. For the completeness, we give it as follow:

Lemma 1. *Assume(5),(6) and* $1 < p < p_{CL} := 1 + \frac{4}{N-\frac{4}{3}}$, *Statement 1 is true.*

Proof of Lemma 1. According to (4), this is equivalently to show:

$$\int_0^T \int_\Omega |\partial_t u(t,x)|^2 dx dt = \infty,$$

if $T < +\infty$. We prove it by contradiction. If not, i.e.,

$$\int_0^T \int_\Omega |\partial_t u(t,x)|^2 dx dt \leq B,$$

where B is a constant, then we are going to prove

$$\sup_{t\in[0,T)} \|u(t,\cdot)\|_\infty \leq C,$$

where C denotes all constants depending on B, M, T, p, Ω and ϕ only, but not on u. Since

$$\int_\Omega u\partial_t u = \int_\Omega u(\Delta u + f(u))$$

$$= -\int_\Omega |\nabla u|^2 + \int_\Omega uf(u)$$

$$\geq -(p+1)E(u(t)) + \frac{p-1}{2}\int_\Omega |\nabla u|^2 - C.$$

we have

$$\frac{p-1}{2}\int_\Omega |\nabla u|^2 \leq (p+1)E(u(t)) + C + \|u(t)\|_2\|\partial_t u(t)\|_2.$$

Now, we turn to estimate $g(t) := \|u(t)\|_2$. Since,

$$g(t)g'(t) = \int_\Omega u(t,x)\partial_t u(t,x)dx$$

$$\leq g(t)\|\partial_t u(t,\cdot)\|_2,$$

we have

$$g'(t)^2 \leq \int_\Omega |\partial_t u(t,x)|^2 dx,$$

and then,

$$\int_0^T |g'(t)|^2 dt \leq B. \tag{7}$$

However,

$$\lambda_1 \int_\Omega |u(t,x)|^2 dx \leq \int_\Omega |\nabla u(t,x)|^2 dx,$$

$$\leq 2\frac{p+1}{p-1}E(u(t)) + \frac{2}{p-1}\|u(t,\cdot)\|_2\|\partial_t u(t,\cdot)\|_2 + C. \tag{8}$$

This implies

$$\int_0^T \int_\Omega |u(t,x)|^2 dxdt \leq C(E(\phi)T + B + 1).$$

that is,

$$\int_0^T g(t)^2 dt = \int_0^T \int_\Omega |u(t,x)|^2 dxdt \leq C. \tag{9}$$

Combining (7) and (9), we obtain

$$\sup_{t\in[0,T)} |g(t)|_\infty \leq C(\int_0^T (|g(t)|^2 + |g'(t)|^2)dt)^{\frac{1}{2}} \leq C. \tag{10}$$

Substituting (10) into (8), it follows

$$\int_\Omega |\nabla u(t,x)|^2 dx \leq C(1 + (\int_\Omega |\partial_t u(t,x)|^2 dx)^{\frac{1}{2}}).$$

Therefore, provided by Sobolev Embedding Theorem,

$$\int_0^T (\int_\Omega |u(t,x)|^{2^*} dx)^{\frac{1}{4}} dt \leq \int_0^T (\int_\Omega |\nabla u(t,x)|^2 dx)^2 dt \leq C.$$

i.e., $u \in L^4([0,t), L^{2^*}(\Omega))$.

Also we know $\partial_t u \in L^2([0,t), L^2(\Omega))$. By the interpolation theorem (see [CL]) , it follows, $u \in L^\infty([0,T), L^r(\Omega))$, where $r = \frac{6N}{3N-4}$. Since then

$$\frac{f(u)}{u} \in L^\infty([0,T), L^{\frac{r}{p-1}}(\Omega)).$$

and $p < 1 + \frac{4}{N-\frac{4}{3}}$, we apply the Maximum Principle,

$$\sup_{t\in[0,T)} \|u(t,\cdot)\|_\infty \le C.$$

This contradicts with the assumption that T is the blowing up time.

Remark 1. In all the previous study, the function f may depend on x and u jointly; i.e., $f = f(x,u)$.

According to Y. Giga [G], the conclusion of lemma 1 is true if the assumption $p < p_{CL}$ is replaced by $f(u) = u^p$ and $\phi \ge 0$, i.e., for the special form of f, Statement 1 holds on the positive cone of $H_0^1(\Omega)$.

Moreover, without the assumption $\phi \ge 0$, we have the following

Lemma 2. *Assume $f = |u|^{p-1}u$ and if Ω is convex, then Statement 1 is true.*

Proof of Lemma 2. $\forall a \in \bar{\Omega}$, define $w(s,y) = (T-t)^{\frac{1}{p-1}} u(t, y(T-t)^{\frac{1}{2}} + a)$, on

$$W = \{(s,y) | s > -\ln T, \exp(-\frac{s}{2})y + a \in \Omega\}.$$

where $s = -\ln(T-t)$, $\quad x = (T-t)^{\frac{1}{2}} y + a$.

Assume that (T,a) is a blowing up point, according to [GK1], [GK2] , we obtain

$$w(s,y) \to \pm\kappa, \tag{11}$$

where $\kappa = \beta^\beta, \beta = \frac{1}{p-1}$, and $a \in int(\Omega)$.

It satisfies the equation:

$$\partial_s w = \Delta w - \frac{1}{2} y \nabla_y w - \beta w + f(\exp(-\frac{s}{2})y, \exp(\beta s) w). \tag{12}$$

We assume the following conclusions at this moment, and shall verify them at the end of this proof: $\exists$ a sequence $s_1 < s_2 < .... < s_j < ... \to +\infty$ such that

(1)$\partial_s w(s + s_j, y) \to 0$, and

(2)$y\nabla_y w(s + s_j, y) \to 0$,

uniformly on $[0,1] \times B_C(a)$ as $j \to \infty$, $\forall C > 0$.

Once the above limits are proved, from (11), $\forall C > 0, \exists \sigma_0 = \sigma_0(C)$ such that $|\partial_s w(s+s_j,y)| < \frac{\beta\kappa}{8}$, $|y\nabla_y w(s+s_j,y)| < \frac{\beta\kappa}{4}$, and $|w(s+s_j,y)| > \frac{3\kappa}{4}$, as $s_j > \sigma_0$, on $[0,1] \times B_C(a)$.

Also, $\exists \sigma_1 = \sigma_1(C)$ such that $B_C(a) \subset W(s) := \{y \in R^N | (s,y) \in W\}$ as $s > \sigma_1$.

Setting $\sigma_2 = Max\{\sigma_0, \sigma_1\}$, then for $s_j > \sigma_2$, we have

$$\int_0^T \int_\Omega |\partial_t u(t,x)|^2 dx dt$$

$$= \int_{-\ln T}^{+\infty} \exp([2\beta - \frac{N}{2} + 1]s) ds \int_{W(s)} [\beta w(s,y) + \frac{1}{2} y \nabla_y w(s,y) + \partial_s w(s,y)]^2 dy$$

$$\geq (\frac{\beta\kappa}{2})^2 \int_{s_j}^{s_j+1} \exp([2\beta - \frac{N}{2} + 1]s) m(B_C(a)) ds \to +\infty \quad \text{as} \quad j \to \infty$$

Thus the proof will be finished if the limits (1) and (2) are verified. We use the argument in [GK1] to verify (1) and (2). Let us consider any positive sequence s_j tending to infinity, and let $w_j(s,y) = w(s+s_j,y)$, according to [GK1], $\forall C > 0, \exists M(C) > 0$, such that $|w_j| \leq M(C)$ on $(s_0(C), +\infty) \times B_C(a)$, for j large, where $M(C)$ denotes constants depending on C, but not on j. Applying the parabolic regularity to the equation (12), first we have

$$||w_j||_\infty, ||\partial_s w_j||_\infty, ||\nabla w_j||_\infty \leq M(C).$$

and then by the Schauder estimates,

$$||\nabla w_j||_\alpha, ||\nabla^2 w_j||_\alpha, ||\partial_s w_j||_{\frac{\alpha}{2}} \leq M(C)$$

for some $\alpha \in (0,1)$ on $[0,1] \times B_C(a)$, where $|| \cdot ||_\alpha$ denotes the Holder norm. They imply that after a subsequence,

$$w_j \to w_\infty, \quad \partial_s w_j \to \partial_s w_\infty, \quad \nabla w_j \to \nabla w_\infty,$$

uniformly on $[0,1] \times B_C(a)$ as $j \to \infty$

Since $w_\infty = \pm\kappa$, The two limits (1),(2) are proved.

Now we turn to study Statement 2 for the heat equation, in which the non linear term satisfies (5).

On one hand, as we shall see in the following that Statement 2 is, in some sense, weaker than the $(PS)_c$ condition,

In fact, let $\lim_{t\to+\infty} E(u(t,\cdot)) = c \in [a,d]$, by (4), we have

$$\int_0^{+\infty} \int_\Omega |\partial_s u(s,x)|^2 dx ds = E(\phi) - c.$$

Thus, there is a sequence $t_j \to +\infty$ such that

$$\int_\Omega |\partial_t u(t_j,x)|^2 dx \to 0.$$

and then, $||(-\Delta)^{-1}\partial_t u(t_j,\cdot)||_{H^1} \to 0$, i.e., $\{u_j\}$ is a $(PS)_c$ sequence. If $(PS)_c$ condition is true, then $\omega(\phi) \neq \emptyset$.

On the other hand, we have,

Assume (5). If $\sup_{t\geq 0}\|u(t)\|_{p+1}\leq M$, *a constant, then Statement 2 is true.*

We assume that $O(\phi)$ globally exists and is in $E^{-1}[a,d]$. From $\|u(t)\|_{p+1}\leq M$, we have

$$\int_\Omega F(u(t))\leq K(M),$$

and then $\|u(t)\|_{H^1}\leq K(M)$, where $K(M)$ is a constant depending on M.

Provided by the regularity of parabolic equations,

$$\sup_{t\geq 0}\|u(t)\|_\infty\leq K(M).$$

By a standard argument, see for instance [CH], $\omega(\phi)\neq\emptyset$.

There is a lot of results concerning $\omega(\phi)$ for the heat equation (1) independent to (PS), see [CH],[H],[P] etc. Also, there are results on the connection between the non-existence of global unbounded solutions of (1) and the validity of the (PS) condition for the corresponding Lyapunov functional, see for instance [Q].

As to Statement 3, it is implied by Statement 2 for equation (1).

Theorem 2 *For the equation (1), in which the non linear term is either of linear growth or satisfying (5) and (6), If $\omega(\phi)$ is finite, then Statement 2 implies Statement 3*

Proof of Theorem 2. Assume Statement 2 i.e., $\omega(\phi)\neq\emptyset$, and that $\omega(\phi)$ is finite. First, we are going to prove that $O(\phi)$ tends to a critical point $v\in K_c$ in H^1- topology. Let $v\in\omega(\phi)$. If the conclusion is not true, then

$$\exists\epsilon_0>0, t_1'<t_1"<...<t_j'<t_j"<t_{j+1}'<...\to\infty$$

such that

$$\|u(t_j')-v\|_2=\frac{\epsilon_0}{2}\leq\|u(t)-v\|_2\leq\epsilon_0=\|u(t_j")-v\|_2\quad\forall t\in[t_j',t_j"].$$

where $\|\cdot\|_2$ denotes the L^2 norm.

This implies that $u(t)$ is bounded in $L^2\quad\forall t\in[t_j',t_j"],\forall j$. In the linear growth case, by parabolic regularity, we have

$$\int_I\|u(t,\cdot)\|_{H^2}^2dt\leq C|I|. \tag{13}$$

where $I\subset[t_j',t_j"]$ is any interval, and $|I|$ is its length. and C denotes constant, which does not depend on the sequence $t_j',t_j"$

In the superlinear case, (5) and (6) imply that $u(t)$ is H^1 bounded. Provided by the L^p estimates and the bootstrap iteration, it follows

$$\|u(t)\|_\infty \leq C, \quad \forall t \in [t'_j, t_j"],$$

Again we obtain (13).

However, we have

$$\int_{t'_j}^{t_j"} \|\partial_t u(t,\cdot)\|_2^2 dt = E(u(t_j")) - E(u(t'_j)) \to 0 \quad \text{as} \quad j \to \infty.$$

Since,

$$\frac{\epsilon_0}{2} \leq \|u(t_j",\cdot) - u(t'_j,\cdot)\|_2 \leq \int_{t'_j}^{t_j"} \|\partial_t u(t,\cdot)\|_2 dt$$

$$\leq (t_j" - t'_j)^{\frac{1}{2}} \Big(\int_{t'_j}^{t_j"} \|\partial_t u(t,\cdot)\|_2^2 dt\Big)^{\frac{1}{2}}.$$

Therefore, $t_j" - t'_j \to \infty$. This implies that $\exists \quad t_j^* \in [t'_j, t_j"]$ such that $\|u(t_j^*)\|_{H^2} \leq C$, and $\|\partial_t u(t_j^*,\cdot)\|_2 \to 0$.

Thus, there is a subsequence of $u(t_j^*)$ convergent in L^2 norm to a point in $\omega(\phi)$. But it is impossible, because there is no point in $\omega(\phi)$ falling into $B_{\epsilon_0}(v) \backslash B_{\frac{\epsilon_0}{2}}(v)$, where $B_r, r > 0$ is the L^2- ball with radius r . This contradiction means that $u(t) \to v$ as $t \to +\infty$ in L^2.

Now, we show $\|u(t) - v\|_{H^1} \to 0$. If not, $\exists \epsilon_0 > 0$ and $t'_j \geq t_j$ such that $\|u(t'_j) - v\|_2 \to 0$, but $\|u(t'_j) - v\|_{H^1} = \epsilon_0$. On one hand, $u(t'_j)$ is bounded in H^1, after a subsequence $u(t'_j) \rightharpoonup v$ in H^1. On the other hand, from

$$\lim_{j\to\infty} \int_\Omega F(u(t'_j)) = \int_\Omega F(v(t)),$$

and

$$\lim_{j\to\infty} E(u(t'_j)) = c = E(v),$$

it follows,

$$\lim_{j\to\infty} \int_\Omega |\nabla u(t'_j)|^2 = \int_\Omega |\nabla v|^2.$$

And then, $u(t'_j) \to v$ in H^1. This is a contradiction.

Remark 3. One may consider the case where $\Omega = R^N$, but the equation (1) is replaced by $\partial_t u(t,x) = \Delta u(t,x) - u(t,x) + f(x,u)$ in R^N. Of course, the elliptic equation and the Lyapunov functional should be modified accordingly.

Remark 4. The semilinear parabolic systems as well as the higher order parabolic equation with variational structures can be studied in the same manner.

In summary, the heat flow does play a role as the gradient flow in the Deformation Principles. Since the heat flow produces a special Palais Smale sequence, i.e., the Statement 2 is weaker than the $(PS)_c$ condition; in principle, the heat method should be applied to more general cases. In fact, there is a lot of results on the ω limit set for parabolic equations, they provide alternative way to approach Statements 2 and 3 without appealing to the (PS) condition. In particular, to the quasilinear elliptic problem:

$$-\sum_{i,j=1}^{N} \partial_i(a_{ij}(x,u)\partial_j u) + \frac{1}{2}\sum_{i,j=1}^{N} \partial_u a_{ij}(x,u)\partial_i u \partial_j u = f(x,u),$$

where we assume that each $a_{ij} : \Omega \times R \to R$ is measurable in $x, \forall u \in R$ and of class C^1 in u for $a.e.x \in \Omega$ with $a_{ij} = a_{ji}$, and that $\exists C, \nu > 0$ such that

$$|a_{ij}(x,u)| \leq C, \quad |\partial_u a_{ij}(x,u)| \leq C,$$

$$\sum_{i,j=1}^{N} a_{ij}(x,u)\xi_i\xi_j \geq \nu \Sigma_{i=1}^{N} \xi^2, \quad \forall (u,\xi) \in R \times R^N.$$

Let us define

$$E(v) = \frac{1}{2}\int_{\Omega} \sum_{i,j=1}^{N} a_{ij}(x, v(x))\partial_i v(x)\partial_j v(x)dx - \int_{\Omega} F(x, v(x))dx$$

If we study the functional on the space $H_0^1(\Omega)$, then E is not differentiable, if on the space $H_0^1 \cap L^\infty(\Omega)$ then the (PS) condition is lost; see M. Struwe [St1], D. Arcoya, L. Boccardo [AB], and A. Canino [Ca] etc. Various modified methods have been used to approach this problem. However,the associate heat flow method can be applied more naturally and directly.

Indeed, the heat flow method in elliptic theory has been introduced in geometric problems, which can be traced back to J. Eells and J.H. Sampson [ES]. It was to find a harmonic map in each component of the space of continuous mappings between two Riemannian manifolds. The harmonic maps satisfy an elliptic system, in order to obtain a solution of the system in a given component, a natural way is to use the associate heat flow with initial data in that component. In [ES], only Statement 2 was applied, because by their assumption, the heat flow globally exists. The method was later extended to study non-minimum solutions for harmonic maps, see for instance, K. C. Chang [Ch2]. Also, M. Struwe [St2] developed the heat method to free boundary problem for constant mean curvature surfaces, and obtained a mountain pass solution.

In the following, we announce results on the Plateau problem for minimal surfaces by the heat method due to J. Q. Liu and the author [ChL1], [ChL2] and [ChL3].

Suppose that (N, h) is a Riemannian manifold, and that D is the unit disk in R^2. Given a simple imbedded curve Γ in N. A map $X : D \to N$ is called a minimal surface (MS) spanned on Γ, if it satisfies the following system:

$$Tr(\nabla dX) = 0,$$

$$h(\partial_x X, \partial_x X) - h(\partial_y X, \partial_y X) = h(\partial_x X, \partial_y) = 0,$$

in D, and $X_{\partial D} := X|_{\partial D} : \partial D \to \Gamma$ is a homeomorphism. This problem has been extensively studied with long history, see for instance, the two volumes book [DHKW] and J. Jost [Jo], M. Struwe [St3].

Let $\gamma : \partial D :\to \Gamma$ be an oriented parametrization of Γ, and let x be an extension of $\gamma^{-1} X_{\partial D}$ satisfying: $x(\theta + 2\pi) = x(\theta) + 2\pi \quad \forall \theta \in R^1$. Define

$$M = \{X \in C \cap H^1(\bar{D}, N) | X_{\partial D} : \partial D \to \Gamma, x : R \to R \text{weakly monotone}\}.$$

Setting $\xi = x - id$, we define $E(X) = \int_D |\nabla X|_h^2$ on the space:

$$M^+ := \{X \in M| \int_0^{2\pi} \xi(\theta) d\theta = \int_0^{2\pi} \xi(\theta) \exp(i\theta) d\theta = 0\}.$$

X is a solution of the Minimal Surface Equations if and only if it is a critical point of E on M^+, i.e.,

$$\int_D \nabla X \nabla \psi \geq 0 \quad \forall \psi \in X^{-1} TM^+.$$

What is the heat flow for (MS)?

Definition. Given $T > 0$ and $X_0 \in H^1(D, N)$, $\quad X \in L^2([0,T], C \cap H^1(\bar{D}, N))$ is called a weak heat flow of (MS), if

(1) $X(t, \cdot) \in M^+$ for a.e.$t \in [0, T]$,

(2) $X(0, \cdot) = X_0$,

(3) $\int_{D^T} \partial_t X \phi + \nabla X \nabla \phi \geq 0, \forall \phi \in L^2([0,T], X^{-1} TM)$.

We have the following results:

Theorem 3(Chang-Liu). *Assume* $\pi_2(N) = 0$. *Given a simple closed curve* $\Gamma \in C^3$, *which is imbedded in* N, *and a component* $\Sigma \subset M^+$, *and let* $X_0 \in \Sigma$. *Suppose*

$$E(X_0) < m_\Sigma + b,$$

where

$$m_\Sigma = \inf\{E(X) | X \in \Sigma\},$$

and

$$b = \inf\{\int_{S^2} |\nabla U|^2 | U : S^2 \to N, \textit{non constant}\}.$$

Then $\forall T > 0, \exists$ *a unique solution of the weak heat flow* $X \in W_2^{1,2}(D^T, N)$ *for (MS). Moreover, the map* $(t, X_0) \to X(t, \cdot)$ *is continuous in* $H^1(D, N)$. *And* $\exists W \in H^2(D, N)$ *satisfying (MS) such that*

$$X(t, \cdot) \to W \quad \textit{in} \quad H^1(D, N),$$

and

$$X(t,\cdot) \rightharpoonup W \quad in \quad H^2(D,N).$$

The limit W is in fact a minimal surface spanned by Γ.

Remark 5. By the Trace Theorem, $W_2^{1,2}(D^T,N) \hookrightarrow C([0,T],H^1(D,N))$ is continuous. This means that W is homotopic to X_0 in H^1 topology; thanks to a result due to Brezis, Y. Y. Li [BL], they are homotopic in C-topology.

Theorem 4(Chang-Liu). *Suppose that $\pi_2(N) = 0$ and $d < m_\Sigma + b$. If there are only finitely many minimal surfaces $X_i \in \Sigma$with energy below d. Then the following Morse relation holds:*

$$\sum_{q=0}^{\infty} M_q t^q = \sum_{q=0}^{\infty} \beta_q t^q + (1+t)Q(t),$$

where $\beta_q = rank H_q(E_d), M_q = \sum_i rank C_q(X_i)$, and Q is a formal series with non negative coefficients.

In the statement, $C_q(.)$ denotes the $q-th$ critical group.

As a corollary, one obtains an extension of a result due to Sacks Uhlenbeck [SU],

Corollary. *Let (N,h) be a Riemannian manifold, with $\pi_2(N) = 0$ and let Γ be a simple, closed, imbedded C^3 curve in N . Then in each component of $\{X \in C(\bar{D},N)| \quad X_{\partial D} : \partial D \to \Gamma \quad$ is a homeomorphism $\}$, $\exists$ a minimal surface spanned on Γ.*

Some multiple solution results are also obtained, see [CL2].

References

[AB] D. Arcoya, L. Boccardo; Critical points for multiple integrals of the calculus of variations, (1994)

[BL] H. Brezis, Y. Y. Li; Topology and Sobolev spaces, preprint, (2001).

[Ca] A. Canino; Multiplicity of solutions for quasilinear elliptic equations,

[CH] T. Cazenave, A. Haraux; An introduction to semilinear evolution equations, Oxford Univ. Press, (1998).

[CL] T. Cazenave, P. L. Lions; Solutions globales d'equations de la chaleur semilineaires, Comm. PDE **9** (1984), 935-978.

[Ch1] K. C. Chang; Infinite dimensional Morse theory and multiple solution problems, Birkhauser, (1993).

[Ch2] K. C. Chang; Heat flow and boundary value problem for harmonic maps, Analyse nanlineaire, **6** (1989), 363-395.

[ChL1] K. C. Chang, J. Q. Liu; Heat flow for the minimal surface with Plateau boundary condition, to appear Acta Mathematica Sinica (English series),

[ChL2] K. C. Chang, J. Q. Liu; An evolution of minimal surfaces with Plateau condition, preprint Peking University, no.

[ChL3] K. C. Chang, J. Q. Liu; Another approach to the heat flow for Plateau problem, to appear JDE,

[DD] E. N. Dancer, Y. Du; The generalized Conley index and multiple solutions of semilinear elliptic problems, Abstract Applied Analysis, **1, 1** (1996), 103-135

[DHKW] U. Dierkes, S. Hildbrandt, A. Küster, O. Wohlrab; Minimal Surfaces, Vol. 1 and 2, Grundlehren der Mathematik Wissenschaften **295 296**, (1992) Springer.

[G] Y. Giga; A bound for global solutions of semilinear heat equations, Comm. Math. Phys. **103** (1986), 415-421.

[GK1] Y. Giga, R. V. Kohn; Characterizing blowup using similarity variables, Indiana U. Math. J. **36, 1** (1987), 1-40.

[GK2] Y. Giga, R. V. Kohn; Removability of blowup points for semilinear heat equations, Lect. Notes in Pure and Appl. Math. **118** Marcel Dakker, (1989), 257-264.

[H] D. Henry; Geometric theory of semilinear parabolic equations, Springer, (1981).

[Jo] J. Jost; Two dimensional geometric variational problems, John Wiley, (1991).

[P] P. Polacik; Parabolic equations: asymptotic behaviour and dynamics on invariant manifolds, Preprint.

[Q] P. Quittner; Boundedness of trajectories of parabolic equations and stationary solutions via dynamical methods, Diff. Integral Equations, **7** (1994), 1547-1556.

[SU] J. Sacks, K. Uhlenbeck; The existence of minimal immersions of 2-spheres, Ann. of Math. **113** (1981), 1-24.

[St1] M. Struwe; Multiple solutions of differential equations without Palais Smale condition, Math. Ann. **261** (1982), 399-412.

[St2] M. Struwe; The existence of surfaces of constant mean curvature with free boundaries, Acta Mathematica **160** (1988), 19-64.

[St3] M. Struwe; Plateau's problem and the calculus of variations, Math. Notes, **35** (1988) Princeton Press.

Some facts and more questions about the Eight

Alain Chenciner

Astronomie et Systèmes Dynamiques, IMCCE, UMR 8028 du CNRS, 77, avenue Denfert-Rochereau, 75014 Paris, France

and

Département de Mathématiques, Université Paris VII-Denis Diderot, 16, rue Clisson, 75013 Paris, France

ABSTRACT. I discuss some properties of the Eight solution of the three-body problem, many of them conjectural. I describe in particular a simple approach to the P_{12} family, proposed by C. Marchal, which is a choreography in the rotating frame with the same 12-fold symmetry as the Eight.

I - Introduction

I-1 The equal mass 3-body problem in $\mathbb{R}^2$ ([C1],[C2]). We consider three bodies of unit mass in $\mathbb{R}^2$. As we are interested in periodic solutions, we suppose from the start that the center of mass is fixed at the origin. Hence, the configuration space is the open subset $\hat{\mathcal{X}}$ of

$$\mathcal{X} = \{x = (\vec{r}_1, \vec{r}_2, \vec{r}_3) \in (\mathbb{R}^2)^3, \sum_{i=1}^{3} \vec{r}_i = 0\}$$

defined by the condition of "no collision" : $\forall i \neq j,\ \vec{r}_i \neq \vec{r}_j$. The vector space $\mathcal{X}$ is endowed with the "mass metric", which coincides here with the standard euclidean metric:

$$I(x) = ||x||^2 = \sum_{i=1}^{3} ||\vec{r}_i||^2.$$

e-mail: chencine@bdl.fr.
Web address: http://www.bdl.fr/Equipes/ASD/person/chenciner/chenciner.html.

Newton-Lagrange equations can now be written $\ddot{x} = \nabla U(x)$, where U is the Newtonian potential

$$U(x) = \sum_{i<j} \frac{1}{||\vec{r}_i - \vec{r}_j||}.$$

It is well known that they are the Euler-Lagrange equations of the *action* $\mathcal{A}$ which associates to a C^1 path $x : [t_0, t_1] \to \hat{\mathcal{X}}$ the real number

$$\mathcal{A}\big(x(t)\big) = \int_{t_0}^{t_1} \left(\frac{||\dot{x}(t)||^2}{2} + U\big(x(t)\big) \right) dt.$$

I-2 The space of oriented triangles ([AC],[CM],[C4]). The squares a, b, c of the lengths of the sides of a triangle are a good set of parameters of the set of all triangles up to isometries. We fix the size by setting $a + b + c = 1$, which amounts to fixing the moment of inertia, with respect to the center of mass, of three equal masses located at the vertices of the triangle. Not all the points of the simplex $a+b+c = 1$, $a \geq 0, b \geq 0, c \geq 0$ represent triangles. The true triangles correspond to the points inside the disc defined by the inequation

$$16S^2 = 2ab + 2bc + 2ca - a^2 - b^2 - c^2 \geq 0,$$

which asserts the positivity of the squared area S of the triangle, a formula going back to Heron of Alexandria. The positivity of S^2 is equivalent to the three triangular inequalities.

Gluing 2 discs together, we represent all similitude classes of *oriented* triangles by a 2-sphere (a more "metric" construction is given in [CM]). This sphere – the so-called "shape sphere" – is represented on figure 1 with the action of its symmetry group, the dihedral group

$$D_6 = \{s, \sigma \mid s^6 = 1,\ \sigma^2 = 1,\ s\sigma = \sigma s^{-1}\}.$$

The North and South poles represent respectively the positively and negatively oriented equilateral triangle, the equator represents the flat triangles and the three vertical planes represent the isosceles triangles while C_1, C_2, C_3 are the collision half-lines.

I-3 The action of the dihedral group D_6 on the loop space ([C2]). The actions of the two generators on a loop $x : \mathbb{R}/T\mathbb{Z} \to \mathcal{X}$ are defined by the following formulas: if $x(t) = \big(\vec{r}_1(t), \vec{r}_2(t), \vec{r}_3(t)\big)$,

$$\begin{aligned}(s \cdot x)(t) &= \big(\Sigma\vec{r}_3(t - T/6), \Sigma\vec{r}_1(t - T/6), \Sigma\vec{r}_2(t - T/6)\big),\\ (\sigma \cdot x)(t) &= \big(-\vec{r}_1(-t), -\vec{r}_3(-t), -\vec{r}_2(-t)\big),\end{aligned}$$

where Σ is the symmetry with respect to the y axis in $\mathbb{R}^2$.

I-4 Choreographies. As the subgroup $\mathbb{Z}/3\mathbb{Z}$ generated by s^2 acts by cyclically permuting the bodies after one-third of a period, each D_6-invariant loop is a *choreography*, that is a loop of the form $(q(t), q(t + T/3), q(t + 2T/3))$, where $q : \mathbb{R}/T\mathbb{Z} \to \mathbb{R}^2$ is a parametrized plane curve. More generally, in [CGMS] and

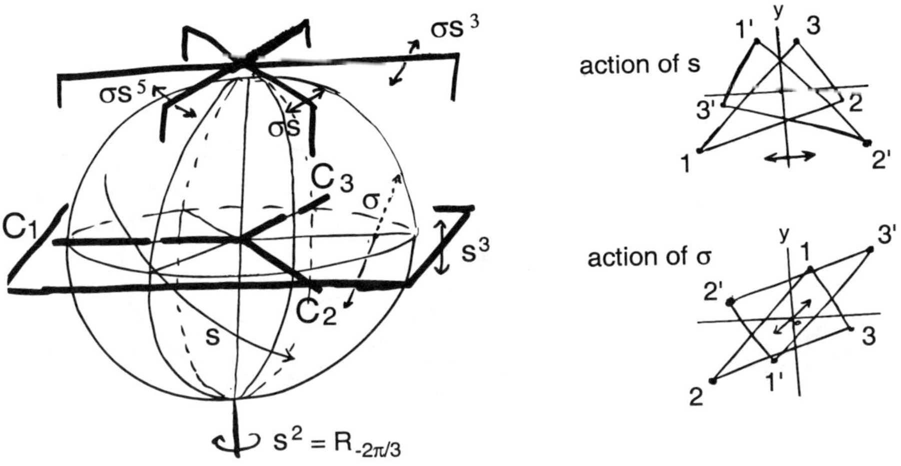

FIGURE 1. (the shape sphere and the action of D_6)

[S2] the authors call (simple) *choreography* a loop in the configuration space of the n-body problem in $\mathbb{R}^k$ of the form

$$\big(q(t), q(t+T/n), \cdots, q(t+(n-1)T/n)\big),$$

where q is a parametrized closed curve in $\mathbb{R}^k$.

II - The Eight ([Mo], [CM], [C2], [Ch], [ZZ2]).

FACT 1: it exists. The existence of a minimizer of the Lagrangian action in the space Λ_{D_6} of D_6-invariant loops in the configuration space is a consequence of Tonelli's theory: coercivity comes from the D_6-invariance which forces the length – and hence the action – of an element of Λ_{D_6} to be big as soon as some part of the loop goes far away ("tied" class in the sense of Gordon, see [M1]). The main point is to show that such a minimizer is collision-free, which implies that it is a solution of the three-body problem and even a choreography supported by an eight-shaped curve. Accurate pictures were obtained numerically by Carles Simó

The proof boils down to showing the inequality $\mathcal{A}_{coll} > \mathcal{A}_{test}$, where $\mathcal{A}_{coll}$ is a lower bound for the action of a D_6-invariant loop undergoing at least one collision and $\mathcal{A}_{test}$ is the action of a collision-free equipotential test path (see [CM]). The best estimate for $\mathcal{A}_{coll}$ (better than the ones in [CM]: $\mathcal{A}_{coll} = 2^{\frac{1}{3}}A_2$, or even in [Ch]) is found in [ZZ2]: thanks to a formula of Leibnitz, the action of a 3-body problem splits into the sum of three terms, each of which is one third of the action of the Kepler problem with attraction constant equal to the total mass $M = 3$ (see

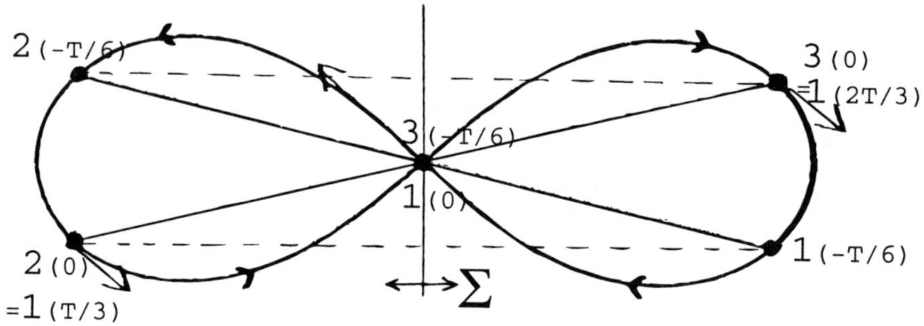

FIGURE 2. (the shape sphere and the action of D_6)

[V1],[ZZ1]):

$$\mathcal{A}(x(t)) = \frac{1}{3}\sum_{i<j}\int_0^T \left[\frac{\|\dot{\vec{r}}_i(t) - \dot{\vec{r}}_j(t)\|^2}{2} + \frac{3}{\|\vec{r}_i(t) - \vec{r}_j(t)\|}\right] dt.$$

As the configurations at t and $t + T/2$ are symmetric with respect to the $0y$ axis (compute the action of s^3), any collision which occurs at t_0 occurs also at $t_0 + T/2$. The lower bound of the Kepler action during a period T is then twice the minimum of the Kepler action of an ejection-collision wih attraction constant 3 and period $T/2$. This is equivalent to $\mathcal{A}_{coll} = \mathcal{A}(2L(T/2))$, where $2L(T/2)$ is the equilateral relative equilibrium which makes two complete turns in one period T. An estimate for $\mathcal{A}_{test}$ is given in [CM].

QUESTION 1: but is it unique? Numerical evidence by Carles Simó in [S1] suggests unicity of the minimizer. From now on we shall indulge in speaking of "the" Eight.

QUESTION 2: is each lobe convex? in [CM] we prove only that each lobe of a minimizer is starshaped (the problem is near the crossing point, convexity outside a neighborhood of this point is easy to prove).

QUESTION 3: is it a transcendental curve? C. Simó ([S1],[S3]) computed numerically truncated Fourier expansions of the components of the Eight. We take the period $T = 2\pi$. Because of the symmetries, the curve $q(t)$ is of the form:

$$q(t) = \left(\sum_{j odd} a_j \sin jt,\ \sum_{j even} b_j \sin jt\right),$$

with $a_j = 0$ if $j \equiv 3$ mod. 6 and $b_j = 0$ if $j \equiv 0$ mod. 6. Because of the missing terms and the quick decrease with j of the non-zero ones, one gets a good approximation of the Eight by the *quartic* $t \mapsto (a_1 \sin t, b_2 \sin 2t + b_4 \sin 4t)$. Indeed, the next terms are $(a_5 \sin 5t, b_8 \sin 8t)$, with a_5 of the order of 2×10^{-2} and b_8 of the order of 10^{-4}. Going to order 30 is enough to have all harmonics larger than 10^{-16}. Finally, fits of the solution by algebraic curves $g(x, y) = 0$ up to degree 12 definitely show disagreements which, according to C. Simó, cannot be due to computational errors. Note that, according to [F.F.O.], the lemniscate $(x^2 + y^2)^2 = x^2 - y^2$ supports a choreographic solution of three equal masses for the potential $\sum_{i<j} \left((1/2)\ln r_{ij} - (\sqrt{3}/24) r_{ij}^2\right)$.

FACT 2: existence of a priori less symmetric Eights ([C3]). The $\mathbb{Z}/3\mathbb{Z}$ symmetry is certainly not enough to characterize the Eight because it follows from [CD] that minimizing among choreographies of three equal masses leads to Lagrange equilateral relative equilibrium. The dihedral group D_6 has three subgroups which contain $\mathbb{Z}/3\mathbb{Z}$. The one generated by s^2 and σs, isomorphic to D_3, is not interesting for our purpose because every element acts on the shape sphere preserving the orientation of the triangles (in particular, the Lagrange equilateral relative equilibrium is symmetric under the action of this subgroup). The two others are respectively generated by s^2 and σ (isomorphic to D_3) and by s (isomorphic to $\mathbb{Z}/6\mathbb{Z}$). It is shown in [C3] that a minimizer of the action among D_3 or $\mathbb{Z}/6\mathbb{Z}$-invariant loops is collision-free.

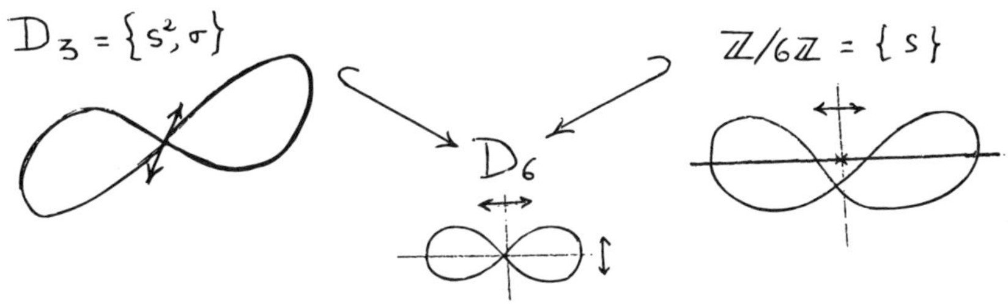

FIGURE 3. (less symmetric Eights?)

QUESTION 4: all the same ? This is of course intimately linked to the unicity question. For more on possible extra-symmetries of action minimizers, see [C2],[C3],[V2].

III - The spatial case

An important feature of the spatial case is that the shape sphere (see I-2) is now replaced by a 2-disk: this is because orientation of a triangle is no more meaningful. The non-collision similitude classes of spatial 3-body configurations form a 2-disk minus three points on its boundary, that is a contractible space. In particular, the loop of configurations described by the Eight solution can now be deformed without collisions to the constant equilateral configuration, leading to the P_{12} family (see III-2).

III-1 Action of D_6 on spatial configurations ([C3]) Let Δ be the axis $0x$ and P, Σ be respectively the coordinate planes $0xz$ and $0xy$ (see figure 4). We extend to loops of spatial configurations of three bodies the action of D_6 by replacing the symmetry with respect to 0 (resp. 0x, resp. Oy) by the symmetry with respect to Δ (resp. P, resp. Σ):

$$\begin{aligned}(s\cdot x)(t) &= \big(\Sigma\vec{r}_3(t-T/6), \Sigma\vec{r}_1(t-T/6), \Sigma\vec{r}_2(t-T/6)\big),\\ (\sigma\cdot x)(t) &= \big(\Delta\vec{r}_1(-t), \Delta\vec{r}_3(-t), \Delta\vec{r}_2(-t)\big),\end{aligned}$$

where Σ (resp. Δ) denotes the symmetry with respect to the horizontal plane Σ (resp. to the line Δ). The Eight in the plane orthogonal to Δ is an obvious example of an invariant loop, but there is also the equilateral relative equilibrium $2L(T/2)$ which makes two full turns in one period in the horizontal plane (figure 4).

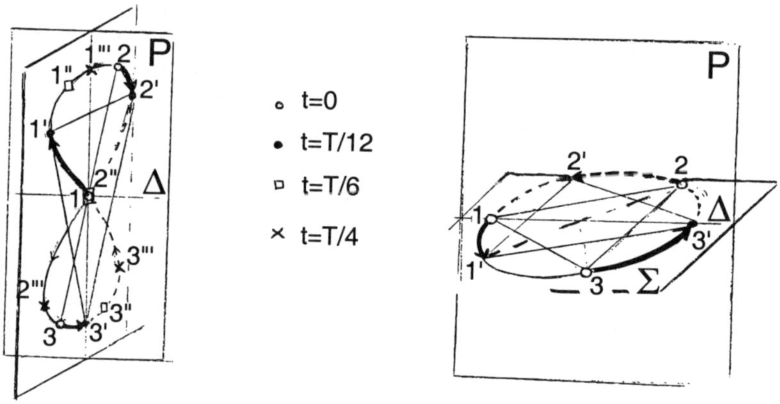

FIGURE 4. (Two examples of D_6-invariant spatial loops)

Moreover the lower bound given for the action D_6-symmetric loops with collision in the plane still holds in space.

QUESTION 5: is a minimizer in $\Lambda_{D_6}^{spatial}$ necessarily planar? i.e. of angular momentum zero, i.e. an Eight?

III-2 The P_{12} family ([Ma],[C3]). This family continues the Eight solution in three-space up to Lagrange equilateral solution, through choreographies in a rotating frame. It is described in detail in [C3]. It is parametrized by an angle u between 0 and $\frac{\pi}{6}$: the solution labeled by u is supposed to minimize the action in fixed time $T/12$ between configurations which are symmetric with respect to the line Δ with $1 \in \Delta$ and configurations which are symmetric with respect to a vertical plane P through the origin which contains body 3 and makes angle u with Δ. In a frame rotating around the vertical axis of an angle $-u$ in time $T/12$, one gets a family of D_6-symmetric choreographies of period T which connects the two examples depicted on figure 4 (the Eight and twice Lagrange $2L(T/2)$) via progressive folding in the direction of Δ (see [C3] fig. 1 and 2, and [N] fig. 5).

FACT 3: existence of a family. For all values of u, minimizers are collision-free

The surprise is that, using as a model the horizontal Lagrange family x_u (which satisfies the symmetry requirements), one can give a simple direct proof of the absence of collisions in a minimizer of the action in $\Lambda_{D_6}^{spatial}$, simpler than in the planar case:

1) the action of an admissible path undergoing a collision is bigger than the action $\hat{A}_2 = 2^{-\frac{5}{3}}3^{\frac{2}{3}}\pi^{\frac{2}{3}}T^{\frac{1}{3}}$ (masses $=1$) of the horizontal relative equilibrium solution x_0 of an equilateral triangle which rotates by $\frac{\pi}{3}$ in the same amount of time $T/12$;

2) this last action is, for any $u \leq \frac{\pi}{3}$, bigger than the one $A(u) = \hat{A}_2\left[\frac{3}{\pi}(\frac{\pi}{3}-u)\right]^{\frac{2}{3}}$ of the horizontal relative equilibrium solution x_u of an equilateral triangle which rotates by an angle $(\frac{\pi}{3} - u)$ during the given amount of time.

Finally, we prove that, for $0 \leq u < \frac{\pi}{6}$, the Lagrange solution x_u is not a minimizer. This is because the value $d^2\mathcal{A}(x_u)(\xi,\xi)$ of the Hessian of the action on the vertical variation

$$\xi = \left(\sin(\frac{2\pi t}{T}),\ \sin(\frac{2\pi t}{T} + \frac{2\pi}{3}),\ \sin(\frac{2\pi t}{T} + \frac{4\pi}{3})\right)$$

which "opens" x_u in the direction of the Eight, is negative for $u < \pi/6$ and positive for $u > \pi/6$. Indeed, the Hessian of x_u is positive when $\pi/6 < u \leq \pi/3$, which supports Marchal's claim that x_u is the minimizer when $\pi/6 \leq u \leq \pi/3$ (notice that its size increases to infinity and its action decreases to 0 when u tends to $\pi/3$). To be sure that this family really connects the Lagrange and Eight solutions, we need answering positively Questions 5 and 6.

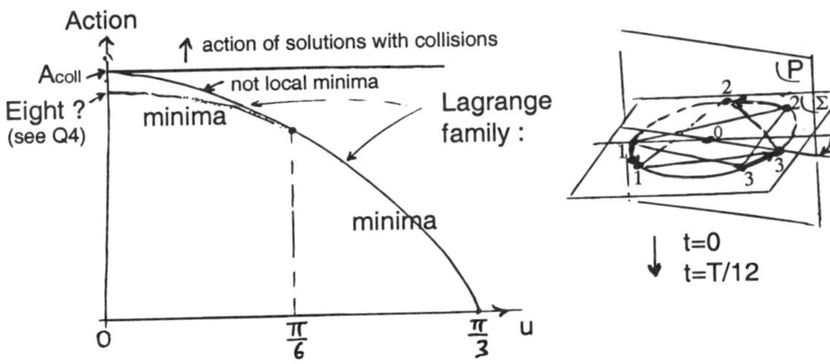

FIGURE 5. (Bifurcation of the P_{12} family from the Lagrange family)

QUESTION 6: unicity of minimizer for any value of the parameter u? or at least continuity of a minimizing family? Such continuity would imply the existence among the family of spatial (non-planar) 3-body choreographies in the fixed frame. Indeed, for well-chosen values of u, the period of the rotating frame and the one of the solution in the rotating frame would be resonant.

III-3 Other continuations in a rotating frame. The first continuation of the Eight into a family of rotating planar choreographies was given – up to the first orbit encountering a collision – by Michel Hénon [CGMS], using the same program as in [H]. The continuation beyond this orbit can be found is [S1]. A third family should exist, rotating around an axis orthogonal to the first two.

IV - Fixing homology

FACT 4: homology class of the Eight is $(0,0,0)$. This means that, during a period, each side of the triangle has zero total rotation. Hence the Eight does not minimize the action in its homology class: the minimum, equal to 0, is attained for still bodies infinitely far from each other (for the case of homology class $(1,1,1)$, see Poincaré [P] 1896 and Venturelli 2001 [V1], [ZZ1]).

QUESTION 7: does the Eight minimize the action among choreographies in its homology class? What makes this question hard is the mixture of topological and symmetry constraints

V - Fixing homotopy. The homotopy class of a loop in the configuration space of the planar n-body problem may be thought of as the *braid* described by the bodies in (periodic) space time $\mathbb{R}^2 \times (\mathbb{R}/T\mathbb{Z})$. Knowing that each lobe of the Eight is starshaped is enough to imply the following:

FACT 5: the braid defined by the Eight is the "Borromean rings" ([Ber],[C3]). This is the signature of a truly triple interaction.

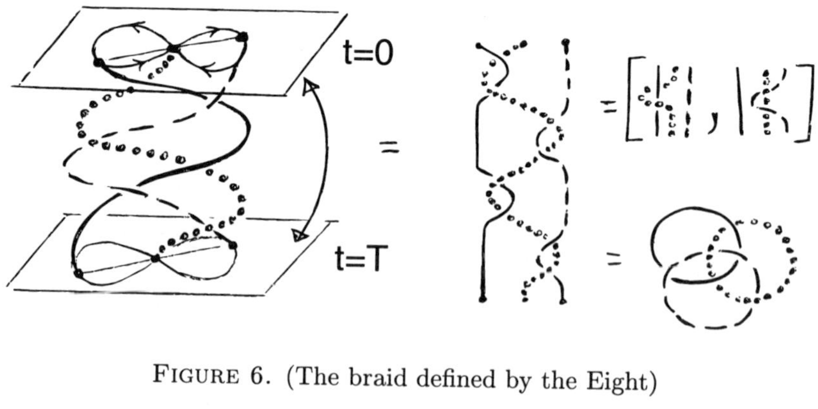

FIGURE 6. (The braid defined by the Eight)

FACT 6 ([C2],[M2]): the Eight does not minimize the action in its homotopy class

QUESTION 8: does the Eight minimize the action among choreographies in its homotopy class? And are there other choreographies in this homotopy class?

Remark. An interesting example of mixed conditions (topology and symmetry) for a minimization problem may be found in [V2] where generalizations of the Hip-Hop lead to spatial choreographies of 4 equal masses. But, as for most choreographies, no proof was found of the existence of Gerver's "supereight" with four equal masses [CGMS], [C2].

VI - Stability

FACT 7: numerically, the Eight is KAM stable. A numerical computation of a Poincaré map to high order around the fixed point corresponding to the Eight and the subsequent computation of the normal form shows that one can apply KAM theorem and, hence, that the Eight is KAM stable on the manifold of zero angular momentum ([S1]). For theoretical works on the stability properties of action minimizers, see [Ar],[O].

QUESTION 9: Give a detailed proof of the KAM stability of the Eight.

VII - Masses ([C5],[C6],[BCS])

FACT 8: a choreography with $n \leq 5$ bodies must have equal masses. This is proved in [C6] using the remark that if a choreography is solution of the n-body problem with masses $m_1, m_2, \ldots, m_n$, it is also solution of the n-body problem with masses $\mu, \mu, \ldots, \mu$, where $\mu = (m_1 + m_2 + \cdots + m_n)/n$.

QUESTION 10: is the same true for any number of bodies? in particular $n = 6$?

VIII - Eights with more bodies and limit when the number $n = 2p+1$ of bodies tends to $+\infty$ ([S2],[C2]). Eight-shaped choreographies exist numerically with any number n of bodies. If n is odd, the Eight-curve has the full $(\mathbb{Z}/2\mathbb{Z}) \times (\mathbb{Z}/2\mathbb{Z})$ symmetry (see figure 7) but if n is even, it has only a $\mathbb{Z}/2\mathbb{Z}$ symmetry (the two lobes are unequal ([CGMS] figure 3a)).

In [C2], the action of D_6 on the loop space of the configuration space of the equal mass three-body problem is extended, for *n odd*, to an action of the dihedral group D_{2n} on the loop space of the configuration space of the equal mass n-body problem. We call $\Lambda_{D_{2n}}$ the subspace of invariant loops under this action.

QUESTION 11: Is an Eight with n (odd) bodies an action minimizer in $\Lambda_{D_{2n}}$?

QUESTION 12: understand the limit of the Eight when n odd tends to $+\infty$. According to C. Simó [S3], the angle at the crossing point tends to $\pi/2$ and, for a given period, the size has a precise scaling law in n.

IX - Other homogeneous potentials ([CGMS]). According to [Mo],[CGMS], the Eight exists for all potentials of the form r^α with $\alpha \in]-\infty, 0[$. When $\alpha = -2$

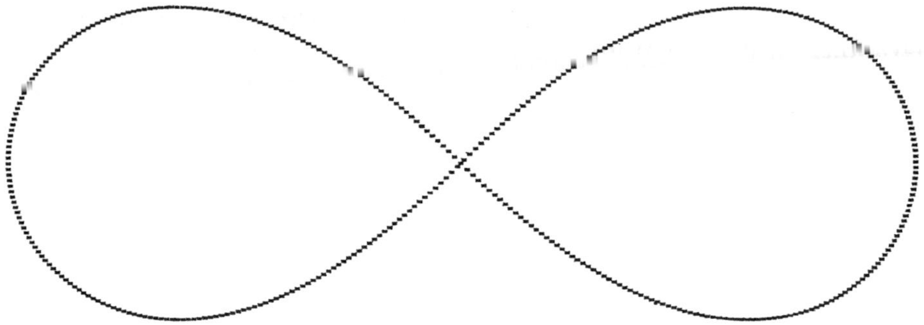

FIGURE 7. (399 bodies on a Eight, computed by C. Simó)

(Jacobi potential), it follows from the Lagrange-Jacobi identity that the energy of any periodic solution is necessarily equal to 0, and its moment of inertia $I = ||x(t)||^2$ is constant. For the Newtonian potential ($\alpha = -1$), it is a conjecture of D. Saari that a solution of the n-body problem can have a constant moment of inertia with respect to the center of mass only in the case it is a *relative equilibrium*, that is ([AC], Proposition 2.5) when the mutual distances stay constant along the motion (*rigid motion*). Numerically, the variations of I for the Newtonian Eight are of the order of 0.5% ([S3]).

QUESTION 13: show that the moment of inertia $I = ||x(t)||^2$ of the Eight stays constant only when $\alpha = -2$.

Two curiosities. Another nice property of the Eight, consequence of its high symmetry, is the shape of its hodograph (figure 8-1); also interesting are the small eight described by the *center of force* (see [W]) of the configuration and the hyperbola-like curve described by the *center of velocity* (which exists because the angular momentum is equal to 0, see [F.F.O.]) (figure 8-2 from figure 13 of [Br] and figure 2 of [F.F.O.]).

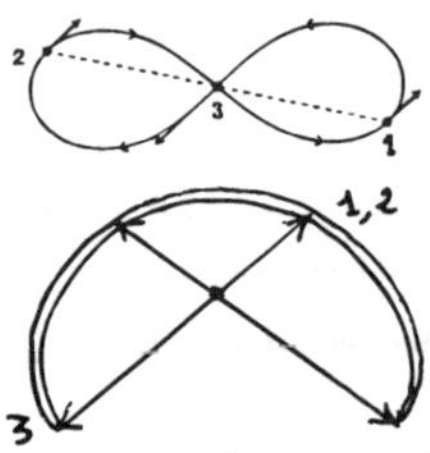

FIGURE 8-1
(the hodograph)

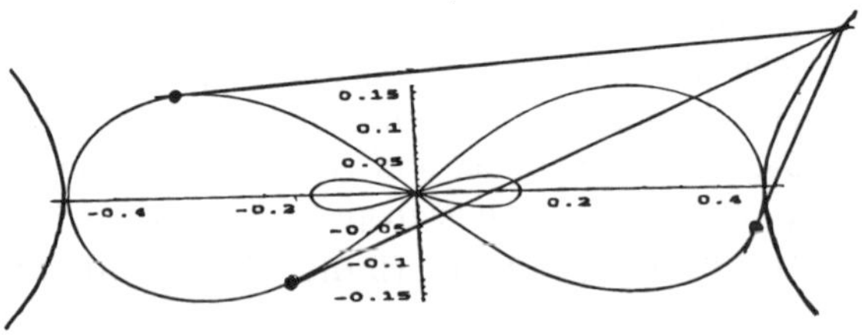

FIGURE 8-2 (motion of the centers
of force and velocity)

Thanks to Carles Simó for numerous and precise comments which I included in the text. Warm thanks to the organizers of the Taiyuan satellite conference of

ICM 2002 for giving me the opportunity to present these questions in the enchanting environment of Jinci.

References

[AC] Albouy A. & Chenciner A., *le problème des n corps et les distances mutuelles, Inventiones Mathematicæ, 131, pp. 151-184 (1998)*

[Ar] Arnaud M.C. *On the type of certain periodic orbits minimizing the Lagrangian action, Nonlinearity 11, pp. 143-150 (1998)*

[BCS] Bang D., Chenciner A. & Simó C. *Truly perverse relative equilibria of the planar n-body problem, in preparation*

[Ber] Berger M.A. *Hamiltonian dynamics generated by Vassiliev invariants, Journal of Physics A: Math. Gen. 34, 1363-1374 (2001)*

[Br] Broucke R., Elipe E. & Riaguas A. *On the Figure-8 Periodic Solutions in the 3-Body Problem, AIAA/AAS Astrodynamics Specialist Conference, Monterey August 2002*

[C1] Chenciner A. *Introduction to the N-body problem, notes of the Ravello summer school, september 1997,* http://www.bdl.fr/Equipes/ASD/person/chenciner/chenciner.htm

[C2] Chenciner A. *Action minimizing periodic solutions of the n-body problem, "Celestial Mechanics, dedicated to Donald Saari for his 60th Birthday", A. Chenciner, R. Cushman, C. Robinson, Z.J. Xia ed., Contemporary Mathematics 292, pp. 71-90 (2002)*

[C3] Chenciner A. *Action minimizing solutions of the n-body problem: from homology to symmetry, Proceedings ICM Beijing 2002, vol. III, pp. 279-294*

[C4] Chenciner A. *La forme de N corps, Proceedings of the conference in honor of Gilles Chatelet (Paris, June 2001), to appear*

[C5] Chenciner A. *Perverse solutions of the planar n-body problem, submitted to the proceedings of the International Conference dedicated to Jacob Palis for his 60th anniversary (July 2000), to appear in Astérisque*

[C6] Chenciner A. *Are there perverse choreographies ?, to appear in the Proceedings of the HAMSYS conference held in Guanajuato in march 2001*

[CD] Chenciner A. & Desolneux N. *Minima de l'intégrale d'action et équilibres relatifs de n corps, C.R. Acad. Sci. Paris. t. 326, Série I (1998), 1209-1212.* Corrections in *C.R. Acad. Sci. Paris. t. 327, Série I (1998), 193 and in [C3].*

[CGMS] Chenciner A., Gerver J., Montgomery R. and Simó C. *Simple choreographies of N bodies: a preliminary study, to appear in Geometry, Mechanics and Dynamics, Springer*

[CM] Chenciner A. and Montgomery R. *A remarkable periodic solution of the three body problem in the case of equal masses, Annals of Math., 152, pp. 881-901 (2000)*

[Ch] Chen K.C. *On Chenciner-Montgomery's orbit in the three-body problem, Disc. Cont. Dyn. Systems 7 (2001)* $n^0 1$*, pp. 85-90*

[F.F.O] Fujiwara T., Fukuda H. & Ozaki H. *Choreographic Three Bodies on the Lemniscate, arXiv:math-ph/0210044 v1 (2002)*

[H] Hénon M. *Families of periodic orbits in the three-body problem, Celestial Mechanics 10, pp. 375-388 (1974)*

[Ma] Marchal C *The family* P_{12} *of the three-body problem. The simplest family of periodic orbits with twelve symmetries per period, Fifth Alexander von Humboldt Colloquium for Celestial Mechanics (2000)*

[M1] Montgomery R. *The N-body problem, the braid group, and action-minimizing periodic solutions, Nonlinearity v. 11, pp. 363-376 (1998)*

[M2] Montgomery R. *Action spectrum and collisions in the three-body problem, "Celestial Mechanics, dedicated to Donald Saari for his 60th Birthday", A. Chenciner, R. Cushman, C. Robinson, Z.J. Xia ed., Contemporary Mathematics 292, pp. 173-184 (2002)*

[Mo] Moore C. *Braids in Classical Dynamics, Physical Review Letters 70* $n^0 24$ *(1993)*

[N] Nauenberg M. *Periodic orbits for three particles with finite angular momentum, Physics Letters A 292, pp. 93-99 (2001)*

[O] Offin D, *Maslov index and instability of periodic orbits in Hamiltonian systems, preprint 2002*

[P] Poincaré H. *Sur les solutions périodiques et le principe de moindre action, C.R.A.S. t. 123, pp. 915-918 (1896)*

[S1] Simó C. *Dynamical properties of the figure eight solution of the three-body problem, "Celestial Mechanics, dedicated to Donald Saari for his 60th Birthday", A. Chenciner, R. Cushman, C. Robinson, Z.J. Xia ed., Contemporary Mathematics 292, pp. 209-228 (2002)*

[S2] Simó C. *New families of Solutions in N–Body Problems, Proceedings of the Third European Congress of Mathematics, C. Casacuberta et al. eds. Progress in Mathematics, 201, pp. 101-115 (2001)*

[S3] Simó C. *Private communication*

[S4] Simó C. *Periodic orbits of the planar N-body problem with equal masses and all bodies on the same path. In B. Steves, J. Maciejewski, editors, The Restless Universe: Applications of N-Body Gravitational Dynamics to Planetary, Stellar and Galactic Systems. Bristol, IOP Publishing, 2001.*

[V1] Venturelli A., *Une caractérisation variationnelle des solutions de Lagrange du problème plan des trois corps, C.R. Acad. Sci. Paris, t. 332, Série I, p. 641-644, (2001)*

[V2] Venturelli A., *Thèse, Paris (to be defended in 2002)*

[W] Wintner A. *The Analytical Foundations of Celestial Mechanics, Princeton University Press 1947, par. 403-404*

[ZZ1] Zhang S. & Zhou Q., *A Minimizing Property of Lagrangian Solutions, Acta Mathematica Sinica, English Series, Vol. 17, No.3, p. 497-500 (2001)*

[ZZ2] Zhang S. & Zhou Q., *Variational method for the choreography solution to the three-body problem, Science in China (Series A), Vol. 45 No. 5, 594-597, (2002)*

Boundary blow-up solutions and their applications

Yihong Du

School of Mathematics, Statistics and Computer Science
University of New England, Armidale, NSW2351, Australia

ABSTRACT. We demonstrate how boundary blow-up solutions of some elliptic equations can be used to obtain better understanding of various well-known nonlinear problems on a bounded domain or over the entire R^N.

1. Introduction

In this paper, we consider several applications of boundary blow-up problem of the standard form

$$\Delta u = f(u) \text{ in } \Omega, \ u|_{\partial\Omega} = \infty, \tag{1.1}$$

and its variations. Here $\Omega \subset R^N$ is a bounded domain with smooth boundary, $u|_{\partial\Omega} = \infty$ means $u(x) \to \infty$ as $d(x) = d(x, \partial\Omega) \to 0$, and $f(s)$ is a C^1 function of $s \in R^1$.

The study of (1.1) has a long history. Motivated by a geometric problem, L. Bieberbach considered, in a 1916 paper [B], the case $N = 2, f(u) = e^u$, and proved that (1.1) has a unique solution $u \in C^2(\Omega)$, and

$$u(x) - \ln d(x)^{-2} = O(1) \text{ as } d(x) \to 0.$$

We list below some other early results we are aware of:

- H. Rademacher [R], 1943 (motivated by a physical problem): Same result holds when $N = 3$.
- J.B. Keller [K], 1957 (also related result by R. Osserman [O], 1957): Let $N \geq 1$, $f(u)$ positive, non-decreasing, $\int_0^\infty F(t)^{-1/2}dt < \infty$, where $F(t) = \int_0^t f(s)ds$. Then (1.1) has at least one solution.

e-mail: ydu@turing.une.edu.au.

- C. Loewner and L. Nirenberg [LN], 1974:
  For $N \geq 3, f(u) = u^{(N+2)/(N-2)}$, (1.1) has a unique positive solution, and

$$u(x)d(x)^{(N-2)/2} \to [N(N-2)/4]^{(N-2)/4} \text{ as } d(x) \to 0.$$

Since the 1990's, the problems regarding existence, uniqueness, asymptotic behaviour and multiplicity of solutions for (1.1) with more general nonlinearity $f(u)$, or for

$$\Delta_p u = f(u), \text{ where } \Delta_p u = div(|\nabla u|^{p-2}\nabla u),\ p > 1,$$

have been extensively studied; see, e.g., [AR], [ACS], [BM], [DL], [GLS], [HMR], [LM], [MV], [M], [MPZ], [MRW], [V], [Z] and the references therein.

Remark: Let (M, g) be a compact Riemannian manifold of dimension $N \geq 2$, and Γ a nonempty closed subset of M. Motivated by the singular Yamabe problem, the problem

$$\Delta_g u = u^p - \lambda u \text{ in } M \setminus \Gamma,\ u|_\Gamma = \infty,$$

where $p = (N+2)/(N-2)$, or more generally $p > 1$, has been studied by many people since the 1980's. In this context, people are interested in knowing what Γ is admissible, and the asymptotic behaviour of u near Γ. For more details in this direction, we refer the interested reader to the recent paper by D. Finn [F] and the references therein.

In this paper, we look at several situations where boundary blow-up problems arise naturally in the study of certain standard nonlinear problems, and demonstrate how boundary blow-up solutions can help to better understand the original problems. The results here are chosen from several recent papers of the author and his collaborators (see, [DD, D1, D2, D3, D4, DG1, DG2, DYG, DH, DL, DM]).

In section 2 we consider the logistic equation and two other standard population models. Boundary blow-up solutions will be used to describe their asymptotic behavior.

In section 3, we demonstrate that by introducing a small parameter in the logistic and competition models, one can obtain stable positive solutions with prescribed patterns. An important role in this analysis will be played by boundary blow-up solutions.

In section 4, we make use of the properties of certain boundary blow-up problems to obtain some Liouville type results; some related problems will also be discussed.

This paper is an expanded version of the talk I gave at the ICM2002 Satellite Conference on Nonlinear Functional Analysis held in Taiyuan in July, 2002. I wish to thank the organizers of the conference for the invitation. I also want to thank my collaborators E.N. Dancer, Yuxia Guo, Zongming Guo, Qingguang Huang, Shujie Li and Li Ma; without their help, this paper would not exist.

2. Asymptotic behaviour of the logistic and other population models

Consider the (degenerate) logistic model

$$\begin{cases} u_t - \Delta u = \lambda u - b(x)u^p, & x \in \Omega, t > 0, \\ u = 0, & x \in \partial\Omega, t > 0, \\ u(x,0) = u_0(x) \geq 0, \not\equiv 0, & x \in \Omega, \end{cases} \tag{2.1}$$

where $p > 1$, $b(x)$ is continuous, nonnegative and

$$b(x) \equiv 0 \text{ in } \Omega_0 \subset\subset \Omega, \; b(x) > 0 \text{ in } \overline{\Omega} \setminus \overline{\Omega}_0.$$

For simplicity, we assume that Ω_0 is connected and has smooth boundary.

It was proved by Ouyang in 1992 [Ou] that (2.1) has a positive steady-state solution if and only if $\lambda \in (\lambda_1^{\Omega}, \lambda_1^{\Omega_0})$; moreover, there is at most one positive steady-state. Here we use λ_1^D to denote the first eigenvalue of the problem

$$-\Delta u = \lambda u, \; u|_{\partial D} = 0.$$

The following result by J.M Fraile et al in 1996 [FKLM] provides a description of the long-time dynamical behavior of (2.1):

(i) If $\lambda \leq \lambda_1^{\Omega}$, then $\lim_{t\to\infty} u(x,t) = 0$,
(ii) if $\lambda \in (\lambda_1^{\Omega}, \lambda_1^{\Omega_0})$, then $\lim_{t\to\infty} u(x,t) = u_\lambda(x)$, the unique positive steady-state of (2.1).
(iii) if $\lambda \geq \lambda_1^{\Omega_0}$, then $\lim_{t\to\infty} \|u(\cdot,t)\|_\infty = \infty$.

A natural question is: What is the point-wise behaviour of $u(x,t)$ as $t \to \infty$ when $\lambda \geq \lambda_1^{\Omega_0}$?

It turns out that one needs a suitable boundary blow-up problem to answer this question, as demonstrated by the following result of Y. Du and Qingguang Huang, 1999 ([DH]):

THEOREM 2.1. (a) *If $\lambda \geq \lambda_1^{\Omega_0}$, then*
$\lim_{t\to\infty} u(x,t) = \infty$ *for* $x \in \overline{\Omega}_0$,
$\underline{\lim}_{t\to\infty} u(x,t) \geq \underline{U}_\lambda(x)$, $\overline{\lim}_{t\to\infty} u(x,t) \leq \overline{U}_\lambda(x)$ *for* $x \in \Omega \setminus \overline{\Omega}_0$.
Here $\underline{U}_\lambda$ and $\overline{U}_\lambda$ are minimal and maximal positive solutions of the boundary blow-up problem

$$-\Delta U = \lambda U - b(x)U^p \text{ in } \Omega \setminus \overline{\Omega}_0, \; U|_{\partial\Omega} = 0, \; U|_{\partial\Omega_0} = \infty. \tag{2.2}$$

(b) *For any $\lambda \in (-\infty, \infty)$, problem (2.2) has a minimal positive solution $\underline{U}_\lambda$ and a maximal positive solution $\overline{U}_\lambda$.*
(c) *If for some positive constants α and c,*

$$\lim_{x\to\partial\Omega_0} b(x)/d(x,\Omega_0)^\alpha = c, \tag{2.3}$$

then (2.2) has a unique positive solution (and hence $\underline{U}_\lambda = \overline{U}_\lambda$).

Remark: In a recent work by Y. Du and Zongming Guo [DG1], parts (b) and (c) have been generalized to the case $-\Delta_p u = \lambda u^{p-1} - b(x)u^q$, $q > p - 1 > 0$, and (2.3) can be relaxed to $c = c(x)$, a positive continuous function on $\partial\Omega_0$.

Boundary blow-up problems also arise naturally in the study of competition models and predator-prey models that have a certain degeneracy. In [D1,D2, DD], we have shown that a certain critical number λ^* exists such that when a parameter λ in the models is less that λ^*, then the models behave as if no degenerecy exists, but when λ is greater than λ^*, then essential changes occur from the non-degenerate case, and boundary blow-up solutions of certain systems of equations arise naturally in this context. These boundary blow-up solutions play an important role in understanding the dynamical behaviour of the model. We refer to [D1,D2,DD] for more details.

3. Realization of prescribed patterns in population models

For a given smooth bounded domain $\Omega \subset R^N$ $(N \geq 2)$, let D_k, $k = 1, 2, ..., m$ be smooth subdomains such that

$$D_k \subset\subset \Omega,\ \overline{D}_i \cap \overline{D}_j = \emptyset \text{ if } i \neq j.$$

In a recent paper by Y. Du and Shujie Li [DL], the following results are proved:

THEOREM 3.1. *Let $b(x)$ be a continuous function, identically zero on $D = \cup_{k-1}^m D_k$ and positive on the rest of Ω, and let $\lambda > \max_{1 \leq k \leq m} \lambda_1^{D_k}$. Let B denote the Dirichlet, Neumann or Robin boundary operator. Then the following conclusions hold.*

(a) *The unique positive solution u_ϵ of*

$$-\Delta u = \lambda u - [b(x) + \epsilon]u^p,\ Bu|_{\partial\Omega} = 0$$

satisfies

$$\lim_{\epsilon \to 0} u_\epsilon = \infty,\ \forall x \in \overline{D},\ \lim_{\epsilon \to 0} u_\epsilon = U(x),\ \forall x \in \Omega \setminus \overline{D},$$

where U is the minimal positive solution of

$$-\Delta U = \lambda U - b(x)U^p \ in\ \Omega \setminus \overline{D},\ BU|_{\partial\Omega} = 0,\ U|_{\partial D} = \infty. \tag{3.1}$$

Note: if $\Omega \setminus \overline{D}$ is not connected, (3.1) represents several boundary blow-up problems.

(b) *Let $v_\epsilon = \epsilon^{p-1} u_\epsilon$. Then v_ϵ is the unique positive solution to*

$$-\Delta v = \lambda v - (\frac{b(x)}{\epsilon} + 1)v^p,\ v|_{\partial\Omega} = 0,$$

and

$$\lim_{\epsilon \to 0} v_\epsilon(x) = 0, \forall x \in \Omega \setminus \overline{D},$$

$$\lim_{\epsilon \to 0} v_\epsilon = \theta_k(x), \forall x \in D_k,$$

where θ_k is the unique positive solution to

$$-\Delta u = \lambda u - u^p \ in\ D_k,\ Bu|_{\partial D_k} = 0.$$

It should be noted that the conclusion in part (b) of Theorem 3.1 is proved as a consequence of part (a) where boundary blow-up solutions play a central role.

We observe that the graph of the solution v_ϵ in Theorem 3.1 develops a clear pattern as $\epsilon \to 0$. It is well known that as a steady-state solution, v_ϵ is globally attractive for all nontrivial nonnegative solutions of the corresponding parabolic problem. Therefore, as a biological interpretation, one can say that for a single species whose growth is governed by the logistic law, it is possible to design a spatially heterogeneous environment (depending on ϵ) so that the population distribution of the species, v_ϵ, concentrates on the desired spatial region $D = \cup_{k=1}^m D_k$.

It is natural to ask whether this is still possible if there are more than one species in the environment. In Y. Du [D3], the two species competition model was investigated and a positive answer was obtained.

THEOREM 3.2. *Let Ω, D and D_k be as before. If c, d are positive constants and the continuous function $b(x)$ is identically zero on $D = D_1 \cup ... \cup D_m$ and positive in the rest of Ω, then for suitable choices of the parameters λ, μ and all small $\epsilon > 0$, the competition model*

$$\begin{cases} u_t(x,t) - \Delta u(x,t) = \lambda u - [\epsilon^{-1}b(x) + 1]u^2 - cuv, \\ v_t(x,t) - \Delta v(x,t) = \mu v - e(x)v^2 - \epsilon^{-1}duv, \end{cases} \quad (x,t) \in \Omega \times [0,\infty)$$

under natural boundary conditions on $\partial\Omega$, possesses an asymptotically stable positive steady-state solution (u_ϵ, v_ϵ) that has pattern D, that is, roughly speaking, as $\epsilon \to 0$, u_ϵ converges to a positive function ϕ over D, while it converges to 0 over the rest of Ω; on the other hand, v_ϵ converges to 0 over D but converges to some positive function ψ in the rest of Ω. In other words, the two competing species u_ϵ and v_ϵ become spatially segregated as $\epsilon \to 0$, with u_ϵ concentrating on D and v_ϵ concentrating on $\Omega \setminus D$.

Remark: In fact, $\phi|_{D_k} = \theta_k$, and ψ is the second component of a positive solution of the boundary blow-up problem

$$-\Delta u = \lambda u - b(x)u^2 - cuv, -\Delta v = \mu v - v^2 - duv, x \in \Omega \setminus D,$$

$$u|_{\partial\Omega} = 0, u|_{\partial D} = \infty; \quad v|_{\partial\Omega\cup\partial D} = 0.$$

4. Loiuville type result and eventual flatness of solutions

For simplicity, we explain the results with the model equation:

$$-\Delta_p u = u^{p-1} - u^q, \; x \in \Omega \subset R^N, \; q > p-1 > 0. \tag{4.1}$$

We first consider the case that $\Omega = R^N$. We have the following Liouville type result (see Y. Du and L. Ma [DM] for $p = 2$, and Y. Du and Z. Guo [DG1] for the general case):

THEOREM 4.1. *If $\Omega = R^N$ and $u \in C^1(R^N)$ is a weak nonnegative solution to (4.1), then $u \equiv$ constant (i.e., $u \equiv 0$ or $u \equiv 1$).*

Idea of proof: Denote $B_R = \{x \in R^N : |x| < R\}$ and consider the auxiliary problems

$$-\Delta_p v = v^{p-1} - v^q \text{ in } B_R,\ v|_{\partial B_R} = 0, \tag{4.2}$$

and

$$-\Delta_p w = w^{p-1} - w^q \text{ in } B_R,\ w|_{\partial B_R} = \infty. \tag{4.3}$$

The following conclusions hold:

(i) For large R, (4.2) and (4.3) have unique positive solutions v_R and w_R, respectively.
(ii) Any positive solution of (4.1) satisfies $v_R \le u \le w_R$ on B_R, $\forall R$.
(iii) $u_R \to 1$, $w_R \to 1$ as $R \to \infty$ locally uniformly in R^N.
(iv) By the strong maximum principle, any nonnegative solution of (4.1) is strictly positive unless it is identically zero.

Now let u be any nonnegative solution of (4.1). By (iv), u is either identically zero, in which case there is nothing to prove, or u is strictly positive on R^N. Therefore we may assume $u > 0$ in R^N. Now it follows from (ii) and (iii) that $u \equiv 1$. This finishes the proof. □

Next we consider the case that Ω is a bounded domain and $p > 2$. In this case, it is well known that for all small $\epsilon > 0$, the problem

$$-\epsilon\Delta_p u = u^{p-1} - u^q,\ u|_{\partial\Omega} = 0$$

has a unique positive solution u_ϵ and $O_\epsilon := \{x \in \Omega : u_\epsilon(x) = 1\} \neq \emptyset$. Moreover, as $\epsilon \to 0$, $O_\epsilon \to \Omega$ in the following sense: For any given compact $D \subset \Omega$, there exists $\epsilon_0 > 0$ such that $O_\epsilon \supset D$ for $\epsilon \in (0, \epsilon_0)$.

In Y. Du and Z. Guo [DG2], by making use of boundary blow-up solutions, we obtained the following result:

THEOREM 4.2. *For any given compact $D \subset \Omega$, there exists $\epsilon_0 > 0$ such that when $\epsilon \in (0, \epsilon_0)$, any positive solution u of*

$$-\epsilon\Delta_p u = u^{p-1} - u^q \text{ in } \Omega,\ (\text{no boundary condition on } \partial\Omega) \tag{4.4}$$

satisfies $u(x) \equiv 1$ on D.

Idea of proof: Consider the boundary blow-up problem

$$-\epsilon\Delta_p v = v^{p-1} - v^q,\ v|_{\partial\Omega} = \infty. \tag{4.5}$$

The following holds:

(i) Problem (4.5) has a unique positive solution v_ϵ for every $\epsilon > 0$.
(ii) $\tilde{O}_\epsilon := \{x \in \Omega : v_\epsilon(x) = 1\} \neq \emptyset$ for small ϵ and $\tilde{O}_\epsilon \to \Omega$ as $\epsilon \to 0$.
(iii) Any positive solution of (4.4) satisfies $u_\epsilon \le u \le v_\epsilon$.

Now the conclusion of Theorem 4.2 follows directly. □

Remarks: (i) If $1 < p \le 2$, then by the strong maximum principle, $O_\epsilon = \emptyset$, $\tilde{O}_\epsilon = \emptyset$. But we still have an eventual flatness result: For any given small $\delta > 0$ and given compact $D \subset \Omega$, there exists $\epsilon_0 > 0$ such that when $\epsilon \in (0, \epsilon_0)$, any positive solution u of (4.4) satisfies

$$|u(x) - 1| < \delta, \ \forall x \in D.$$

(ii) The ideas in the above outlined proofs are very intuitive. However, they are difficult to apply for more general nonlinearities of a similar nature. In [DG2], a variant of Serrin's sweeping principle was used to deal with more general nonlinearities.

The ideas here can be used to deal with many other problems. In [D4] and [DYG], we considered an indefinite superlinear problem of the following form:

$$-\Delta u = \lambda u + a(x)u^p, \ u > 0 \ \text{ in } R^N, \tag{4.6}$$

where $p > 1$ and $a(x)$ is a sign-changing continuous function satisfying

$$\lim_{|x|\to\infty} a(x) = a_\infty < 0.$$

THEOREM 4.3. (a) Existence and nonexistence: *There exists $\Lambda > 0$ such that (4.6) has at least one solution when $\lambda \in (0, \Lambda)$, and there is no solution when $\lambda > \Lambda$.*

(b) Multiplicity and further existence result: *Suppose further that*

$$\text{(i) } \nabla b(x) \neq 0 \ \textit{ whenever } \ b(x) = 0, \quad \textit{and} \text{ (ii) } p < N^*$$

where $N^* = \infty$ when $N = 1, 2$, $N^* = (N+2)/(N-2)$ when $N \geq 3$. Then (4.6) has at least one solution for $\lambda = \Lambda$ and $\lambda \leq 0$, at least two solutions when $\lambda \in (0, \Lambda)$.

Here one crucial point in proving the multiplicity result is to use boundary blow up solutions and zero boundary solutions of suitable auxiliary elliptic equations to obtain accurate estimates for the positive solution $u(x)$ of (4.6) near ∞. One rough estimate is the following:

$$\lim_{|x|\to\infty} u(x) = (\lambda^+/|a_\infty|)^{1/(p-1)}.$$

Some finer estimates are also needed.

References

[AR] A. Aftalion and W. Reichel, *Existence of two boundary blow-up solutions for semilinear elliptic equations*, J. Differential Equations 141 (1997), 400–421.

[ACS] V. Anuadha, C. Brown and R. Shivaji, *Explosive nonnegative solutions to two point boundary value problems*, Nonlinear Anal., 26(1996), 613-630.

[BM] C. Bandle and M. Marcus, *'Large' solutions of semilinear elliptic equations: existence, uniqueness and asymptotic behaviour*, J. Anal. Math., 58(1992), 9-24.

[B] L. Bieberbach, *$\Delta u = e^u$ und die automorphen funktionen*, Math. Ann., 77(1916), 173-212.

[DD] E.N. Dancer and Y. Du, *Effects of certain degeneracies in the predator-prey model*, to appear in SIAM J Math. Anal.

[DL] G. Diaz and R. Letelier, *Explosive solutions of quasilinear elliptic equations: existence and uniqueness*, Nonlinear Anal., 20(1993), 97-125.

[D1] Y. Du, *Effects of a degeneracy in the competition model, part I: classical and generalized steady-state solutions*, J. Diff. Eqns, 181(2002), 92-132.

[D2] Y. Du, *Effects of a degeneracy in the competition model, part II: perturbation and dynamical behaviour*, J. Diff. Eqns, 181(2002), 133-164.

[D3] Y. Du, *Realization of prescribed patterns in the competition model*, Research Report No. 02-204, Univ of New England, 2002, submitted for publication.

[D4] Y. Du, *Multiplicity of positive solutions for an indefinite superlinear elliptic problem in R^N*, Research Report No. 02-198, Univ. of New England, 2002, submitted for publication.

[DG1] Y. Du and Z. Guo *Boundary blow-up solutions and their applications in quasilinear elliptic problems*, to appear in J. d'Analyse Math.

[DG2] Y. Du and Z. Guo *Liouville type results and eventual flatness of positive solutions for p-Laplacian equations*, Adv. Diff. Eqns., 7(2002), 1479-1512.

[DYG] Y.Du and Y. Guo, *Mountain pass solutions and an indefinite superlinear elliptic problem on R^N*, Research Report No. 02-201, Univ. of New England, 2002, submitted for publication.

[DH] Y. Du, Q. Huang, *Blow-up solutions for a class of semilinear elliptic and parabolic equations*, SIAM J. Math. Anal., 31(1999), 1-18.

[DL] Y. Du and S. Li, *Positive solutions with prescribed patterns in some simple semilinear equations*, Diff. Integral Eqns. 15(2002), 805-822.

[DM] Y. Du and L. Ma, *Logistic type equations on R^N by a squeezing method involving boundary blow-up solutions*, J. London Math. Soc., 64(2001), 107-124.

[F] D. Finn, *Behavior of positive solutions to $\Delta_g u = u^q + Su$ with prescribed singularities*, Indiana Univ. Math. J., 49(2000), 177-219.

[FKLM] J.M. Fraile, P. Koch Medina, J. Lopez-Gomez, and S. Merino, *Elliptic eigenvalue problems and unbounded continua of positive solutions of a semilinear elliptic equation*, J. Diff. Eqns., 127(1996), 295-319.

[GLS] J. Garcia-Melian, R. Letelier-Albornoz and J. Sabina de Lis, *Uniqueness and asymptotic behaviour for solutions of semilinear problems with boundary blow-up*, Proc. Amer. Math. Soc., 129(2001), 3593-3602.

[HMR] S. Hill, K.S. Moore and W. Reichel, *Existence and uniqueness theorems for singular anisotropic quasilinear elliptic boundary value problems*, Proc. Amer. Math. Soc., 128(2000), 1673-1683.

[K] J.B. Keller, *On solutions of $\Delta u = f(u)$*, Comm. Pure Appl. Math., 10(1957), 503-510.

[LM] A.C. Lazer and P.J. McKenna, *On a problem of Bieberbach and Rademacher*, Nonlinear Anal., 21(1993), 327-335.

[LN] C. Loewner and L. Nirenberg, *Partial differential equations invariant under conformal or projective transformations*, in Contributions to Analysis (a collection of papers dedicated to Lipman Bers), pp. 245-272. Academic Press, New York, 1974.

[MV] M. Markus and L. Veron, *Uniqueness and asymptotic behavior of solutions with boundary blow-up for a class of nonlinear elliptic equations*, Ann. Inst. H. Poincare Anal. Non Lineaire, 14(1997), 237-274.

[M] J. Matero, *Quasilinear elliptic equations with boundary blow-up*, J. Anal. Math., 69(1996), 229-247.

[MPZ] J. Mawhin, D. Papini and F. Zanolin, *Boundary blow-up for differential equations with indefinite weight*, preprint, 2001.

[MRW] P.J. McKenna, W. Reichel and W. Walter, *Symmetry and multiplicity for nonlinear elliptic differential equations with boundary blow-up*, Nonlinear Anal., 28(1997), 1213-1225.

[O] R. Osserman, *On the inequality $\Delta u \geq f(u)$*, Pacific J. Math., 7(1957), 1641-1647.

[Ou] T. Ouyang, *On the positive solutions of semilinear equation $\Delta u + \lambda u - hu^p = 0$ on the compact manifolds*, Trans. Amer. Math. Soc., 331(1992), 503-527.

[R] H. Rademacher, *Einige becondere probleme partieller differential gleichungen*, in Die Differential-und Integral- gleichungen der Mechanik und Physik I, 2nd edition, pp. 838-845. Rosenberg, New York, 1943.

[V] L. Veron, *Semilinear elliptic equations with uniform blow-up on the boundary*, J. Anal. Math., 59(1992), 231-250.

[Z] Z. Zhang, *A remark on the existence of explosive solutions for a class of semilinear elliptic equations*, Nonlinear Anal., 41(2000), 143-148.

On periodic solutions for forced Van der Pol type equations

Chikahiro Egami and Norimichi Hirano

Department of Mathematics, Graduate School of Environment and Information Sciences, Yokohama National University, Tokiwadai, Hodogayaku, Yokohama Japan

1. Introduction

Let $f, g, e : \mathbb{R} \to \mathbb{R}$ be continuous functions. The Liénard type equation with forcing term

$$u_{tt} + f(u)u_t + g(u) = e(t), \qquad t \in \mathbb{R} \tag{L}$$

has been studied by many authors due to its adoption to a wide variety of mechanical, electrical, biological and economical systems. The equation (L) is usually called Duffing type when $f(u) = \zeta$, or Van der Pol type when $g(u) = \eta u$, where ζ and η are constants.

In economics, there have been a number of elegant mathematical treatments of some of the traditional business cycle theories e.g. the treatment of Kaldor's model by Chang and Smyth (1971) and Schinasi (1982) and of a complete Keynesian system by Torre (1977). Because of one-dimensional relaxation oscillator just like Van der Pol type without an forcing term, their treatments concentrated on the question of the existence of limit cycles and consequently made use of the planar properties such as Poincaré-Bendixson theory and Jordan curve (cf. C. Chiarella [14, §2, §3, §7]. N-dimensional extension of the equation (L) has also been studied by many authors. In the present paper, we consider the existence and stability of periodic solutions of N-dimensional Van del Pol type equations. The equation

$$u_{tt} + \varepsilon(u^2 - 1)u_t + u = e(t) \qquad t \in \mathbb{R} \tag{1.1}$$

is called Van der Pol equation, where $\varepsilon > 0$ and $e : R \to R$ is periodic function with period $T > 0$. Van der Pol equation has been studied these 50 years and the behavior of the solutions is now well understood(cf. Guchenheimer and Holmes[9]). On the

e-mail: `hirano@hiranolab.jks.ynu.ac.jp`.

other hand, few seems to be known about N-dimensional Van der Pol equation of the form

$$u_{tt} + \frac{d}{dt}F(u) + g(u) = e(t) \qquad t \in \mathbb{R}, \tag{V}$$

where $u(t) = (u_1(t), u_2(t), \dots, u_N(t))$, and $u_i(t) \in \mathbb{R}^N$ for each $t \in \mathbb{R}$ and each integer $i \in [1, N]$, and $F, g : \mathbb{R}^N \to \mathbb{R}^N$ are continuous functions, and $e : \mathbb{R} \to \mathbb{R}^N$ is a periodic function with period T. As in one dimensional case, it is natural to expect under a suitable condition on F and g that equation (V) has a nontrivial limit cycle in case $e = 0$, and that equation (V) has periodic solutions when $e \neq 0$. However, it is not easy for $N \geq 2$ to obtain similar results as in one-dimensional cases, becuase we can not rely on Poincaré-Bendixson theory or properties of Jordan curve in case $N \geq 2$. In fact, N-dimensional existence results for periodic solutions of the forced Van der Pol type are not yet established(cf. J. Mawhin[1] for forced Duffing type). In this paper, we consider the existence and stablity of periodic solutions for the forced Van del Pol equation in the case that $N \geq 2$.

2. Statement of Main Results

To state our results, we need some preliminaries. In the following, $|\cdot|$ and $\langle\cdot,\cdot\rangle$ stands for the norm and the inner product of $\mathbb{R}^N$, respectively. In the rest of this paper, we identify each T-periodic function in $L^2_{loc}(\mathbb{R};\mathbb{R}^N)$ with the element of $L^2([0,T];\mathbb{R}^N)$. For each $u \in L^2([0,T];\mathbb{R}^N)$, we put $\|u\| = \left(\int_0^T |u|^2\right)^{1/2}$. We put

$$H = \left\{ u \in C([0,T];\mathbb{R}^N) : u(0) = u(T), \int_0^T |u|^2\,dt < \infty, \int_0^T |u_t|^2\,dt < \infty \right\}.$$

The norm of H is defined by $\|u\|_H = \left(\|u\|^2 + \|u_t\|^2\right)^{1/2}$ for each $u \in H$. We also put

$$\widetilde{H} = \left\{ u \in H : \int_0^{2\pi} u(t)dt = 0 \right\}.$$

We consider the case that $g = A$ and F has the form

$$F(x_1, x_2, \cdots, x_n) = \begin{pmatrix} F_1(x_1) \\ F_2(x_2) \\ \cdots \\ F_N(x_N) \end{pmatrix} \tag{F1}$$

where A is (n,n)-matrix and $F_i : \mathbb{R} \to \mathbb{R}$ is continuous mapping. We put $f_i = F_i'$ for $1 \leq i \leq N$. We assume that each f_i satisfies

$$f_i(0) < 0 \qquad \text{and} \qquad \frac{f_i(s) - f_i(0)}{s^2} > 0 \qquad \text{for } s \neq 0. \tag{F2}$$

We put

$$\mu = \min\left\{\frac{f_i(s)-f_i(0)}{s^2} : s \neq 0, 1 \le i \le N\right\} \text{ and } \nu = \min\{|f_i(0)| : 1 \le i \le N\}.$$

We also assume that

$$\langle Au, u\rangle \le |u|^2 \qquad \text{for all } u \in \mathbb{R}^N. \tag{A1}$$

Then we have

THEOREM 1. *(cf.[2]) Suppose that (F1), (F2) and (A1) hold. Let $e \in \widetilde{H}$. Then problem (V) has a T -periodic solution. Moreover if F is odd function and $e(t+T/2) = -e(t)$ for each $t \in \mathbb{R}$, then problem (V) has a T-perodic solution $u \in \widetilde{H}$ such that $u(t+T/2) = -u(t)$ for all $t \in \mathbb{R}$.*

The proof of Theorem 1 is based on the degree theory. We next consider the stability of solutions of (V) when the forcing term is sufficiently large. From the proof of Theorem 1 given in [2] and the numerical simulations, we expect that if the forcing term is sufficiently small, then (V) has a periodic solution which is a repeller(see [2]). On the other hand, if e is sufficiently large, (V) has a periodic solution which is an attractor. Here we restrict the problem (V) to the case that $N = 2$,

$$F(x_1,x_2) = \begin{pmatrix} \varepsilon_1(\frac{x_1^3}{3} - x_1) \\ \varepsilon_2(\frac{x_2^3}{3} - x_2) \end{pmatrix} \quad , \quad A = \begin{pmatrix} 1 & c_1 \\ c_2 & 1 \end{pmatrix}$$

and

$$e(t) = d\begin{pmatrix} \cos t \\ \cos t \end{pmatrix},$$

where ε_1,ε_2, c_1, c_2 and d are positive constants. Then we have

THEOREM 2. *Let $\varepsilon_1, \varepsilon_2 > 0$. There exists $c_0 > 0$ and $d_0 > 0$ such that if $c_1, c_2 \in (0, c_0)$ and $d \ge d_0$, then problem (V) has a stable 2π-periodic solution.*

3. Proof of Theorem 2.

Our purpose in this section is to give a sketch of the proof of Theorem 2. That is we show that if the forcing term is sufficiently large, the problem (V) has a stable solution under the assumption of Theorem 2. From the assumption, problem (V) has the form

$$\begin{pmatrix} \ddot{u}_1 \\ \\ \ddot{u}_2 \end{pmatrix} + \begin{pmatrix} \varepsilon_1(u_1^2-1)\dot{u}_1 \\ \\ \varepsilon_2(u_2^2-1)\dot{u}_2 \end{pmatrix} + \begin{pmatrix} u_1 + c_1u_2 \\ \\ c_2u_1 + u_2 \end{pmatrix} = d\begin{pmatrix} \cos t \\ \\ \cos t \end{pmatrix} \tag{V2}$$

where $\varepsilon_1, \varepsilon_2 > 0$ and $c_1, c_2 \in (0,1)$. Throughout the rest of this section, we denote by C_0, C_1, $\cdots$,C_0', C_1',$\cdots$ various constants independent of d. For each $i \in \{1,2\}$, we define $\bar{\imath} \in \{1,2\}$ by $\bar{\imath} = 1$ if $i = 2$ and $\bar{\imath} = 2$ if $i = 1$. Under our assumption, we have that there exists a 2π-periodic solution $u \in \widetilde{H}$ such that $u(t+\pi) = -u(t)$ for all $t \in \mathbb{R}$. In the following, we assume that $u \in \widetilde{H}$ is a solution of (V2) satisfying the above property. Then the following lemma holds(cf. [2]).

LEMMA 1. *There exists $C_0, C_1 > 0$ such*

$$\|\dot{u}_1\| + \|\dot{u}_2\| \le C_0 + C_1 d. \tag{3.1}$$

We also have

LEMMA 2. *There exist $C_2, C_3 > 0$ such that*

$$\sup\left\{|\dot{u}_i(t)| : t \in [0,2\pi], i = 1,2\right\} < C_2 + C_3 d. \tag{3.2}$$

PROOF. Let $u = (u_1, u_2)$ be a solution of (V2). First we note that from Lemma 1, there exists C_0' and $C_1' > 0$ such that

$$\|u_1\| + \|u_2\| < C_0' + C_1' d. \tag{3.3}$$

Multiplying (V2) by $\dot{u}$ and integrating over $[t, s]$ with $t > s$, we have

$$\left\{\begin{aligned} &|\dot{u}_1(t)|^2 + \varepsilon_1 \textstyle\int_s^t (u_1^2 - 1)\dot{u}_1^2 + |u_1(t)|^2 + c_1 \int_s^t \dot{u}_1 u_2 \\ &\qquad = |\dot{u}_1(s)|^2 + |u_1(s)|^2 + \textstyle\int_s^t e\dot{u}_1, \\ &|\dot{u}_2(t)|^2 + \varepsilon_2 \textstyle\int_s^t (u_2^2 - 1)\dot{u}_2^2 + |u_2(t)|^2 + c_2 \int_s^t \dot{u}_2 u_1 \\ &\qquad = |\dot{u}_2(s)|^2 + |u_2(s)|^2 + \textstyle\int_s^t e\dot{u}_2. \end{aligned}\right.$$

Then noting that

$$\left|\int_s^t \dot{u}_1 u_2\right| \le \|\dot{u}_1\| \, \|u_2\| \qquad \text{and} \qquad \left|\int_s^t e\dot{u}_1\right| \le \|e\| \, \|\dot{u}_1\|,$$

we find

$$|\dot{u}_1(t)|^2 \le |\dot{u}_1(s)|^2 + |u_1(s)|^2 + \varepsilon_1 \|\dot{u}_1\|^2 + c_1 \|\dot{u}_1\| \, \|u_2\| + \|e\| \, \|\dot{u}_1\|.$$

Since (3.1) and (3.3) holds, we can choose $s \in [0, 2\pi)$ such that

$$|\dot{u}_1(s)|^2 + |u_1(s)|^2 \le C_0' + C_1' d^2.$$

Then again by using (3.1) and (3.3), we have from the inequalities above that there exists C_2', $C_3' > 0$ such that $|\dot{u}_1(t)| \le C_2' + C_3' d$ for all $t \in [0, 2\pi]$. We can get the same estimate for $|\dot{u}_2(t)|$. This completes the proof. ∎

LEMMA 3. *There exists $C_4, C_5 > 0$ such that*

$$\sup_{t\in[0,2\pi]} |u_i(t)| \le C_4 + C_5 d^{1/3} \qquad i = 1, 2.$$

PROOF. Integrating (V2) over $[t, s]$, we find

$$\begin{aligned} \dot{u}_i(t) + \varepsilon_i\left(\frac{u_i^3(t)}{3} - u_i(t)\right) + \int_s^t (u_i + c_i u_{\bar{i}})dt \qquad \qquad \qquad \qquad \qquad \qquad (3.4) \\ = \dot{u}_i(s) + \varepsilon_i\left(\frac{u_i^3(s)}{3} - u_i(s)\right) + d\int_{s_i}^t \cos t \end{aligned}$$

for $i = 1, 2$. Recalling that u satisfies $u(t+\pi) = -u(t)$ for each $t \in \mathbb{R}$, we have that

$$\int_0^{2\pi} (u_i(s) - \varepsilon_i(\frac{u_i^3}{3} - u_i)(s)) - 0.$$

Then there exists $s_i \in [0, 2\pi]$, $i = 1, 2$ such that

$$\dot{u}_i(s_i) - \varepsilon_i(\frac{u_i^3}{3} - u_i)(s_i) = 0, \qquad i = 1, 2.$$

Then we have

$$\varepsilon_i \left| \frac{u_i^3(t)}{3} \right| \le |\dot{u}_i(t)| + \varepsilon_i |u_i(t)| + \left| \int_{s_i}^t (u_i + c_i u_{\bar{i}}) dt \right| + d \left| \int_{s_i}^t \cos t dt \right|$$

for $i = 1, 2$. Then by using (3.1) and (3.3), we find that there exist C_0' and C_1' such that

$$|u_i^3(t)| \le C_0' + C_1' d \qquad \text{for all } t \in [0, 2\pi], i = 1, 2.$$

Then one can see that the assertion holds. ∎

LEMMA 4. *There exists $\rho_0 > 0$ such that for each $\rho \in (0, \rho_0)$ and $\varepsilon > 0$, there exists $d_0 > 0$ and an interval $[a_i, b_i] \subset [0, 2\pi]$ satisfying that $|[a_i, b_i]| \ge \pi - \varepsilon$ and*

$$|u_i(t)| \ge \rho d^{1/3} \qquad \text{for } d \ge d_0 \text{ and } t \in [a_i, b_i], \qquad i = 1, 2.$$

PROOF. From (3.4), we find

$$\dot{u}_i(t) + \varepsilon_i(\frac{u_i^3}{3} - u_i)(t) + \int_{s_i}^t u_i + c_i u_{\bar{i}} = d \int_{s_i}^t \cos t \qquad \text{for } i = 1, 2. \tag{3.5}$$

First we claim that there exists $\rho_0 > 0$ such that if d is sufficiently large

$$\max\{|u(t)| : t \in [0, 2\pi]\} > \rho_0 d^{1/3}. \tag{3.6}$$

Fix $\varepsilon > 0$. We choose $\delta > 0$ and an interval $A_i \subset [0, 2\pi]$, $i = 1, 2$ such that $|A_i| \ge \pi - \varepsilon/2$ and

$$d \int_{s_i}^t \cos t \ge \delta d \qquad \text{for all } t \in A_i.$$

Then we have by (3.5) and Lemma 3 that there exists $C > 0$ and

$$\begin{aligned} |\dot{u}_i(t)| &\ge - \left| \varepsilon_i(\frac{u_i^3}{3} - u_i)(t) + \int_{s_i}^t u_i + c_i u_j \right| + d \int_{s_i}^t \cos t \\ &\ge - \left| \varepsilon_i(\frac{u_i^3}{3} - u_i)(t) \right| - C(1 + d^{1/3}) + \delta d \qquad \text{on } A_i. \end{aligned}$$

Here we choose $\rho_0 > 0$ so small that for d sufficiently large,

$$\max\{\varepsilon_1, \varepsilon_2\} |t^3/3 - t| \le \delta d/2 \qquad \text{for all } |t| \le \rho_0 d^{1/3}.$$

Now suppose that $|u_i(t)| \le \rho_0 d^{1/3}$ on A_i. Then from the inequality above, we have

$$|\dot{u}_i(t)| \ge \delta d/2 - C(1 + d^{1/3}) \qquad \text{on } A_i, \qquad i = 1, 2.$$

Then by choosing d sufficiently large, we have $|\dot{u}_i(t)| \geq \delta d/3$ on A_i. Let $t, s \in A_i$ such that $|t-s| \geq \pi - \varepsilon/2$. Then we have

$$|u_i(t) - u_i(s)| \geq (2\pi - \varepsilon)\delta d/3.$$

Then we have $\rho_0 d^{1/3} \geq |u_i(t)| \geq (2\pi - \varepsilon/2)\delta d/6$. Therefore we reach to a contradiction by choosing d sufficiently large. Thus we find that (3.6) holds as claimed. We now choose $\rho \in (0, \rho_0)$. Then by the same argument as above, we have that for sufficiently large d,

$$|\dot{u}_i(t)| \geq \delta d/3 \qquad \text{on } A_i, \qquad\qquad i = 1, 2.$$

This implies that if we choose d sufficiently large, the assertion holds for some interval $[a_i, b_i]$ with $|[a_i, b_i]| \geq \pi - \varepsilon$, $i = 1, 2$. ∎

By a similar argument as in the proof of lemma above, we have

LEMMA 5. *There exists $\delta_0 > 0$ such that for d sufficiently large,*

$$\{t \in [0, 2\pi] : |u_i(t)| \leq 1\} = B_{i,1} \cup B_{i,2}, \qquad\qquad i = 1, 2$$

where $B_{i,1}$, $B_{i,2}$ are intervals such that $|B_{i,j}| \leq \delta_0 d^{-1}$, $i, j \in \{1, 2\}$ and

$$B_{i,2} = \{t + \pi : t \in B_{i,1}\} \qquad i = 1.2.$$

Here we define $t_{i,j} \in [0, \pi]$, $i, j \in \{1, 2\}$ by

$$B_{i,1} = [t_{i,1}, t_{i,2}] \qquad\qquad \text{for } i = 1, 2. \tag{3.7}$$

Now we assume that d is so large that the assertions of Lemmas above hold. Let consider the linearlized problem

$$\begin{pmatrix} \ddot{v}_1 \\ \\ \ddot{v}_2 \end{pmatrix} + \begin{pmatrix} \varepsilon_1(u_1^2 - 1)\dot{v}_1 \\ \\ \varepsilon_2(u_2^2 - 1)\dot{v}_2 \end{pmatrix} + \begin{pmatrix} (2u_1\dot{u} + 1)v_1 + c_1 v_2 \\ \\ c_1 v_1 + (2u_2 + 1)v_2 \end{pmatrix} = 0. \tag{3.8}$$

To prove the stability of the solution u, it is sufficient to show that the linearized problem (3.8) is stable. Let $s_1, s_2 \in [0, 2\pi]$ with $s_1 < s_2$ such that

$$|u_i(s_j)| \neq 1 \qquad \text{for } i, j \in \{1, 2\}.$$

We integrate the equations above on $[s_i, t], i = 1, 2$. Then we have

$$\begin{aligned} &\begin{pmatrix} \dot{v}_1(t) \\ \\ \dot{v}_2(t) \end{pmatrix} + \begin{pmatrix} \varepsilon_1(u_1^2 - 1)v_1(t) \\ \\ \varepsilon_2(u_2^2 - 1)v_2(t) \end{pmatrix} + \begin{pmatrix} \int_{s_i}^t v_1 + c_1 \int_{s_i}^t v_2 \\ \\ c_2 \int_{s_i}^t v_1 + \int_{s_i}^t v_2 \end{pmatrix} \\ &\qquad = \begin{pmatrix} \dot{v}_1(s_i) + \varepsilon_1(u_1^2 - 1)v_1(s_i) \\ \\ \dot{v}_2(s_i) + \varepsilon_2(u_2^2 - 1)v_2(s_i) \end{pmatrix}. \end{aligned} \tag{3.9}$$

We choose $(x_{i,1}, y_{i,1})$ and $(x_{i,2}, y_{i,2})$ such that $y_{i,1} + \varepsilon_1(u_1^2(s_i) - 1)x_{i,1} = 0$ and $y_{i,2} + \varepsilon_2(u_2^2(s_i) - 1)x_{i,2} = 0$, respectively. By choosing s_1, s_2 appropriatively, we may assume that

$$\dot{v}_i(s_2) + \varepsilon_i(u_i^2(s_2) - 1)v_i(s_2) \neq 0 \qquad i = 1, 2,$$

where $v = (v_1, v_2)$ is the solution of (3.8) with $(\dot{v}_1(s_1), v_1(s_1)) = (y_{1,1}, x_{1,1})$ and $(\dot{v}_2(s_1), v_2(s_1)) = (y_{1,2}, x_{1,2})$. Then we have that $(x_{2,1}, y_{2,1}, 0, 0)$, $(0, 0, y_{2,2}, x_{2,2})$, $(\dot{v}_1(s_2), v_1(s_2), 0, 0)$ and $(0, 0, \dot{v}_2(s_2), v_2(s_2))$ form the basis of $\mathbb{R}^4$. Now let $i \in \{1, 2\}$ and $v = (v_1, v_2)$ be a solution of (3.8) satisfying

$$\dot{v}_1(s_i) + \varepsilon_1(u_1^2 - 1)v_1(s_i) = \dot{v}_2(s_i) + \varepsilon_2(u_2^2 - 1)v_2(s_i) = 0.$$

We put $w_1(t) = \int_{s_i}^t v_1 dt$ and $w_2(t) = \int_{s_i}^t v_2 dt$ for $t \in \mathbb{R}$. Then we have that

$$\begin{pmatrix} \ddot{w}_1(t) \\ \\ \ddot{w}_2(t) \end{pmatrix} + \begin{pmatrix} \varepsilon_1(u_1^2 - 1)\dot{w}_1(t) \\ \\ \varepsilon_2(u_2^2 - 1)\dot{w}_2(t) \end{pmatrix} + \begin{pmatrix} w_1 + c_1 w_2 \\ \\ c_2 w_1 + w_2 \end{pmatrix} = 0. \tag{3.10}$$

We can rewrite the equation above as

$$\begin{pmatrix} \dot{w}_1 \\ \dot{z}_1 \\ \dot{w}_2 \\ \dot{z}_2 \end{pmatrix} = \begin{pmatrix} 0 & 1 & 0 & 0 \\ -1 & -\varepsilon_1(u_1^2 - 1) & -c_1 & 0 \\ 0 & 0 & 0 & 1 \\ -c_2 & 0 & -1 & -\varepsilon_2(u_2^2 - 1) \end{pmatrix} \begin{pmatrix} w_1 \\ z_1 \\ w_2 \\ z_2 \end{pmatrix}$$

Then by Froque's theorem, fundamental matrix $\Phi(t)$ can be written as

$$\Phi(t) = P(t)e^{Rt}$$

where $P(t)$ is a matrix such that each element of $P(t)$ is a 2π -periodic function, and R is a constant matrix. Then from the observation above, to prove the stability of solutions of (3.8), it is sufficient to prove that $\lim_{t\to\infty} \left|e^{Rt}\right| = 0$. To prove this property, we first consider the linear problem

$$\begin{pmatrix} \ddot{w}_1(t) \\ \\ \ddot{w}_2(t) \end{pmatrix} + \begin{pmatrix} \varepsilon_1(u_1^2 - 1)\dot{w}_1(t) \\ \\ \varepsilon_2(u_2^2 - 1)\dot{w}_2(t) \end{pmatrix} + \begin{pmatrix} w_1 \\ \\ w_2 \end{pmatrix} = 0. \tag{3.11}$$

Fix a solution (w_1, w_2) of (3.11). Then by multiplying (3.11) by $(\dot{w}_1, \dot{w}_2)$ and integrating over $[s, t]$, we find

$$\left|\dot{w}_i(t)\right|^2 + |w_i(t)|^2 + \int_s^t \varepsilon_i(u_i^2 - 1)\left|\dot{w}_i\right|^2 = \left|\dot{w}_i(0)\right|^2 + |w_i(0)|^2, \qquad i = 1.2.$$

From the inequality above and Lemma 5, we have the following lemma by a standard argument.

LEMMA 6. *There exists $\delta_1 > 0$ such that*

$$\left|\dot{w}_i(t_{i,2})\right|^2 + |w_i(t_{i,2})|^2 \leq \left(\left|\dot{w}_i(t_{i,1})\right|^2 + |w_i(t_{i,1})|^2\right) e^{\delta_1 d^{-1}} \qquad i = 1, 2.$$

We also have by Lemma 4 that

LEMMA 7. *There exists $\delta_2 > 0$ such that*

$$\left|\dot{w}_i(t_{i,1}+\pi)\right|^2 + |w_i(t_{i,1}+\pi)|^2 \le \left(\left|\dot{w}_i(t_{i,2})\right|^2 + |w_i(t_{i,2})|^2\right)(1-\delta_2 d^{-2/3})$$

for $i = 1, 2$.

Combining two lemmas above, we find that that there exists $\alpha \in (0,1)$ such that

$$\left|\dot{w}_i(t)\right|^2 + |w_i(t)|^2 \le \alpha(\left|\dot{w}_i(0)\right|^2 + |w_i(0)|^2, \qquad i = 1.2.$$

Since (w_1, w_2) is arbitrary, we find that each solution (w_1, w_2) of (3.11) is stable. Then by choosing c_1, c_2 sufficiently small, we obtain that each solution (w_1, w_2) of (3.10) is stable and then $\lim_{t\to\infty}\left|e^{Rt}\right| = 0$. Then from the definition of w_1 and w_2, it follows that each solution (v_1, v_2) of (3.9) is stable. This completes the proof.

References

[1] Jean Mawhin, *An Extension of a Theorem of A.C.Lazer on Forced Nonlinear Oscillations*, Journal of Mathematical Analysis and Applications, Vol.40, pp.20-29, (1972).

[2] C. Egami and N. Hirano, Existence of periodic solutions for N-dimensional Van der Pol equations, in preparation.

[3] J.E.Flaherty, F.C.Hoppensteadt, *Frequency Entrainment of a Forced van der Pol Oscillator*, Studies in Applied Mathematics, Vol.58, pp.5-15, (1978).

[4] Duane W.Storti, Per G.Reinhall, *Stability of in-phase and out-of-phase modes for a pair of linearly coupled van der Pol oscillators*, Nonlinear dynamics, pp.1-23, Series on Stability, Vibration and Control Systems. Series B, Vol.2, World Scientific Publishing, River Edge, NJ, (1997).

[5] Nguyen Phuong Các, *Periodic Solutions of a Liénard equation with forcing term*, Nonlinear Analysis, Vol.43, pp.403-415, (2001).

[6] Chaitan P.Gupta, Juan J.Nieto, Luis Sanchez, *Periodic Solutions of Some Lienard and Duffing Equations*, Journal of Mathematical Analysis and Applications, Vol.140, pp.67-82, (1989).

[7] Norman Levinson, *A Second Order Differential Equation with Singular Solutions*, Annals of Mathematics, Vol.50, No.1, pp.127-153, (1949).

[8] M.L.Cartwright, J.E.Littlewood, *Some Fixed Point Theorems*, Annals of Mathematics, Vol.54, No.1, pp.1-37, (1951).

[9] John Guckenheimer, Philip Holmes, "Nonlinear Oscillations, Dynamical Systems, and Bifurcations of Vector Fields", Applied Mathematical Sciences, 42, Springer-Verlag, (1983).

[10] Frank C. Hoppensteadt, "Analysis and Simulation of Chaotic Systems", Applied Mathematical Sciences, 94, Springer-Verlag, (2000).

[11] Ferdinand Verhulst "Nonlinear Differential Equations and Dynamical Systems", Springer-Verlag, (1990).

[12] Wolfgang Walter "Ordinary Differential Equations ", Springer-Verlag, (1998).

[13] N.G.Lloyd, "Degree Theory", Cambridge University Press, (1978).

[14] G.Hardy, J.E.Littlewood, G.Pólya, "INEQUALITIES", Cambridge University Press, (1934).

[15] Carl Chiarella, "The Elements of a Nonlinear Theory of Economic Dynamics", LectureNotes in Economics and Mathematical Systems, 343, Springer-Verlag, (1990).

[16] Toru Maruyama, Wataru Takahashi, "Nonlinear and Convex Analysis in Economic Theory", LectureNotes in Economics and Mathematical Systems, 419, Springer-Verlag, (1995).

[17] Gianni Ricci, Kumaraswamy Velupillai, "Growth Cycles and Multisectoral Economics: the Goodwin Tradition", LectureNotes in Economics and Mathematical Systems, 309, Springer-Verlag, (1988).

A nonlinear Lyapunov-Schmidt reduction and multiple solutions for some elliptic problem

Pierpaolo Esposito[1] and Gianni Mancini[2]

[1] *Università di Roma Tor Vergata, Dipartimento di Matematica, via della Ricerca Scientifica, Roma 00133 Italy.*

[2] *Università di Roma Tre, Dipartimento di Matematica, Largo S. Leonardo Murialdo 1, Roma 00146 Italy.*

ABSTRACT. We present a finite dimensional reduction for perturbed variational functionals and discuss some nonlinear elliptic PDE with Sobolev critical growth in bounded domains.

1. Introduction

We will present in this talk a general variational principle for perturbative problems in presence of a manifold of "quasi critical points" for the unperturbed energy functional. A model problem is the following:

$$(P) \qquad \begin{cases} -\triangle u = u^p + f(\delta, x, u) & \text{in}\,\Omega \\ u = 0 & \text{on}\,\partial\Omega \\ u > 0 & \text{in}\,\Omega \end{cases}$$

where Ω is an open bounded set in $\mathbf{R}^N$, $N \geq 3$, $p = \frac{N+2}{N-2}$ is the limiting Sobolev exponent. Here $f(\delta, x, u)$ is a perturbation term, small if δ is small, satisfying the growth condition

$$\exists c > 0 : \quad |f(\delta, x, u)| \leq c(1 + |u|^p)$$

The authors' research is supported by M.U.R.S.T. under the national project *Variational methods and nonlinear differential equations*.

[2]Corresponding author.

e-mail: `pesposit@axp.mat.uniroma2.it` `mancini@mat.uniroma3.it`.

For $f(\delta, x, u) = \delta u$ and $0 < \delta < \lambda_1(\Omega)$, precise existence results for (P) were established in [**11**] (see also [**1**] for sharp conditions in higher dimensions and general nonlinearities); existence of multiple solutions and asymptotic behaviour for $\delta \to 0^+$ were discussed in [**17**], [**19**] .
Problem (P) can be seen as the stationary equation for some functional $E_\delta(u) := E(u) - G(\delta, u)$, where, for δ small, $G(\delta, u)$ is a small perturbation of the unperturbed energy $E(u)$: it fits into a general framework which we will describe in the next section (see [**15**]- [**16**] for details).

2. A finite dimensional reduction

We consider a functional $E_\delta(u) = E(u) - G(\delta, u)$, u in some Hilbert space H. We assume for $E(u)$ the existence of a smooth manifold $Z \subset H$ in the form

$$Z = \{z(\epsilon, y) : \epsilon > 0, y \in U \subset \mathbf{R}^N\},$$

z a smooth parametrization, such that

$$(A1) \qquad \|\nabla E\,(z(\epsilon, y))\,\| \to_{\epsilon \to 0} 0 \text{ uniformly on compact subsets of } U.$$

We will refer to Z as "an almost critical manifold" for $E(u)$. We require for Z some nondegeneracy property. Denoted by T_z the tangent space to Z at $z \in Z$, let $\pi_z : H \to T_z$ and $\pi_z^\perp := Id - \pi_z$ be orthogonal projections, and set $L_z := \pi_z^\perp E''(z)\,|_{T_z^\perp}$. We assume

$$(A2) \qquad L_z \in Iso(T_z^\perp, T_z^\perp)) \qquad \text{and} \qquad \sup_{z \in Z} \|L_z^{-1}\| < \infty$$

We will also need a smallness assumption on $G(\delta, u)$

$$(A3) \quad G(\delta, u) \to_{\delta \to 0} 0 \text{ in } C^2_{loc}$$

Looking for critical points of E_δ close to Z, one can perform a nonlinear Lyapunov-Schmidt reduction : given $U_0 \subset\subset U$, there exist $\bar{\epsilon} > 0$, $\bar{\delta} > 0$ and a smooth map $w(\delta, z(\epsilon, y))$ such that

$$(i) \;\; w(\delta, z) \in T_z^\perp \quad \text{and} \quad \|w\| = O(\|\nabla E_\delta(z)\|), \qquad (z = z(\epsilon, y), \epsilon \leq \bar{\epsilon}, \delta \leq \bar{\delta}, y \in U_0)$$

$$(ii) \quad \pi_z^\perp \nabla E_\delta\,(z + w(\delta, z)) = 0, \qquad (z = z(\epsilon, y),\ \epsilon \leq \bar{\epsilon},\ \delta \leq \bar{\delta},\ y \in U_0)$$

It remains to solve the "bifurcation equation"

$$(b) \quad \pi_z\,\nabla E_\delta\,(z + w(\delta, z)) = 0\,,\ z \in Z_0.$$

If Z is linear, equation (b) is known to have a variational structure: it is equivalent to

$$(b)' \quad \frac{\partial}{\partial z}\; E_\delta\,(z + w(\delta, z)) = 0$$

because $\frac{\partial w}{\partial z} h \in Z^\perp \ \forall h \in Z$, in such a case. Actually, to have equivalence between (b) and $(b)'$ it is enough that $||\pi_z \frac{\partial w}{\partial z}|| = O(||w||)$. The argument goes as follows. Let $z(t)$ be a smooth curve on Z_0, with $z(0) = z_0$ and $\dot{z}(0) = \pi_{z_0} \nabla E_\delta(z_0 + w(\delta, z_0))$. By assumption,

$$0 = \frac{d}{dt} E_\delta \left(z(t) + w(\delta, z(t)) \right) |_{t=0} =< \nabla E_\delta(z_0 + w(\delta, z_0)), \dot{z}(0) + \frac{\partial w}{\partial z}(\delta, z_0)\dot{z}(0) >$$

Since $\pi_{z_0}^\perp \nabla E_\delta(z_0 + w(\delta, z_0)) = 0$, using the estimate for $||w||$ we get

$$\|\dot{z}(0)\|^2 \leq \|\dot{z}(0)\|^2 \|\pi_{z_0} \frac{\partial w}{\partial z}(\delta, z_0)\| \leq c\, \|w(\delta, z_0)\| \|\dot{z}(0)\|^2 \leq \tilde{c}\, \|\dot{z}(0)\|^2 \|\nabla E_\delta(z)\|$$

and hence $\dot{z}(0) = 0$ because $\|\nabla E_\delta(z)\| << 1$ if ϵ, δ are small, by (A2)-(A3).
In turn, the estimate for $||\pi_z \frac{\partial w}{\partial z}||$ involves the variation of T_z: it holds true assuming

$$(A4) \qquad \exists\, c > 0 : \quad \|\pi_z \frac{\partial}{\partial z} \left(\pi_z^\perp v\right) \| \leq c \|\pi_z^\perp v\| , \ \forall v \in H$$

Let us derive from $(A4)$ the estimate for $||\pi_z \frac{\partial w}{\partial z}||$. Let $\bar{w} = w(\delta, \bar{z})$ for some $\bar{z} \in Z_0$, δ fixed.

From $\pi_z w(\delta, z) \equiv 0$ it follows $\pi_z \frac{\partial w}{\partial z} = -\frac{\partial}{\partial z}(\pi_z \bar{w})$ at $z = \bar{z}$. Since $-\frac{\partial}{\partial z}(\pi_z \bar{w}) = \frac{\partial}{\partial z}(\pi_z^\perp \bar{w})$, we have, by (A4),

$$\|\pi_{\bar{z}} \frac{\partial w}{\partial z}(\delta, \bar{z})\| \leq c\, \|\pi_{\bar{z}}^\perp \bar{w}\|$$

This gives the desired estimate, because $\pi_{\bar{z}}^\perp \bar{w} = \bar{w}$

Relevant informations can be derived by the variational structure of (b). After a Taylor expansion, (b)' rewrites as

$$stat_z \left[E\left(z(\epsilon, y) \right) - G\left(\delta, z(\epsilon, y) \right) + O(\|\nabla E_\delta(z)\|^2) \right]$$

and this leads to look for "stable" critical points of the "Melnikov function"

$$(\epsilon, y) \to E\left(z(\epsilon, y) \right) - G\left(\delta, z(\epsilon, y) \right).$$

Of course, estimates for the "error term" $\|\nabla E_\delta(z)\|^2$, either in L^∞ or in C^1, are crucial.

REMARK 2.1. *1. Similar (somehow different) procedure has been used by Rey* [**19**], *and then by many others for (P) in the perturbative case (see, to quote a few,* [**2**], [**9**], [**10**] *)*
2. Similar principle, but in case Z is a critical manifold (i.e. $\nabla E(z) = 0$ for any $z \in Z$), goes back to Ambrosetti, Coti Zelati and Ekeland in [**5**] *(for compact Z) and to Ambrosetti and Badiale in* [**4**] *(for non compact Z). Recent advances in more*

complicated situations are due to Ambrosetti, Malchiodi and Ni in [**6**] *(solutions concentrating on codimension-1 manifolds for singularly perturbed problems) See also* [**18**], [**7**]

3. Applications to problem (P)

We consider the functional $E_\delta(u) = E(u) - G(\delta, u)$, $u \in H_0^1(\Omega)$, associated to problem (P), where

$$E(u) = \frac{1}{2}\int_\Omega |\nabla u|^2 - \frac{1}{p+1}\int_\Omega |u|^{p+1}.$$

Here, the manifold Z is given by

$$Z = \{PU_{\epsilon,y} : \epsilon > 0, y \in \Omega\},$$

where $P : D^{1,2}(\mathbf{R}^N) \to H_0^1(\Omega)$ is the orthogonal projection and

$$U_{\epsilon,y}(x) = \epsilon^{-\frac{N-2}{2}} U(\frac{x-y}{\epsilon}) \qquad U(x) = \frac{c_N}{(1+|x|^2)^{\frac{N-2}{2}}}$$

$c_N = [N(N-2)]^{\frac{N-2}{4}}$, are the extremal functions for the Sobolev inequality on $\mathbf{R}^N$. Assumptions (A1)-(A2)-(A3) are satisfied (see [**19**] and [**16**] for detailed proofs). We will consider two distinguished problems of type (P) : to find $u > 0$ in Ω such that

$$(NHD) \qquad \begin{cases} -\triangle u = u^p & \text{in}\,\Omega \\ u = \delta\varphi & \text{on}\,\partial\Omega \qquad \varphi \in C^1(\partial\Omega) \text{ positive somewhere} \end{cases}$$

$$(PSCE) \qquad \begin{cases} -\triangle u = (1+\delta a(x))u^p & \text{in}\,\Omega \\ u = 0 & \text{on}\,\partial\Omega \end{cases}$$

For problem (NHD), the existence of a small solution ($u_\delta \to 0$ in $H^1(\Omega)$ as $\delta \to 0$) holds in much greater generality (see [**14**]). The existence of a "large" solution for $\delta < \delta^*$, δ^* some positive number, is due, among other things, to Caffarelli and Spruck [**12**]. Some multiplicity results for problem (NHD) with δ small have been obtained by Rey [**20**]. We present here some improvements (a detailed proof is in [**16**]). Let $h(x)$ be the harmonic extension of $\varphi(x)$ and let $\psi(x, y)$ be the regular part of the Green function of Ω. Set

$$K(x) = \frac{h^2(x)}{H(x)} \quad , \quad H(x) = \psi(x, x),$$

THEOREM 3.1. *(i) Let $N \geq 3$. Let C be a compact subset of Ω with $h(x) > 0$ on C and $\max_{\partial C} K < \max_C K$. Then, for $\delta_j \to 0$ there exist u_j solutions and $x \in C$ such that*

$$|\nabla u_j|^2 \rightharpoonup c_N^2 S^{\frac{N}{2}} \delta_x \qquad (S := \textit{best Sobolev constant})$$

(ii) Let $N \geq 4$. Then the number of "large" solutions is greater (or equal) than the number of non degenerate critical points of K with $h > 0$.

According to (i), highly oscillating boundary datas should give rise to a large number of solutions:

EXAMPLE 3.1. *If $\varphi = \sum_{j=1}^{m} \frac{1}{|x - tx_j|^{N-2}}, x \in \Omega = B_1$ with $\{x_1, .., x_m\} \subset \partial B_1$ and $1 < t < t(N, x_1, .., x_m)$, problem (NHD) with boundary data $\delta\varphi$ admits for δ small at least m solutions.*

As for (PSCE), if $\Omega = \mathbf{R}^N$, it is the prescribed scalar curvature problem for S^N. Known results, (see [**13**]) recently improved by Ambrosetti, Azorero and Peral [**3**], are variations of the following one: assume $a \in C^2$ has a finite number of non degenerate critical points x_j ; assume in addition

$$\text{(c.c.)} \qquad \Delta a(x_j) \neq 0 \ \forall j, \qquad \sum_{\Delta a(x_j) > 0} i(\nabla a, x_j) \neq 0$$

$$(*) \qquad < \nabla a(x), x > < 0 \text{ for } |x| >> 1$$

Then (PSCE) is solvable. The "counting condition" (c.c.) was introduced by Bahri-Coron in [**8**].
If Ω is a bounded domain in $\mathbf{R}^N$, different kind of results can be obtained (see, e.g. [**16**]).

THEOREM 3.2. *Let $N > 6$. Then, for δ small, the number of solutions for (PSCE) is greater (or equal) than the number of isolated critical points of a with positive laplacian and non zero topological index.*

REMARK 3.3. *(i) If $< \nabla a(x), x > \leq 0$ for any $x \in \Omega$ and Ω is starshaped, no solution can exist.*
(ii) There are no "ground state" solutions (our solutions have Morse index ≥ 2).

The above results are obtained by an analysis of the corresponding "Melnikov functions" . For the unperturbed energy along the manifold Z, we have that, for $y \in K \subset \Omega$ compact:

$$E(PU_{\epsilon,y}) = \frac{S^{\frac{N}{2}}}{N} + \alpha_N H(y)\epsilon^{N-2} + O(\epsilon^{N-1}), \quad \alpha_N = \frac{c_N^{p+1}}{2} \int_{\mathbf{R}^N} \frac{dx}{(1 + |x|^2)^{\frac{N+2}{2}}}$$

We first consider (NHD) , for which the functional $G(\delta, u)$ turns out to be equal to

$$G(\delta, u) = \frac{1}{p+1} \int_{\Omega} |u + \delta h|^{p+1} - |u|^{p+1}$$

By Taylor formula, and for every $N \geq 3$, we have

$$E_\delta(PU_{\epsilon,y}+w)=\tfrac{1}{N}S^{\frac{N}{2}}+\alpha_N H(y)\epsilon^{N-2}-\tfrac{2\alpha_N}{c_N}h(y)\delta\epsilon^{\frac{N-2}{2}}\ +$$
$$+O(\epsilon^{N-1}+\delta\epsilon^{\frac{N}{2}}+\delta^2\epsilon+\delta^{p+1})$$

After a change of variables $\epsilon=(\theta\delta)^{\frac{N-2}{2}}$, we are lead to look for critical points of

$$M_\delta(\theta,y)=c_N\theta^2H(y)-2\theta h(y)+\delta^{\frac{2}{N-2}}O(1+\theta^{\frac{N}{N-2}})$$

As for part i) in Theorem 3.1, we just remark that the above L^∞ estimate implies stability of strict local minima of $M(\theta,y):=c_N\theta^2H(y)-2\theta h(y)+\delta^{\frac{2}{N-2}}$; on the other hand

$$\max_{\partial C}\frac{h^2}{H}<\max_{C}\frac{h^2}{H}\Rightarrow\min_{[\theta_1,\,\theta_2]\times C}M<\min_{\partial([\theta_1,\,\theta_2]\times C)}M$$

where $0<\theta_1<\min_C\frac{h}{c_NH},\ \theta_2>\max_C\frac{h}{c_NH}$. We prove this claim. First, if $\theta(y):=\frac{h(y)}{c_nH(y)},\ \theta(y)\in(\theta_1,\,\theta_2)$ and

$$M(\theta,\,y)\ge M(\theta(y),y)\ =\ -\frac{1}{c_N}K(y)\ \ \forall\,y\in\Omega$$

So, by using the assumption on K, we can find $\gamma>0$ such that

$$\forall(\theta,y)\in[\theta_1,\theta_2]\times\partial C: M(\theta,y)\ge\min_{\partial C}M(\theta(y),y)\ge\gamma+(\min_{C}M(\theta(y),y))\ge\gamma+\min_{[\theta_1,\theta_2]\times C}M$$

Finally

$$\text{for } i=1,2:\quad \min_{C}M(\theta_i,y)=:M(\theta_i,y_i)>M(\theta(y_i),y_i)\ \ge\min_{[\theta_1,\theta_2]\times C}M$$

Part ii) in Theorem 3.1, deals with critical points of M_δ other than minima, so we also need C^1 estimates on the error terms. The following estimates hold true. Let $N\ge4$. Then

$$\tfrac{\partial}{\partial y_i}E_\delta(PU_{\epsilon,y}+w)\ =\ \alpha_N\tfrac{\partial H}{\partial y_i}(y)\epsilon^{N-2}-\tfrac{2\alpha_N}{c_N}\tfrac{\partial h}{\partial y_i}(y)\delta\epsilon^{\frac{N-2}{2}}\ +$$
$$+O(\epsilon^{N-1}\log\frac{1}{\epsilon}+\delta\epsilon^{\frac{N}{2}}\log\frac{1}{\epsilon}+\delta^2\epsilon\log\frac{1}{\epsilon}+\frac{\delta^{2p}}{\epsilon})$$
$$\tfrac{\partial}{\partial\epsilon}E_\delta(PU_{\epsilon,y}+w)\ =\ (N-2)\alpha_NH(y)\epsilon^{N-3}-(N-2)\tfrac{\alpha_N}{c_N}h(y)\delta\epsilon^{\frac{N-4}{2}}\ +$$
$$+O(\epsilon^{N-2}+\delta\epsilon^{\frac{N-2}{2}}+\delta^2\epsilon\log\frac{1}{\epsilon}+\delta^p\epsilon^{\frac{N-4}{2}}+\frac{\delta^{2p}}{\epsilon})$$

Part ii) in Theorem 3.1 follows from

$$\nabla K(y_0)=0,\ \ D^2K(y_0)\in Gl_n(\mathbf{R}),\ h(y_0)>0\ \Rightarrow$$

$$\nabla M(\theta(y_0), y_0) = 0 \text{ and } D^2M(\theta(y_0), y_0) \in Gl_{n+1}(\mathbf{R})$$

Let us prove prove this fact. As noticed above,

$\frac{\partial M}{\partial \theta}(\theta(y), y) = 0 \quad \forall y \in C$, while $\frac{\partial M}{\partial y}(\theta(y), y) = -\frac{1}{c_n}\nabla K(y)$. Thus $(\theta(y_0), y_0)$ is a critical point of M because $\nabla K(y_0) = 0$. We just have to check that it is nondegenerate.

Using $\frac{\partial H}{\partial y_j}(y_0) = 2\frac{H(y_0)}{h(y_0)}\frac{\partial h}{\partial y_j}(y_0)$ and the above relations, we find

$D_y^2 M(\theta(y_0), y_0) = \frac{2}{c_n H(y_0)}\nabla h^t(y_0)\nabla h(y_0) - \frac{1}{c_n}D^2K(y_0)$, and hence

$$D^2M(\theta(y_0), y_0) = \begin{pmatrix} 2c_nH(y_0) & 2\nabla h(y_0) \\ 2\nabla h^t(y_0) & \frac{2}{c_nH(y_0)}\nabla h^t(y_0)\nabla h(y_0) - \frac{1}{c_n}D^2K(y_0) \end{pmatrix}$$

Thus $0 = D^2M(\theta_0, y_0)(\tau, \eta)$, $\tau \in \mathbf{R}$, $\eta \in \mathbf{R}^n$ implies

$$2\tau c_n H(y_0) + 2 < \nabla h(y_0), \eta >= 0$$

$$2\tau\nabla h(y_0) + \frac{2}{c_nH(y_0)}\nabla h(y_0) < \nabla h(y_0), \eta > -\frac{1}{c_n}D^2K(y_0)\eta = 0.$$

Inserting the first relation in the second one, we find

$$H(y_0)D^2K(y_0)\eta = \nabla h(y_0)[2c_nH(y_0)\tau + 2 < \nabla h(y_0), \eta >] = 0$$

and hence $\eta = 0$ because $D^2K(y_0)$ is invertible. This implies $D^2M(\theta_0, y_0)$ is invertible.

Similarly, for problem (PSCE), we have that

$$G(\delta, u) = \frac{\delta}{p+1}\int_\Omega a(x)|u|^{p+1} \qquad G(\delta, PU_{\epsilon,y}) = \frac{S^{\frac{N}{2}}}{p+1}a(y)\delta + \frac{S^{\frac{N}{2}}}{4N}\Delta a(y)\delta\epsilon^2 + h.o.t.$$

and the corresponding "Melnikov function" is

$$(\epsilon, y) \to \frac{S^{\frac{N}{2}}}{N} + \alpha_N H(y)\epsilon^{N-2} - \frac{S^{\frac{N}{2}}}{p+1}a(y)\delta - \frac{S^{\frac{N}{2}}}{4N}\Delta a(y)\delta\epsilon^2 + h.o.t..$$

As above, we can define $\theta := \epsilon\delta^{\frac{1}{N-4}}$ and, after a change of variable, we look for "stable" critical points of

$$(\theta, y) \to \delta^{\frac{2}{N-4}}[4N\alpha_N H(y)\theta^{N-2} - S^{\frac{N}{2}}\Delta a(y)\theta^2].$$

(since the error term is of order $O(\delta^{\frac{3}{N-4}})$, the assumption $N > 6$ is needed to ensure that it is a fortiori of order $o(\delta^{\frac{2}{N-4}})$). Equivalently, we may look for zeros of the vector field

$$\frac{\partial M_\delta}{\partial \theta} = -\frac{\delta^{\frac{2}{n-4}}}{4n}\left[2n(n-2)DH(y)\theta^{n-3} - 2S^{\frac{n}{2}}\Delta a(y)\theta + o(1)\right]$$

$$\frac{\partial M_\delta}{\partial y} = \frac{S^{\frac{n}{2}}}{p+1}\nabla a(y) - \frac{\delta^{\frac{2}{n-4}}}{4n}\left[2nD\nabla H(y)\theta^{n-2} - S^{\frac{n}{2}}\nabla\Delta a(y)\theta^2 + o(1)\right].$$

We want to show that, if y_0 is an isolated critical point of a with positive laplacian and non zero topological index, then, for δ small, ∇M_δ has a zero close to (θ_0, y_0), where

$\theta_0 := \theta(y_0) = \left(\frac{S^{\frac{n}{2}}\Delta a(y_0)}{n(n-2)DH(y_0)}\right)^{\frac{1}{n-4}}$. To see this, we just consider the homothopy $\Phi = (\Phi_1, \Phi_2)$,

$$\Phi_1(t;\theta,y) := -t\frac{\delta^{\frac{2}{n-4}}}{4n}\left[2n(n-2)DH(y_0)\theta^{n-3} - 2S^{\frac{n}{2}}\Delta a(y_0)\theta\right] + (1-t)\frac{\partial M_\delta}{\partial\theta}$$

$$\Phi_2(t;\theta,y) := \frac{S^{\frac{n}{2}}}{p+1}\nabla a(y) - (1-t)\frac{\delta^{\frac{2}{n-4}}}{4n}\left[2nD\nabla H(y)\theta^{n-2} - S^{\frac{n}{2}}\nabla\Delta a(y)\theta^2 + o(1)\right].$$

Easily, Φ does not vanish, for δ small, on $\partial([\underline{\theta},\ \overline{\theta}]\times U_{y_0})$ for suitable $0 < \underline{\theta} < \theta_{y_0} < \overline{\theta}$, U_{y_0} being a neighborhood of y_0 with $\|\nabla a\| \geq c > 0$ on ∂U_{y_0}. Thus we conclude that

$$\deg(\nabla M_\delta, (\underline{\theta},\ \overline{\theta})\times U_{y_0}, 0) = \deg(\Phi(1;\theta,y), (\underline{\theta},\ \overline{\theta})\times U_{y_0}, 0) = -\deg(\nabla a, U_{y_0}, 0)$$

REMARK 3.4. *If $\Omega = B_R$, the finite dimensional reduction requires the bound $\delta << \frac{1}{R^{N-2}}$; hence, we cannot obtain solutions for (PSCE) on $\mathbf{R}^N$ sending R to infinity: one needs uniform size of the perturbation as $R \to \infty$.*

THEOREM 3.5. *Let $N > 6$, $a \in C_b^3(\mathbf{R}^N)$ and assume (c.c.) and (*). Then (PSCE) has on B_R, for $R \geq \bar{R}$, $\delta \leq \bar{\delta}$, at least one solution.*

Thus the result follows from the fact that, for $\delta \leq \bar{\delta}$, $r << R$,

$$\deg(-\nabla E_\delta(\epsilon,y), \{\epsilon > \epsilon_\delta\}\times\{|y| < r\}, 0) = -\sum_{\nabla a(x)=0\,\Delta a(x)>0} i(\nabla a, x) \neq 0$$

for some $\epsilon_\delta \to 0$ as $\delta \to 0$. The right hand side is the degree for δ small, whose computation relies on the local analysis of ∇M_δ around its critical points. Assumption $*$ provides the a priori bounds required to perform a continuation argument: property (c.c.) plays the role of an obstruction to the complete collapse of the bifurcating solutions given by Theorem 3.2

REMARK 3.2. *Assuming instead of (*) the opposite condition* $< \nabla a(x), x > >$ 0 *for* $|x| >> 1$ *and*

$$l := \sum_{\nabla a(x)=0 \Delta a(x)<0} i(\nabla a, x) \neq 0,$$

we get the same result, but with different argument: for $\delta \leq \bar{\delta}$ *and* $\delta >> \frac{1}{R^{N-2}}$, *we get that*

$$deg(-\nabla E_\delta(\epsilon, y), \{\epsilon > \epsilon_\delta\} \times \{|y| < r\}, 0) = l \neq 0.$$

Notice that, for $\delta << \frac{1}{R^{N-2}}$, *it still holds true that*

$$deg(-\nabla E_\delta(\epsilon, y), \{\epsilon > \epsilon_\delta\} \times \{|y| < r\}, 0) = - \sum_{\nabla a(x)=0 \Delta a(x)>0} i(\nabla a, x)$$

and this quantity is different from l; in particular, no a priori bounds are available in this case.

References

[1] Adimurthi , G. Mancini and K. Sandeep *A sharp solvability condition in higher dimensions for some Brezis-Nirenberg type equation, Calc. Var.* 14, 275-317 (2002)

[2] Adimurthi, G. Mancini and Yadava *The role of the mean curvature in semilinear Neumann problem involving critical exponent, Comm. in Partial Differential equations* **20** (1995), 591-631

[3] Ambrosetti A., J. G. Azorero and I. Peral, *Perturbation of* $\Delta u + u^{\frac{n+2}{n-2}} = 0$, *the scalar curvature problem in* R^n *and related topics* , *J. Funct. Analysis* **165** (1999), 117-149.

[4] Ambrosetti A., M. Badiale, *Homoclinics: Poincaré-Melnikov type results via a varational approach* , *Ann. Inst. H. Poincaré Anal. Non Linèaire* **15** (1998), no. 2, 233-252.

[5] Ambrosetti A., V. Coti Zelati and I. Ekeland, *Symmetry breaking in Hamiltonian systems, J. Differential Equations* **67** (1987), no. 2, 165-184.

[6] Ambrosetti A., A. Malchiodi and W.-M. Ni, *Solutions, concentrating on spheres, to symmetric singularly perturbed elliptic problems, C.R.Acad.Sci.Paris* **Ser.I335** (2002), 145-150.

[7] Ambrosetti A., A. Malchiodi and S. Secchi, *Multiplicity results for some nonlinear Schroedinger equations with potential, Arch. Rat. Mech. Anal.* **159** (2001), 253-271

[8] Bahri A. and J. M. Coron, *On a nonlinear elliptic equation involving the critical Sobolev exponent: the effect of the topology of the domain, Comm. Pure Appl. Math.* **41** (1988), 253-290.

[9] Berti M. and P. Bolle, *Homoclinics and chaotic behaviour for perturbed second order systems, Ann. Mat. Pura Appl.* **176** (1999), no. 4, 323-378.

[10] Berti M. and A. Malchiodi, *Non-compactness and multiplicity results for the Yamabe problem on* S^n , *J. Funct. Analysis* **180** (2001), no. 1, 210-241.

[11] Brezis H. and L. Nirenberg, *Positive solutions of nonlinear equations involving critical Sobolev exponents, Comm. Pure Appl. Math.* **36** (1983), 437-477.

[12] Caffarelli L. A. and J. Spruck, *Variational problems with critical Sobolev growth and positive Dirichlet data, Indiana Univ. Math. J.* **39** (1990), no. 1, 1-18.

[13] Chang S. A. and P. Yang, *A perturbation result in prescribing scalar curvature on* S^n, *Duke Math. J.* **64** (1991), no. 1, 27-69.

[14] Crandall M. G. and P.H. Rabinowitz, *Some continuation and variational methods for positive solutions of nonlinear elliptic eigenvalue problems,* Arch. Ration. Mech. Anal. **58** (1975), 207-218.

[15] P. Esposito, *Tesi di Laurea*

[16] Esposito P.P. and G. Mancini, *On some nonlinear elliptic BVP: almost critical manifolds and multiple solutions*, preprint.

[17] Han Z. *Asymptotic approach to singular solutions for nonlinear elliptic equations involving critical Sobolev exponent,* Ann. Inst. H. Poincaré Analyse non linéaire **8** no. 2 (1991), 159-174.

[18] Malchiodi A. and Montenegro, it Boundary concentration phenomena for a singularly perturbed elliptic problem, preprint

[19] Rey O., *The role of the Green's function in a non-linear elliptic equation involving the critical Sobolev exponent,* J. Funct. Analysis **89** (1990), 1-52.

[20] Rey O. *Concentration of solutions to elliptic equations with critical nonlinearity,* Ann. Inst. H. Poincaré Analyse non linéaire **9** no. 2 (1992), 201-218.

$p(x)-$Laplacian equations

Xian-Ling Fan

Department of Mathematics, Lanzhou University, Lanzhou 730000, P. R. China

ABSTRACT. We discuss the $p(x)-$Laplacian equation of the form

$$-div(|\nabla u|^{p(x)-2}\nabla u) = f(x,u), x \in \Omega \subset R^N,$$

where $p(x)$ is a function. This equation is a generalization of the p-Laplacian equation where p is a constant . In this paper we summarize our research results about the $p(x)$-Laplacian equations. The main contents contain: Lavrentiev phenomenon, generalized Sobolev space $W^{1,p(x)}(\Omega)$, regularity of weak solutions, strong maximum principle, the "first eigenvalue", existence and multiplicity of solutions.

1. Introduction

We consider the following $p(x)-$Laplacian equation

$$-div(|\nabla u|^{p(x)-2}\nabla u) = f(x,u), \quad x \in \Omega \subset R^N, \tag{1.1}$$

where Ω is an open domain in R^N, $p : \Omega \to R$ is a measurable real-value function such that

$$1 < p_- := ess\inf\{p(x) : x \in \Omega\} \le p_+ := ess\sup\{p(x) : x \in \Omega\} < +\infty, \tag{1.2}$$

$f : \Omega \times R \to R$ satisfies some suitable conditions. The operator $-\Delta_{p(x)}u := -div(|\nabla u|^{p(x)-2}\nabla u)$ is called $p(x)-$Laplacian. It is a generalization of the well-known $p-$Laplacian. In the $p-$Laplacian p is a constant, but in the $p(x)-$Laplacian $p(x)$ is a function. The $p(x)-$Laplacian can be applied to describe the physical phenomenon with "pointwise different properties" which earliest arose from the nonlinear elasticity theory (see[16,17]) . The variational functional corresponding to the $p(x)-$Laplacian is $\int_\Omega \frac{1}{p(x)}|\nabla u|^{p(x)}\,dx$. In general, for an integral functional

This reseach is supported by the National Science Foundation of China (19971036).

$J(u) = \int_\Omega F(x, \nabla u)dx$, we say that J (or F) satisfies the $p-$standard growth condition if

$$c_1 |\xi|^p - c_0 \le |F(x,\xi)| \le c_2 |\xi|^p + c_0, \tag{1.3}$$

and that J (or F) satisfies the $(p,q)-$nonstandard growth condition if

$$c_1 |\xi|^p - c_0 \le |F(x,\xi)| \le c_2 |\xi|^q + c_0, \tag{1.4}$$

where $p < q$, c_0, c_1 and c_2 are positive constants. The study of the variational problems and elliptic equations with nonstandard growth conditions is a new and interesting topic, see e.g. Zhikov [16,17], Marcellini [12], Acerbi and Fusco [1,2]. It is clear that the integral functional $\int_\Omega \frac{1}{p(x)} |\nabla u|^{p(x)} dx$ satisfies the so-called $p(x)-$growth condition:

$$c_1 |\xi|^{p(x)} - c_0 \le |F(x,\xi)| \le c_2 |\xi|^{p(x)} + c_0, \tag{1.5}$$

hence it satisfies the $(p_-, p_+)-$nonstandard growth condition.

The $p(x)-$Laplacian possesses more complicated nonlinearity than the $p-$Laplacian, for example, it does not possess the homogeneity. In order to study the $p(x)-$Laplacian equations we must use some new methods and tools.

In this paper we summarize our research results about the $p(x)-$Laplacian equations, such as the Lavrentiev phenomenon, generalized Sobolev space $W^{1,p(x)}(\Omega)$, regularity and existence of the weak solutions, strong maximum principle, the "first eigenvalue",etc.

2. Lavrentiev phenomenon

Consider the integral functional $J(u) = \int_\Omega F(x, \nabla u)dx$. We say that J (or F) is regular if

$$\inf\{J(u) : u \in W_0^{1,1}(\Omega)\} = \inf\{J(u) : u \in C_0^\infty(\Omega)\}. \tag{2.1}$$

Otherwise we say that J (or F) is nonregular or say that the Lavrentiev phenomenon happened.

It is well known that when J satisfies the $p-$standard growth condition the Lavrentiev phenomenon can not happen. When J satisfies the $p(x)-$growth condition, Zhikov [16] gave some examples of nonregularity and gave the following conjecture:

Zhikov's conjecture: If $p(x)$ is a smooth Morse function with saddle points, then the Lavrentiev phenomenon will happen.

In [3] we proved that the Lavrentiev phenomenon can not happen if the function $p(x)$ is Hölder continuous. This result gave a negative answer to the Zhikov's conjecture. Further results can be found in [4].

3. The spaces $L^{p(x)}(\Omega)$ and $W^{k,p(x)}(\Omega)$

In order to study the $p(x)-$Laplacian equations we need generalize the theory of the classical Lebesgue-Sobolev spaces. We established the theory of the generalized Lebesgue spaces $L^{p(x)}(\Omega)$ and the generalized Sobolev spaces $W^{k,p(x)}(\Omega)$ (see [5,9,11]).

The space $L^{p(x)}(\Omega)$ is defined by

$$L^{p(x)}(\Omega) := \{u : u : \Omega \to R \quad \text{is measurable and} \int_\Omega |u(x)|^{p(x)}\, dx < \infty\} \tag{3.1}$$

with the norm

$$|u|_{p(x)} := |u|_{L^{p(x)}(\Omega)} := \inf\{\lambda > 0 : \int_\Omega \left|\frac{u(x)}{\lambda}\right|^{p(x)} dx \le 1\}. \tag{3.2}$$

The space $L^{p(x)}(\Omega)$ is a separable and uniformly convex Banach space. The conjugate space of $L^{p(x)}(\Omega)$ is the space $L^{q(x)}(\Omega)$ where $\frac{1}{p(x)} + \frac{1}{q(x)} = 1$.

For brevity, here we only discuss the space $W^{1,p(x)}(\Omega)$, for the general space $W^{k,p(x)}(\Omega)$ see [5,9,11].

Define

$$W^{1,p(x)}(\Omega) := \{u \in L^{p(x)}(\Omega) : |\nabla u| \in L^{p(x)}(\Omega)\} \tag{3.3}$$

and

$$\|u\|_{1,p(x)} := \|u\|_{W^{1,p(x)}(\Omega)} := |u|_{p(x)} + |\nabla u|_{p(x)}\,. \tag{3.4}$$

The space $W^{1,p(x)}(\Omega)$ is a separable and reflexive Banach space.

We denote by $p^*(x)$ the Sobolev critical exponent of $p(x)$, that is

$$p^*(x) := \begin{cases} \frac{Np(x)}{N-p(x)} & \text{if} \quad p(x) < N \\ \infty & \text{if} \quad p(x) \ge N. \end{cases} \tag{3.5}$$

For the space $W^{1,p(x)}(\Omega)$ we have the following imbedding theorems of Sobolev type, where the open domain $\Omega \subset R^N$ is assumed to possess the cone property.

Theorem 1 (see [5,9]) (Imbedding theorem of Sobolev type)
1). *If $p : \overline{\Omega} \to R$ is Lipschitz continuous, $q \in L^\infty(\Omega)$ and*

$$p(x) \le q(x) \le p^*(x) \quad \textit{for a.e.} \quad x \in \Omega, \tag{3.6}$$

then there is a continuous imbedding $W^{1,p(x)}(\Omega) \to L^{q(x)}(\Omega)$.
2). *If Ω is bounded, $p : \overline{\Omega} \to R$ is continuous, $q \in L^\infty(\Omega)$, and*

$$\inf\{p^*(x) - q(x) : x \in \Omega\} > 0, \tag{3.7}$$

then there is a compact imbedding $W^{1,p(x)}(\Omega) \to L^{q(x)}(\Omega)$.

Define

$$W_r^{1,p(x)}(R^N) := \{u \in W^{1,p(x)}(R^N) : u \text{ is radially symmetric}\}. \tag{3.8}$$

Theorem 2 (see [11]) (Compact imbedding theorem of Strauss-Lions type) *Suppose that $p : R^N \to R$ is uniformly continuous and radially symmetric, $q \in L^\infty(R^N)$, and*

$$\lim_{x \in R^N} \inf\{q(x) - p(x)\} > 0 \textit{ and } \lim_{x \in R^N} \inf\{p^*(x) - q(x)\} > 0. \tag{3.9}$$

Then there is a compact imbedding $W_r^{1,p(x)}(R^N) \to L^{q(x)}(R^N)$.

Define

$$W_0^{1,p(x)}(\Omega) := \text{ the closure of } C_0^\infty(\Omega) \text{ in } W^{1,p(x)}(\Omega). \tag{3.10}$$

Theorem 3 (see [9]) (Inequality of Poincaré type) *If Ω is bounded, then there is a positive constant c such that*

$$|u|_{p(x)} \le c\,|\nabla u|_{p(x)} \ \textit{for } u \in W_0^{1,p(x)}(\Omega). \tag{3.11}$$

In general, $C^\infty(\Omega)$ may not be dense in $W^{1,p(x)}(\Omega)$.

Theorem 4 (see [9]) *Let Ω be a bounded domain with a Lipschitz boundary. If there is a positive constant L such that*

$$(Z) \qquad -\log|x-y| \cdot |p(x)-p(y)| \le L \ \textit{for } x, y \in \overline{\Omega},$$

then $C^\infty(\Omega)$ is dense in $W^{1,p(x)}(\Omega)$ and there holds

$$W_0^{1,p(x)}(\Omega) = W^{1,p(x)}(\Omega) \cap W_0^{1,1}(\Omega). \tag{3.12}$$

Remark 1 The condition (Z) is satisfied if p is Hölder continuous on $\overline{\Omega}$.

4. Regularity of weak solutions

$u \in W^{1,p(x)}(\Omega)$ is said to be a weak solution of the equation (1.1) if

$$\int_\Omega |\nabla u|^{p(x)-2} \nabla u \nabla \varphi dx = \int_\Omega f(x,u)\varphi dx, \forall \varphi \in C_0^\infty(\Omega). \tag{4.1}$$

Theorem 5 *Suppose $f \in C(\Omega \times R)$ and*

$$|f(x,t)| \le c_1 + c_2 |t|^{q(x)-1}, \forall (x,t) \in \Omega \times R, \tag{4.2}$$

where c_1 and c_2 are positive constants, $q \in C(\Omega)$ and $1 < q(x) < p^*(x)$ for $x \in \Omega$. Suppose $u \in W^{1,p(x)}(\Omega)$ is a weak solution of the equation (1.1). Then

1) (see [7,8]) $u \in C^{0,\alpha}_{loc}(\Omega)$ if p satisfies the condition (Z) locally in Ω.

2) (see [10]) $u \in C^{1,\alpha}_{loc}(\Omega)$ if $p \in C^1(\Omega)$, where $\alpha \in (0,1)$ is a constant.

For the regularity of minimizers of integral functionals with $p(x)-$growth conditions see [6,7,8].

5. A strong maximum Principle

We generalized the strong maximum principle for the $p-$Laplacian equations to the case of $p(x)-$Laplacian, this provides a base to study the positive solutions for the $p(x)-$Laplacian equations.

Consider the following $p(x)-$Laplacian equation

$$-div(|\nabla u|^{p(x)-2} \nabla u) + d(x)\,|u|^{q(x)-2} u = 0 \text{ in } \Omega \subset R^N, \tag{5.1}$$

where $p \in C^1(\overline{\Omega}), q \in C(\overline{\Omega}), p(x) \le q(x) < p^*(x)$ for $x \in \overline{\Omega}, d \in L^\infty(\Omega)$ and $d(x) \ge 0$ for a.e. $x \in \overline{\Omega}$.

Theorem 6 (see [14]) *Let $u \in W^{1,p(x)}(\Omega)$ is a weak super-solution of the equation (5.1) and $u(x) \ge 0$ for a.e. $x \in \Omega$. If $u(x)$ is not identical with 0 in Ω, then, for any nonempty compact subset $K \subset \Omega$, there exists a positive constant c such that $u(x) \ge c$ for a.e. $x \in K$.*

Theorem 7 (see [14]) *Let u be as stated in above Theorem. Suppose that $x_1 \in \partial\Omega$, $u \in C^1(\Omega \cup \{x_1\})$ and $u(x_1) = 0$. If Ω satisfies the interior-ball condition at x_1, then $\frac{\partial u(x_1)}{\partial \gamma} > 0$, where γ is the inward unit normal vector of $\partial\Omega$ on x_1.*

6. On the "first eigenvalue" of $p(x)$–Laplacian

Consider the eigenvalue problem of $p(x)$–Laplacian

$$-div(|\nabla u|^{p(x)-2}\nabla u) = \lambda |u|^{p(x)-2} u \text{ in } \Omega, \ u = 0 \text{ on } \partial\Omega, \tag{6.1}$$

where Ω is a bounded domain in R^N.

Set

$$\lambda_1 = \lambda_{1,p(x)} = \inf\{\lambda \in R : \lambda \text{ is a eigenvalue of } (6.1)\}.$$

It is well known that, when $p(x) \equiv$ a constant p, $\lambda_{1,p}$, the first eigenvalue of p–Laplacian, is positive. However, in general, for example, when the function $p(x)$ has a strictly local minimum (or maximum) point in Ω, we have $\lambda_{1,p(x)} = 0$ (see [14]). In [14] we gave some sufficient conditions for $\lambda_{1,p(x)} > 0$, in particular, when $N = 1$ we have $\lambda_{1,p(x)} > 0$ if and only if the function $p(x)$ is monotone.

The fact $\lambda_{1,p(x)} = 0$ adds heavily to complexity of the $p(x)$–Laplacian equations.

7. Existence and multiplicity of solutions

Consider the Dirichlet problem for $p(x)$–Laplacian equation:

$$-div(|\nabla u|^{p(x)-2}\nabla u) = f(x,u) \text{ in } \Omega, \ u = 0 \text{ on } \partial\Omega. \tag{7.1}$$

The corresponding integral functional is

$$J(u) = \int_\Omega \frac{1}{p(x)} |\nabla u|^{p(x)} dx - \int_\Omega F(x,u)dx, \ u \in W_0^{1,p(x)}(\Omega), \tag{7.2}$$

where $F(x,t) = \int_0^t f(x,s)ds$. The critical points of J are exactly the weak solutions of the problem (7.1).

By using the critical point theory, in [14,15] we obtained some results on the existence and multiplicity for the problem (7.1), which generalized the corresponding results in the p–Laplacian case.

We also considered the periodic solutions of the so-called second-order $p(t)$–Hamiltonian systems

$$\left(|\dot{u}|^{p(t)-2}\,\dot{u}\right)' = \nabla F(x,u),\ u(0)=u(T),\ \ \dot{u}\,(0) = \dot{u}\,(T), \tag{7.3}$$

and obtained some existence results (see [12])

8. Some open problems

Many fine results have been obtained in the $p-$Laplacian case have not been generalized to the $p(x)-$Laplacian case. For example, the following problems are open.

1). Under what conditions the problem (7.1) has a unique positive solution? For example:

a). Consider the following problem

$$-div(|\nabla u|^{p(x)-2}\,\nabla u) = |u|^{q(x)-2}\,u \text{ in } \Omega,\ \ u=0 \text{ on } \partial\Omega, \tag{8.1}$$

where Ω is bounded and $1 < q_- \leq q_+ < p_-$. Dose the problem (8.1) have a unique positive solution?

b). Consider the constrained minimization problem

$$\inf\left\{\int_\Omega \frac{1}{p(x)}\,|\nabla u|^{p(x)}\,dx : \int_\Omega \frac{1}{p(x)}\,|u|^{p(x)}\,dx = 1\right\}\ , \tag{8.2}$$

where Ω is bounded. It is clear that the problem (8.2) has a positive minimizer. Is the positive minimizer unique?

2). Under what conditions the solutions of the problem (7.1) must be radially symmetric?

3). On the nodal solutions.

For a radially symmetric function $u(x)$, a node of u is a radius r such that $u(x) = 0$ when $|x| = r$. Consider the nodal solution of the following $p(x)-$Laplacian equation

$$-div(|\nabla u|^{p(x)-2}\,\nabla u) + |u|^{p(x)-2}\,u = |u|^{q(x)-2}\,u \text{ in } R^N,\ \ u \in W_0^{1,p(x)}(R^N), \tag{8.3}$$

where $p \in C^1\left(R^N\right), q \in C\left(R^N\right)$, p and q are radially symmetric. When

$$\inf\left\{p^*(x) - q(x) : x \in R^N\right\} > 0$$

and $p_+ < q_-$, in [15] we have proved that, for every positive integer k, the problem (8.3) has a radially symmetric solution having exactly k nodes. Is the similar result true when $1 < q_- < q_+ < p_-$?

References

[1] E. Acerbi and N. Fusco, Partial regularity under anisotropic (p,q) growth conditions, J. Differential Equations, 107(1994), 46-67.

[2] E. Acerbi and N. Fusco, A transmission problem in the calculus of variations, Calc. Var. 2(1994), 1-16.

[3] X-L. Fan, The regularity of Lagrangians $f(x,\xi) = |\xi|^{\alpha(x)}$ with Hölder exponents $\alpha(x)$, Acta Math. Sinica (New Series) 12(3)(1996),254-261.

[4] X-L. Fan, Regularity of nonstandard Lagrangians $f(x,\xi)$, Nonlinear Anal. 27(6)(1996),669-678.

[5] X-L. Fan, J-S. Shen and D. Zhao, Sobolev embedding theorems for spaces $W^{k,p(x)}(\Omega)$, J. Math. Anal. Appl. 262(2001), 749-760.

[6] X-L. Fan and D. Zhao, Regularity of minimizers of variational integrals with p(x)-growth conditions, Chin. J. Contemp.. Math., 17(4) (1996), 327-336.

[7] X-L. Fan and D. Zhao, A class of De Giorgi type and Hölder continuity, Nonlinear Anal., 36(1999), 295-318.

[8] X-L. Fan and D. Zhao, The quasi-minimizer of integral functionals with m(x) growth conditions, Nonlinear Anal., 39(2000), 807-816.

[9] X-L. Fan and D. Zhao, On the spaces $L^{p(x)}(\Omega)$ and $W^{m,p(x)}(\Omega)$, J. Math. Anal. Appl., 263(2001), 424-446.

[10] X-L. Fan and D. Zhao, Local $C^{1,\alpha}$ regularity of weak solutions for p(x)-Laplacian equations, J. Gansu Education College, 15(2)(2001), 1-5.

[11] X-L. Fan, Y-Z. Zhao and D. Zhao, Compact imbedding theorems with symmetry of Strass-Lions type for the space $W^{k,p(x)}(\Omega)$, J. Math. Anal. Appl., 255(2001),333-348.

[12] P. Marcellini, Regularity and existence of solutions of elliptic equations with (p,q)-growth conditions, J. Differential Equations, 90(1991), 1-30.

[13] J-S. Shen, Periodic solutions of second order p(t)-Hamiltonian systems, Doctoral Dissertation, Lanzhou University, 2001.

[14] Q-H. Zhang, Some studies on p(x)-Laplace equations, Doctoral Dissertation, Lanzhou University, 2000.

[15] Y-Z. Zhao, Existence of positive and nodal solutions for p(x)-Laplacian equations, Doctoral Dissertation, Lanzhou University, 2001.

[16] V.V. Zhikov, Averaging of functionals in the variations and elasticity, Math USSR Izvestija, 50(4) (1986), 675-711.

[17] V.V. Zhikov, On passing to the limit in nonlinear variational problem, Math. Sbornik, 183(8)(1992), 47-84.

Morse theory for strongly indefinite functional and its applications

Yuxia Guo

Department of Mathematical Sciences, Tsinghua University, Beijing 100084, P. R. China

We say that a functional is strongly indefinite if both of the positive and negative spaces of its Hessian operator are of infinite dimension. In this talk, by using Gelerkin approximation method, we developed a new Morse theory for a class of strongly indefinite functional on its dynamically isolated critical sets. Here, the so called strongly indefinite functional we are considering is the functional with the form:

$$f(x) = \frac{1}{2}\langle Ax, x\rangle + G(x),$$

where A is a linear bounded self-adjoint operator and $G'(x)$ is compact. As applications we study the existence and multiplicity of nontrivial periodic solutions of asymptotically linear Hamiltonian systems, and non-cooperative elliptic systems.

(I) **Purpose and significance**

It is well known that within a variational frame, the existence of the solution for an ordinary or a partial differential equation can be deduced to finding the critical point of a functional. Morse theory has been proved to be a very useful tool to find critical points of a functional. By using Morse theory, we can not only prove the existence of solutions and estimate the number of the solutions of the equations, but also distinguish the difference of the solutions by a series of homology groups. However, since only non-degenerate functional or non-degenerate critical points were well studied, and the non-degeneracy of a solution to a partial differential equation occurring in practice is difficult to verify in advance (sometimes it is even impossible). For this reason, the applications of the classical Morse theory to partial differential equations were very limited.

To extend the applications of classical Morse theory, there have been two directions: one is Gromoll-Meyer theory on isolated critical point which may be degenerate ([1, 2]); the other is Conley's Morse index theory on isolated invariant sets ([3]). However, the Conley index is difficult to compute, In this sense, Gromoll Meyer theory is easier than Conley index theory, and among other things, Gromoll-Meyer theory is sufficient for many applications ([4]).

On the other hand, in a variational setting, Chang and Ghoussoub [5] investigated the connection between the two theories. More precisely , they studied the relationship between the Gromoll-Meyer pair of a dynamically isolated critical set and the Conley index pair of its isolated invariant neighborhoods, and proved that the Gromoll-Meyer pairs are Conley index pairs; and Conley index in Conley index theory coincides with the critical groups in Gromoll-Meyer theory. Hence the information given by the critical groups of such a dynamically isolated critical set is equivalent to that given by the Conley index of its isolated invariant neighborhoods. Thus, for many problems, we can use Gromoll-Meyer theory only to resolve problem, moreover the proof is also relatively easier.

Now, we consider the following functional $\Phi(x) = \frac{1}{2}\|x^+\|^2 - \frac{1}{2}\|x^-\|^2$ (functional of such type are called strongly indefinite) defined on a Hilbert space $E = E^+ + E^-$, where E^+, E^- are orthogonal decomposition of E, and both of them are infinite dimensional subspace. Clearly, zero element θ in space E is the only critical point of Φ. According to Morse theory, the critical groups of functional Φ at θ: $C_q(\Phi, \theta) = \theta, q = 0, 1, 2, \ldots$ This example explains why the previous definition for critical group (both in Morse theory and in Gromeoll-meyer theory)is not applicable to strongly indefinite functionals.

In applications, to find the critical points of a strongly indefinite functional, there are at least two methods that have been developed: (i) Morse theory and its generalizations (Gromoll-Meyer theory, Conley index theory). (ii) Minimax theory. And there have been much work based on minimax theory (see [6-11]), however all these methods have some disadvantages, e.g., requiring strong hypotheses , such as uniform boundedness, convexity etc. In 1992, Szukin ([12]) construct a different theory for an isolated critical point which turned out to be useful for strongly indefinite functional; On the other hand, in 1997, K.Geba [13] presented a new generalization of classical Conley index theory, and gave the definition of homotopy Conley index about an isolated neighborhood of an isolated invariant set.

Motivated by the ideas of papers [5] and [13], based on the Gromoll-Meyer theory, we construct a different Morse theory for strongly indefinite functional on its dynamically isolated set. This new Morse theory provides us a powerful tool in studying the existence problem involving strongly indefinite functional.

Morse theory for strongly indefinite functional[GL]

We consider a functional defined on Hilbert space H with the form

$$f(x) = \frac{1}{2} = \langle Ax, x\rangle + G(x)$$

where A is a bounded self adjoint operator, $G'(x)$ is nonlinear compact mapping.

First, we discuss the stability of the dynamically isolated critical set and its isolating triplet of the functional f with respect to a family of flow associated to f . That is, if S is a dynamically isolated critical set of f associated with flow η_{λ_0}(generated by df), then S is also a dynamically isolated set about to the flow η_λ generated by a vector field related to f, when the parameter λ is sufficiently near to λ_0. i.e.

Theorem 1 Suppose that $f \in C^2(H, R)$ satisfies (P.S.), $\eta_\lambda(t,x) = \eta(t,x,\lambda) = e^{tA}x + U(t,x,\lambda)$ is a family of LS flows generated by F_λ, the zero eigenvalue of A is isolated in $\sigma(A)$, Assume η_{λ_0} is the negative gradient flow generated by $df = A + K_{\lambda_0}$ and S is a dynamicaly isolated critical set of f associated with flow η_{λ_0} . If K_{λ_0} is global Lipschitz continuous on any bounded set and $\|F_\lambda(x) - df(x)\| \to 0$, uniformly in bounded x. Then there exists an isolating triplet (O, α, β) of S and a positive number $\delta > 0$ such that $\forall \lambda \in \Lambda,\ \ |\lambda - \lambda_0| < \delta$ we have

$$cl(\tilde{O}_\lambda) \cap K \cap f^{-1}[\alpha,\beta] = S \tag{2.1}$$

$$cl(\tilde{O}_\lambda) \cap K_\lambda \cap f^{-1}[\alpha,\beta] \subset \text{int}O \tag{2.2}$$

where $K_\lambda = \{x \mid F_\lambda(x) = 0\}$

Moreover, if (O, α, β) is a isolating triplet of S associated with flow η_{λ_0}, then when the parameter λ is sufficiently near to λ_0, (O, α, β) is also an isolating triplet for the invariant set associated with the flow η_λ. i.e.

Theorem 2 Suppose that f satisfies the assumes in Theorem 3.1, η_{λ_0} is negative LS gradient flow of f, S is a dynamically isolated critical set of f associated with flow η_{λ_0} . $\eta_\lambda(\lambda \in \Lambda)$ is a family of approximation (λ_0) LS flows , then there exists an isolate triplet (O, α, β) of S and a real number $\delta > 0$ such that

$$cl(\tilde{O}_\lambda) \cap K \cap f^{-1}[\alpha,\beta] = S \tag{2.3}$$

$$cl(\tilde{O}_\lambda) \cap K_\lambda \cap f^{-1}[\alpha,\beta] \subset \text{int}O \tag{2.4}$$

for $\forall \lambda \in \Lambda,\ |\lambda - \lambda_0| < \delta$.

Secondly, by using Gelerkin approximation method, we prove that: for the considered functional f and its dynamically isolated set S, there always exist an isolating triplet (O, α, β) such that $(O \cap H^n, \alpha, \beta)$ is an isolating triplet of S_n, which is the critical set of the restricted functional $f_n := f \mid_{H^n}$ in $O \cap H^n$ (for largely n). And with the difference of n (large enough), the relative homology group of (W_n, W_{n-}) has some kinds of properties for translating invariant, that is

Theorem 3 For the considered functional $f(x) = \frac{1}{2}\langle Ax, x\rangle + G(x)$, finite dimensional subspace $H^n = \oplus_1^n H_i$ of H and the approximation scheme $\Gamma = \{P_n\}_1^\infty$ w.r.t. A, suppose that (W_n, W_{n-}) is a G-M pair for critical set S_n associated with the flow generated by $df_n = P_n(A+K)P_n : H^n \to H^n$, then for n large enough

$$H^{*+m(P_n(A+P)P_n)}(W_n, W_{n-}) \cong H^{*+m(P_{n+1}(A+P)P_{n+1})}(W_{n+1}, W_{n+1}) \tag{2.5}$$

where $m(\cdot)$ denotes the Morse index of operator $\cdot$.

Thus, we get a series of isotropic homology groups, and such a series of groups is unique in the sense of isotropic. So we can define the critical group of f on its dynamically isolated critical set S as follows:

Definition 4 Let S be a dynamically isolated set for f associated with its negative gradient flow. we define (for n large enough)

$$C_*(f,S) := H^{*+m(P_n(A+P)P_n)}(W_n, W_{n-})$$

and call it the qth critical group of S with respect to flow η generated by negative gradient vector field df. where, (W_n, W_{n-}) is a Gromoll- Meyer pair of S_n the critical set of the restriction functional f_n, $\{P_n|n = 1,2...\}$ is an approximation scheme with respect to operator A, P is the project operator from H onto the kernel space of A.

Remark 1. Our definition is independent the choice of isolating triplet and the choice of the approximation scheme.

At last, we give the corresponding Morse type inequalities.

$$\sum_{i=1}^{n} P(t, C_*(f,S_i)) = P(t, C_*(f,S)) + (1+t)Q(t)$$

where,$\cup_i^n S_i = S, S_i \cap S_j = \emptyset, i \neq j$, S_i is a dynamically isolated critical set.

(III) **Computation of the critical groups**[G3]

First of all, we define the critical groups of functional f at ∞ as follows:

Definition 5 If S is the set containing all of the critical points of f, and f has finite critical values , then we define the critical group at ∞ by

$$C_*(f,\infty) = C_*(f,S).$$

We know that, the computation of the degenerate critical point is very difficult even in finite dimensional space. In this work, a complete description the degenerate critical critical of functional of Ladersman-Lazer type resonance is obtained. Our main results are as follows:

Theorem 6 Assume that

A is a bounded self-adjoint operator

θ is an isolating critical point of f, 0 is the isolating critical value, $G'(x)$ is compact and global Lipschitz continuous on any bounded set.

$(3^{\pm})$ there exists a linear, compact, symmetric operator B_0, such that $||G'(x) - B_0(x)|| \leq c(||x_0||^\alpha + ||x_+ + x_-||^\beta)$, and $(G(x_0) - \frac{1}{2}\langle B_0 x, x\rangle)||x_0||^{-2\alpha} \to \pm\infty$ as $||x_0|| \to 0$, where, $x = x_+ + x_0 + x_- \in H_+ \oplus H_0 \oplus H_-$, which is the orthogonal decomposition corresponding to the spectrum of operator A.

$(4^{\pm})$ there exists a linear, compact, symmetric operator B_∞, such that $||G'(x) - B_\infty(x)|| \leq c||x||^\alpha, \alpha \in (0,1), c > 0$, and $(G(P_0x) - \frac{1}{2}\langle B_\infty P_0 x, P_0 x\rangle)||P_0 x||^{-2\alpha} \to \pm\infty$, as $||P_0 x|| \to \infty$, where, P_0 is the orthogonal projection from H to the kernel space of A. Then

$$C_q(f,\theta) = \begin{cases} \delta_{q, I_-(B_0)+N(B_0)}G & \text{if}(3^-)\text{holds} \\ \delta_{q, I_-(B_0)}G & \text{if}(3^+)\text{holds} \end{cases}$$

and

$$C_q(f,\infty) = \begin{cases} \delta_{q,I_-(B_\infty)+N(B_\infty)}G, & \text{if}(4^-)\text{holds} \\ \delta_{q,I_-(B_\infty)}G, & \text{if}(4^+)\text{holds} \end{cases}$$

Where, $(I_-(\cdot), N(\cdot))$ is the abstract Maslove index of the operator ·

(IV) **Applications**

(a) Nontrivial periodic solution for asymptotically linear Hamiltonian system with resonance[G2].

Consider the following Hamiltonian system

$$\dot{z} = JH'(t,z). \tag{4.1}$$

Suppose that $H \in C^2([0,1] \times R^{2N}, R)$ and there exist periodic continuously symmetric matrix functions $B_0(t)$ and $B_\infty(t)$ such that

$(H_1^\pm)$ there exist $M_0 > 0, c_0 > 0$, and $\sigma \in (1,\infty)$ such that

$$|F_0'(t,z)| \le M_0|z|^\sigma, \quad z \in R^{2N}, \|z\| \le c_0,$$

$$\lim_{|z|\to 0} \frac{F_0(t,z)}{|z|^{2\sigma}} = \pm\infty \quad \text{uniformly in t.}$$

$(H_2^\pm)$ there exist $M_\infty > 0, \alpha \in (0,1)$ such that

$$|F_\infty'(t,z)| \le M_\infty(1+|z|^\alpha) \quad z \in R^{2N},$$

and

$$\lim_{|z|\to\infty} \frac{F_\infty(t,z)}{|z|^{2\alpha}} = \pm\infty \quad \text{uniformly in t}$$

where

$$F(t,z) = H(t,z) - \frac{1}{2}(B_0(t)z,z) \quad \forall (t,z) \in R \times R^{2N},$$

$$F(t,z) = H(t,z) - \frac{1}{2}(B_\infty(t)z,z) \quad \forall (t,z) \in R \times R^{2N}.$$

By the abstract results obtained in section (iii), we know that the critical groups both at the origin and at infinity read as follows

$$C_q(f,\theta) = \begin{cases} \delta_{q,I_-(B_0)+N(B_0)}G, & \text{if } H_1^- \text{ holds} \\ \delta_{q,I_-(B_0)}G, & \text{if } H_1^+ \text{ holds} \end{cases}$$

$$C_q(f,\infty) = \begin{cases} \delta_{q,I_-(B_\infty)+N(B_\infty)}G, & \text{if } H_4^- \text{ holds} \\ \delta_{q,I_-(B_\infty)}G, & \text{if } H_4^+ \text{ holds} \end{cases}$$

where G is field of co-homology , $I_-(B_i) := i_i, N(B_i) := n_i (i = 0, \infty)$ are the Morse type index and the nullity of the quadratic functional respectively.

Thus, the system (1) has at least one nontrivial periodic solution, if one of the following conditions holds.

(A_1) (H_1^-) and (H_2^-) hold, and $i_0 \neq i_\infty$

$(A_2)$$(H_1^+)$ and (H_2^-) hold, and $i_0 + n_0 \neq i_\infty$

$(A_3)(H_1^-)$ and (H_2^+) hold, and $i_0 \neq i_\infty + n_\infty$

$(A_4)(H_1^+)$ and (H_2^+) hold, and $i_0 + n_0 \neq i_\infty + n_\infty$

In addition, if we assume that $x_0 \neq 0$ is a non-degenerate periodic solution, then the system has a second solution $x_1 \neq \theta, x_0$.

(b)Boundary value problem of non-cooperate elliptic system with resonance[G1].

We Consider

$$\begin{cases} \Delta u = f(x,u,v), & x \in \Omega \\ \Delta v = -g(x,u,v), & x \in \Omega \\ u = v = 0, & x \in \partial\Omega \end{cases} \tag{4.2}$$

By introducing a new space, we deduce the problem of finding the nontrivial solution of the system to the problem finding a critical point of a functional defined on a suitable function space. For this reason, we assume that:

$(H_3^\pm)$ $\lim_{|U|\to 0} \dfrac{F_0(x,U)}{|U|^{2\beta}} = \pm\infty$, for a.e.$x \in \Omega$uniformly

$(H_4^\pm)$ $\lim_{|U|\to \infty} \dfrac{F_\infty(x,U)}{|U|^{2\alpha}} = \pm\infty$, for a.e.$x \in \Omega$uniformly

Suppose that $G \in C^2(\bar{\Omega} \times R^2, R)$, then the system (2) has at least one nontrivial solution if one of the following conditions holds.

$(\bar{A}_1)(H_3^+),(H_4^+)$ hold, and $i^0(B_0) + i^-(B_0) \neq i^0(B_\infty) + i^-(B_\infty)$

$(\bar{A}_2)(H_3^+),(H_4^-)$ hold, and $i^0(B_0) + i^-(B_0) \neq i^-(B_\infty)$

$(\bar{A}_3)(H_3^-),(H_4^+)$ hold, and $i^-(B_0) \neq i^0(B_\infty) + i^-(B_\infty)$

$(\bar{A}_4)(H_3^-),(H_4^-)$ hold, and $i^-(B_0) \neq i^-(B_\infty)$

where $i^0(B_i), i^-(B_i)(i = 0, \infty)$ is the abstract Morse index of operator $B_i(i = 1,2)$.

Again, if we assume that $x_0 \neq \theta$ is a non-degenerate solution ,then systems (2) has another solution .

Relevant Publications: [GL] Yuxia Guo& Jiaquan Liu, Morse theory for strongly indefinite functional, Nonlinear Analysis TMA, 2002 (48), pp. 831-851

[G_1] Yuxia Guo, nontrivial solutions for resonant noncooperative elliptic system , Comm. Pure. Appl. Math, Vol. LIII, PP.1335-1349.

[G_2] Yuxia Guo, Nontrivial periodic solutions for asymptotically linear Hamiltonian system with resonance, J. Differential Equations 2001(175) , no. 1 pp. 71-87.

[G_3] Yuxia Guo, Computations of critical groups at degenerate critical point and applications to nonlinear differential equations with resonance, Journal of Math. Anal. Appl., 2001(256), no. 2, pp. 462-477.

References

(1) K.C.Chang, Infinite dimensional Morse theory and its applications, CRM , Universite' de Montreal, SMS, 97(1985).

(2) K.C.Chang, Infinitc dimensional Morse theory and multiple solution problems, Birkhawer Boston, 1993.

(3) C.C.Conley, Isolated invariant sets and the Morse index, CBMS Regional Conference Series 38, American Mathematical Society, Providence, 1978.

(4) K.C.Chang, Solutions of asymptotically linear operator equations via Morse theory, Comm. Pure. Appl. Math., 34(1981), 691-712.

(5) K.C.Chang and N.Ghoussoub, The Conley index and the critical groups via an extension of Gromoll-Meyer theory, Topology Method in Nonlinear Analysis, Vo.7, 1996, 77-93.

(6) P.H.Rabinowitz, Periodic solutions of Hamiltonian systems, Comm. Puer . Appl. Math. 31(1978), 157-184.

(7) I.Ekeland and J.M.Lasry, On the number of periodic trajectories for a Hamiltonian flow on a convex energy surface, Ann. of Math. (2)112 (1980), 283-319.

(8) Y.M.Long, Conley index theory and Hamiltonian systerm, Sencience Publication (in chinese), 1993

(9) D.V.Benci and P.H.Rabinowitz, Critical ponit theorems for indefinite functionals, Inventions Math, 52(1979), 241-273.

(10) V.Benci, The direct method in the study of periodic solutions of Hamiltonian systems with prescribed perid, In "Advances in Hamiltonian Systems", P.Aubin, A, Bensoussan, and I.Ekeland, Birkhauser, Basel, 1983

(11) H.Hofer, On strongly indefinite functionals with applications, J.A.M. S., Vol.275, No.1, 1983.

(12) A.Szulkin, Cohomology and Morse theory for strongly indefinite functionals, Math.Z., Vol.209, (1992), 375-418.

(13) K.Geba, M.Izydorek and A.Pruszko, The Conley index in Hilbert spaces and its applications, Preprint, 1997.

On stability estimates for the regularity property of maps

A.D. Ioffe

Department of Mathematics, Technion–Israel Institute of Technology, Haifa 32000, Israel

1. Introduction

Whenever we have a good and useful property, the question of whether and how stable is this property with respect to perturbations of its bearers appears often and quite naturally. In this note we shall discuss some questions related to stability of the regularity property of locally Lipschitz mappings between Banach spaces.

Recall, to begin with, that a continuously differentiable mapping F from a Banach space X into another Banach space Y is *regular* at $\bar{x} \in X$ if $\mathrm{Im}\ F'(\bar{x}) = Y$. This is certainly one of the most central concepts of the classical nonlinear analysis. The "phenomenological" characterization of this property is given by the Ljusternik-Graves theorem (containing as its consequences the implicit function theorem in the most general form and various Lagrange multiplier rules including those that cannot be obtained with the help of the implicit function theorem) which can now be stated as follows.

Theorem (Ljusternik-Graves). *Let $F: X \to Y$ be continuously Fréchet differentiable at $\bar{x}$. Then the following statements are equivalent:*

(i) F is regular at $\bar{x}$;

(ii) There are $\varepsilon > 0$, $K > 0$ such that the inequality

$$d(x, F^{-1}(y)) \leq K\|y - F(x)\|$$

The reasearch was supported by the USA-Israel Binational Science Foundation under the grant 2000157.

holds whenever $\|x-\bar{x}\|<\varepsilon$, $\|y-F(\bar{x})\|<\varepsilon$ *(where* $d(x,C)$ *stands for the distance from* x *to* C*);*

(iii) There are $\varepsilon>0$, $K>0$ *such that for any* x *and* y *with* $\|x-\bar{x}\|<\varepsilon$, $\|y-F(\bar{x})\|<t\leq\varepsilon$ *there is a* $u\in X$ *such that* $F(u)=y$ *and* $\|u-x\|\leq Kt$.

The property (ii) is usually called *metric regularity* of F near $\bar{x}$. As to (iii), it can be interpreted in two ways: as a statement that for all x close to $\bar{x}$ and all $t>0$ sufficiently close to zero the ball $B(F(x),t)\subset Y$ is covered by the F-image of the ball $B(x,Kt)\subset X$, and also as the statement that for every (x,y) sufficiently close to $(\bar{x},F(\bar{x}))$ the equation $F(u)=y$ has a solution which differers from x by at most $K\|y-F(x)\|$. The first of the two last properties is often called *openness at a linear rate*; the second – *Lipschitz stability* of solutions to the equation $F(u)=y$ at $(\bar{x},F(\bar{x}))$, or the *pseudo-Lipschitz*, or *Aubin* property of the inverse (generally set-valued) mapping F^{-1} at $(F(\bar{x}),\bar{x})$.

The equivalence of (ii) and (iii) is easily seen. The important point is, however, that neither the statements of (ii) and (iii) nor the formal arguments showing their equivalence need differentiability, even continuity of F. Moreover, both (ii) and (iii) as well as the fact that they are equivalent can be naturally extended as far as to set-valued mappings between metric spaces (see e.g. [**6**]). We shall not need such a general setting in this note, so the following definition will be sufficient for our purpose.

Definition 1. We say that a mapping F from a Banach space X into another Banach space Y defined in a neighborhood of an $\bar{x}\in X$ is *regular* near $\bar{x}$ if the equivalent properties (ii), (iii) of the Ljusternik-Graves theorem are satisfied for F.

As our interest lies in studying perturbation stability of the regularity property, we need quantitative measures of regularity.

Definition 2. For a mapping $F:\ X\to Y$ defined in a neighborhood of $\bar{x}$ we define the *constant*, or *rate* of regularity near $\bar{x}$ by

$$\mathrm{Reg}F(\bar{x})=\inf\{K>0:\ \text{(ii) holds for some}\ \varepsilon>0\}$$

As always, we set $\inf\emptyset=\infty$. The quantity

$$\mathrm{Sur}F(\bar{x})=(\mathrm{Reg}f(\bar{x}))^{-1}$$

is called the *constant* (or *rate*) of openness (or surjection) of F near X. It is clear that $\mathrm{Sur}F(\bar{x})$ is the upper bound of $L\geq 0$ with the following property: there is an $\varepsilon>0$ such that $B(F(x),Lt)\subset F(B(x,t))$ if $\|x-\bar{x}\|<\varepsilon$, $0\leq t<\varepsilon$.

The first, and for a long time the only, result on perturbation stability of the regularity property was a theorem of Miljutin published in [**3**]. Below we state the local version of his theorem proved in [**5**].

Theorem (Miljutin). *Suppose that* F *is continuous in a neighborhood of* $\bar{x}$, *and let* $H:\ X\to Y$ *be defined and locally Lipschitz in the neighborhood. Then*

$$\mathrm{Sur}(F+H)(\bar{x})\geq\mathrm{Sur}F(\bar{x})-\mathrm{Lip}H(\bar{x}).$$

Here $\mathrm{Lip}H(\bar{x})$ stands for the Lipschitz constant of H at $\bar{x}$:

$$\mathrm{Lip}H(\bar{x}) = \limsup_{\substack{x,u\to\bar{x}\\ x\neq u}} \frac{\|H(x) - H(u)\|}{\|x - u\|}.$$

The question of how good is the estimate provided by Miljutin's theorem was not discussed till very recently. The recent theorem of by Dontchev-Lewis-Rockafellar **[4]** (which seems to be the first result of that sort[1]) shows that the estimate is exact for mappings between finite dimensional spaces

The central in their approach is the concept of *radius* of regularity:

$$\mathrm{Rad}F(\bar{x}) = \inf\{\|A\| : A \in \mathcal{L}(X,Y) : \mathrm{Sur}(F + A)(\bar{x}) = 0\},$$

where $\mathcal{L}(X,Y)$ is the space of bounded linear operators from X into Y.

Theorem (Dontchev-Lewis-Rockafellar). *Suppose that both X and Y are finite dimensional Banach spaces. Assume that F is regular near $\bar{x}$. Then* $\mathrm{Rad}F(\bar{x}) = \mathrm{Sur}F(\bar{x})$. *Moreover, the infimum in the definition of the radius of regularity is provided by a linear operator of rank one.*

Observe that both theorems, the theorem of Miljutin and the theorem of Dontchev-Lewis-Rockafellar, have been proved for set-valued mappings with closed graphs (the first in **[5]**).[2]

The purpose of this note is to show that mappings between infinite dimensional Banach spaces are in a sense even more robust. Namely, we give an example of a Lipschitz (and even having good differentiability properties) mapping from a Hilbert space into itself for which the radius of regularity at certain points is strictly greater than the constant of surjection.

The analysis of the example is based on the formula for computation of the surjection constant recently obtained by the author **[6]**. The key element of the formula is the concept of the slope of a real valued function introduced by De Giorgi, Marino and Tosques **[2]**. In the next section we have collected all necessary information regarding these matters. The last section contains the construction of the example and all necessary proofs.

---

[1]With the exception of the 1938 theorem of Eskart-Young on non-singularity of linear operators in $\mathbb{R}^n$. But in this case the problem is easily solvable thanks to the fact that such an operator is regular iff it is one-to-one.

[2]There were some infinite extensions of this result. In **[4]** a similar theorem was proved for homogeneous set-valued mappings with closed graph. In a very recent paper **[8]** Mordukhovich proved another extension of the theorem to certain (set-valued) mappings from an Asplund space to $\mathbb{R}^n$. But the condition imposed to the mappings seems to be very restrictive and it is not clear whether it leaves room for reasonable assumptions beyond $\dim X < \infty$ and sufficiently simple cases of F being either continuously differentiable single-valued or set-valued with convex graph.

2. Calculation of the surjection constant

Definition 3. Let f be an extended-real-valued function defined in a neighborhood of $\bar{x}$ and finite at $\bar{x}$. The quantity

$$|\nabla f|(\bar{x}) = \liminf_{\substack{x \to \bar{x} \\ x \neq \bar{x}}} \frac{(f(\bar{x}) - f(x))^+}{\|\bar{x} - x\|}$$

is called the *slope* (or *strong slope*) of f at $\bar{x}$.

The notation chosen by the authors of the definition is due to the fact that for a function which is Fréchet differentiable at $\bar{x}$ the slope at $\bar{x}$ coincides with the norm of its derivative at the point. The notation is a bit confusing when we have to work with functions on a Hilbert space and their gradients. One just have to pay attention to the vertical modulus-type lines which appear in the notation for slope but not in the notation for gradient. The following result is an immediate consequence of Theorem 2b of [**6**]. Given a mapping F from X into Y, we set

$$\psi_y(x) = \|y - F(x)\|$$

Theorem 1. *Assume that* $F: X \to Y$ *is defined and continuous in a neighborhood of* $\bar{x}$. *Then*

$$\mathrm{Sur}F(\bar{x}) = \liminf_{\substack{x \to \bar{x} \\ y \to F(\bar{x}) \\ y \neq F(\bar{x})}} |\nabla \psi_y|(x).$$

The quoted theorems of Miljutin and Dontchev-Lewis-Rockafellar are of course corollaries of this theorem. Consider for instance F and H as in Miljutin's theorem and set $\varphi_y(x) = \|y - F(x) - H(x)\|$ and $v = y - H(x)$. Then

$$\begin{aligned}
|\nabla \varphi_y|(x) &= \limsup_{\substack{u \to x \\ u \neq x}} \|x - u\|^{-1}(\varphi_y(x) - \varphi_y(u))^+ \\
&= \limsup_{\substack{u \to x \\ u \neq x}} \|x - u\|^{-1}(\|y - F(x) - H(x)\| - \|y - F(u) - H(u)\|) \\
&\geq \limsup_{\substack{u \to x \\ u \neq x}} \|x - u\|^{-1}(\|v - F(x)\| - \|v - F(u)\| - \|H(u) - H(x)\| \\
&\geq |\nabla \varphi_v|(x) - \mathrm{Lip}H(x).
\end{aligned}$$

Applying Theorem 1, we immediately get Miljutin's theorem.

If Y a Hilbert (or Euclidean) space and F is locally Lipschitz, it is possible to give a simpler formula for calculation of $\mathrm{Sur}F$.

Proposition 1. *Let* Y *be a Hilbert space with the inner product* $(\cdot|\cdot)$, *and let* $F: X \to Y$ *be defined and Lipschitz in a neighborhood of* $\bar{x}$. *Set* $f_y(x) = (y|F(x))$. *Then*

$$\mathrm{Sur}F(\bar{x}) = \liminf_{x \to \bar{x}} \inf_{\|y\|=1} |\nabla f_y|(x).$$

PROOF. Take an x sufficiently close to $\bar{x}$ and fix a $z \neq F(x)$. We show first that

$$|\nabla \psi_z|(x) = |\nabla f_y|(x), \tag{1}$$

where

$$y = \frac{F(x) - z}{\|F(x) - z\|}. \tag{2}$$

We have

$$\begin{aligned} \psi_z^2(x) - \psi_z^2(u) &= \|F(x) - z\|^2 - \|F(u) - z\|^2 \\ &= (F(x) - z|F(x) - z) - (F(u) \pm F(x) - z|F(u) \pm F(x) - z) \\ &= -2(F(u) - F(x)|F(x) - z) + \|F(u) - F(x)\|^2. \end{aligned}$$

Therefore

$$\begin{aligned} \psi_z(x) - \psi_z(u) &= \frac{2(F(x) - z|F(x) - F(u)) + \|F(x) - F(u)\|^2}{\|F(x) - z\| + \|F(u) - z\|} \\ &= \frac{2(y|F(x) - F(u)) - \dfrac{\|F(x) - F(u)\|^2}{\|F(x) - z\|}}{1 + \dfrac{\|F(u) - z\|}{\|F(x) - z\|}}, \end{aligned}$$

and we get the desired equality (1) as $z \neq F(x)$ and F satisfies the Lipschitz condition near x (so that $\|F(x) - F(u)\|^2 = o(\|x - u\|)$).

To complete the proof of the proposition it is sufficient to note that for any y with $\|y\| = 1$, we can find a z arbitrarily close to $F(x)$ such that (2) holds. ∎

3. The main result

Theorem 2. *Let X be a separable real Hilbert space. Than there is a Lipschitz and weakly continuously Fréchet differentiable mapping $F : X \to X$ which is regular at zero and such that*

$$\text{Rad}\, F(0) > \text{Sur} F(0).$$

In other words, there is a positive γ such that the norm of any linear bounded operator A for which $\text{Sur}(F - A) = 0$ is not smaller than $\text{Sur}F(0) + \gamma$. (By weak continuous Fréchet differentiability we mean continuous Fréchet differentiability of the functions $f_y(x) = (y|F(x))$, $y \in X$.)

The proof of the theorem will be carried out in several steps.

Step 1. Consider the problem:

$$\textit{minimize } (\eta + \xi)^2 + \eta^2, \qquad \textit{s.t. } \eta^2 + \xi^2 = 1. \tag{3}$$

A straightforward calculation based on the Lagrange multiplier rule shows that the minimal value of the cost function in the problem is

$$c = (1/2)(3 - \sqrt{5}) < 1,$$

so that the components of the solution are both distinct from zero.

Consider now a slightly more general problem:

$$(4) \qquad minimize\ (\lambda+\xi)^2+\eta^2, \qquad s.t.\ \eta^2+\xi^2=1, \quad |\lambda|\le|\eta|$$

Let $(\overline{\xi},\overline{\eta},\overline{\lambda})$ be a solution of the problem (which obviously exists). We claim that

$$(\overline{\lambda}+\overline{\xi})^2+\overline{\eta}^2=c, \qquad |\overline{\lambda}|=|\overline{\eta}|.$$

Clearly, the minimal value of the cost function in (2) cannot be greater than c. It is also clear that $\overline{\lambda}\neq 0$ as $c<1$. On the other hand, if $0<|\overline{\lambda}|<|\overline{\eta}|$, then by the Lagrange multiplier rule there must be a multiplier ρ such that

$$\overline{\lambda}+\overline{\xi}=0,\ \overline{\lambda}+\overline{\xi}-\rho\overline{\xi}=0,\ \overline{\eta}-\rho\overline{\eta}=0.$$

The first two equations imply that either $\rho=0$ in which case $\overline{\eta}=0$ contrary to the assumption, or that $\overline{\xi}=0$ in which case the value of the cost function would be greater than one. Thus the only remaining possibility is $|\overline{\lambda}|=|\overline{\eta}|$. We finally notice that changing simultaneously the signs of $\overline{\xi}$ and $\overline{\lambda}$ does not change the result, so it is possible to assume that $\overline{\lambda}=\overline{\eta}$ and (4) reduces to (3). It follows in particular that c is the minimal value also in the problem

$$(4a) \qquad minimize\ (\lambda+\xi)^2+\eta^2, \qquad s.t.\ \eta^2+\xi^2=1, \quad |\lambda|\le|\eta|,\ \lambda\eta\ge 0.$$

Step 2. We next consider an extension of the last problem to X, namely

$$(5) \qquad minimize\ \|y+\lambda e_1\|^2, \quad s.t.\ \|y\|=1,\ 0<|\lambda|\le|\eta_k|,\ \lambda\eta\ge 0$$

where $y=(\eta_1,\eta_2,...)$ and $e_1,e_2,..$ is a certain orthonormal basis in X, fixed throughout the proof, $k\ge 2$ and $\eta_k=(y|e_k)$.

We claim that the minimal value of the cost function in the problem is the same c and it is attained at vectors of the form

$$(6) \qquad \bar{y}=(\overline{\eta}_1,0,...,0,\overline{\eta}_k,0...).$$

This is almost immediate from what we have seen. Indeed, for a $y=(\eta_1,\eta_2,...)$ set $v=y-\eta_1e_1-\eta_ke_k$. Then

$$\|y+\lambda e_1\|^2=\|v\|^2+(\lambda+\eta_1)^2+\eta_k^2\ge\|v\|^2+c(1-\|v\|^2)>c$$

if (y,λ) is admissible in (5) and $v\neq 0$. On the other hand, we take precisely c if we take y as in (6) with $\eta_1=\overline{\xi}$, $\eta_2=\overline{\eta}$, where $(\overline{\xi},\overline{\eta})$ is the solution of (4a).

Step 3. This is the central step in the proof. Namely we shall show that *there is a* $\overline{\gamma}>0$ *such that for any* $B\in\mathcal{L}(X)$ *with*

$$(7) \qquad \liminf_{k\to\infty}\inf\{\|(I-B)y+\lambda e_1\|:\ \|y\|=1,\ |\lambda|\le|\eta_k|\}=0$$

we have $\|B\|\ge\sqrt{c}+\overline{\gamma}$.

So take a B satisfying (7). This means that there are sequences (y_k), (λ_k) such that $\|y_k\|=1$, $|\lambda_k|\le\eta_{kk}$ and

$$(8) \qquad By_k=y_k+\lambda_ke_1+z_k$$

where $z_k \to 0$. Then

$$\|B\| \geq \limsup_{k\to\infty} \|By_k\| = \limsup_{k\to\infty} \|y_k + \lambda_k e_1\| \geq \sqrt{c} \tag{9}$$

(by Step 2). We can therefore set $\|B\| = \sqrt{c} + \gamma$ with $\gamma \geq 0$. Set further

$$v_k = y_k - \eta_{k1}e_1 - \eta_{kk}e_k.$$

Then $\|v_k\|^2 + \eta_{k1}^2 + \eta_{kk}^2 = 1$ and, as above, we conclude that

(10)

$$\|y_k + \lambda_k e_1\|^2 = \|v_k\|^2 + (\eta_{k1} + \lambda_k)^2 + \eta_{kk}^2 \geq \|v_k\|^2 + c(1 - \|v_k\|^2) = c + \|v_k\|^2(1-c).$$

Along with (9), this implies

$$(\sqrt{c} + \gamma)^2 \geq c + \limsup_{k\to\infty} \|v_k\|^2(1-c)$$

and consequently that

$$\limsup_{k\to\infty} \|v_k\|^2 = \delta^2 \leq \frac{(\sqrt{c}+\gamma)^2 - c}{1-c} = \frac{\gamma^2 + 2\sqrt{c}\gamma}{1-c}.$$

By passing to a subsequence, we can assume that $\|v_k\| \to \delta$, $\eta_{k1} \to \xi$, $\eta_{kk} \to \eta$, $\lambda_k \to \lambda$ as $k \to \infty$, and the limit numbers clearly satisfy $\xi^2 + \eta^2 = 1 - \delta^2$, $|\lambda| \leq |\eta|$.

Choose a γ_0 such that $\sqrt{c} + \gamma_0 < 1$ and set $\rho = 1 - (\sqrt{c} + \gamma_0)^2$. Assume now that $\|B\| = \sqrt{c} + \gamma$ and $\gamma < \gamma_0$. In this case we have by (10)

$$\begin{aligned}(\sqrt{c}+\gamma)^2 &\geq \limsup_{k\to\infty} \|By_k\|^2 \\ &= \limsup_{k\to\infty} \|y_k + \lambda_k e_1\|^2 \\ &= \limsup_{k\to\infty} (\|v_k\|^2 + (\eta_{k1} + \lambda_k)^2 + \eta_{kk}^2) \geq 1 + 2\xi\lambda,\end{aligned}$$

from which we get

$$-2\xi\lambda \geq 1 - (\sqrt{c}+\gamma)^2 \geq \rho.$$

As the absolute values of λ and ξ cannot be greater than one, we have

$$|\xi| \geq \sigma = \rho/2, \quad |\lambda| \geq \sigma, \quad |\eta| \geq \sigma.$$

Set $b_{ij} = (Be_i|e_j)$. Then

$$(By_k|e_k) = (Bv_k|e_k) + \eta_{kk}b_{kk} + \eta_{k1}b_{1k}.$$

On the other hand by (8) $(By_k|e_k) = \eta_{kk} + (z_k|e_k)$. Comparing these equalities and taking into account that $|(Bv_k|e_k)| \leq \|B\|\|v_k\| \leq (\sqrt{c}+\gamma)\delta + o(1)$, we get

$$|\eta_{kk}b_{kk} + \eta_{k1}b_{1k} - \eta_{kk}| \leq (\sqrt{c}+\gamma)\delta + o(1).$$

Obviously, $b_{1k} \to 0$. Assuming that b_{kk} converge to a certain β and passing to the limit in the last inequality, we get $|\eta(\beta - 1)| \leq (\sqrt{c}+\gamma)\delta$. If $\gamma \leq \gamma_0$, it follows that

$$|\beta - 1| \leq \delta\frac{\sqrt{c}+\gamma}{\sigma}.$$

Recalling the definitions of σ and δ, we can get from here, taking into account that $\|B\| \geq |\beta|$,

$$\sqrt{c} + \gamma = \|B\| \geq 1 - \frac{\sqrt{\gamma^2 + 2\sqrt{c}\gamma}}{\sqrt{1-c} \cdot \frac{\sqrt{c}+\gamma}{\sigma}} \geq 1 - 2\frac{\sqrt{\gamma}(2\sqrt{c} + \gamma)^{3/2}}{\sqrt{(1-c)}(1 - (\sqrt{c} + \gamma)^2)}$$

The function on the right is strictly decreasing, so we have

$$\sqrt{c} + \gamma \geq 1 - 2\frac{\sqrt{\gamma_0}(2\sqrt{c} + \gamma_0)^{3/2}}{\sqrt{(1-c)}(1 - (\sqrt{c} + \gamma_0)^2)} \tag{11}$$

Now define $\overline{\gamma}$ by the equality

$$\sqrt{c} + \overline{\gamma} =\geq 1 - 2\frac{\sqrt{\overline{\gamma}}(2\sqrt{c} + \overline{\gamma})^{3/2}}{\sqrt{(1-c)}(1 - (\sqrt{c} + \overline{\gamma})^2)} \tag{12}$$

The equation has a unique solution which is strictly positive because the function on the left is strictly increasing and its value at $\overline{\gamma} = 0$ is strictly smaller than one.

If we now choose $\gamma_0 = \overline{\gamma}$, then the assumption that $\|B\| = \sqrt{c} + \gamma$ with $\gamma < \gamma_0$ will lead to a contradiction in view of (11), (12). Thus $\|B\| \geq \sqrt{c} + \overline{\gamma}$ which proves the claim.

Step 4. We are now ready to describe the construction of F. Let $\psi(t)$ be a smooth nonnegative even function on $\mathbb{R}$ supported on $[-1, 1]$ and such that

$$\psi(0) = \max \psi = 1; \qquad \int_{-1}^{1} \psi(t)dt = 1.$$

Set

$$\varphi(t) = \int_0^t \psi(\tau)d\tau,$$

so that $\varphi(t)$ is a nondecreasing function satisfying

$$\varphi(t) = -1/2, \text{ if } t \leq 1, \quad \varphi(t) = 1/2, \text{ if } t \geq 1; \quad 0 \leq \varphi'(t) \leq 1, \quad \varphi'(0) = 1,$$

and $\varphi'(t)$ assumes all values between zero and one on $[-1, 1]$.

We next define a mapping $\pi(t)$ from $\mathbb{R}$ into X as follows: first for every natural k set

$$\pi_k(t) = 2^{-(k+1)}[\varphi(2^{k+1}(t - 3 \cdot 2^{-(k+1)})) + (1/2)]$$

and then set

$$\pi(t) = \sum_{k=1}^{\infty} \pi_k(t) e_{k+1}.$$

It is an easy matter to see that π is a Lipschitz mapping with Lipschitz constant one. As every π_k is supported on $\Delta_k = [2^{-k}, 2^{-(k-1)}]$, we have, setting $k(t) = k$ if $2^{-k} < t \leq 2^{-(k-1)}$, $\alpha_k = \sum_{i=k+1}^{\infty} 2^{-i} e_{i+1}$

$$\pi(t) = \alpha_{k(t)} + \pi_{k(t)}(t) e_{k(t)+1}.$$

Then π is continuously (Fréchet) differentiable at every nonzero t, $\pi(t) = 0$ if $t \le 0$ and for $t > 0$

$$\pi'(t) = \varphi'(2^{k(t)+1}(t - 3 \cdot 2^{-(k(t)+1)}))e_{k(t)+1},$$

in particular, $\pi'(3 \cdot 2^{-k}) = e_k$ for every $k = 2, 3, \ldots$.

Finally, for every $x = (\xi_1, \xi_2, \ldots) \in X$ (where ξ_i stand for the coordinates of x in the chosen basis) we set

$$F(x) = x + \pi(\xi_1).$$

Clearly, F is continuously Fréchet differentiable everywhere except at the points of the hyperplane $H_0 = \{x : (x|e_1) = 0\}$. Consider now for every $y = (\eta_1, \eta_2, \ldots) \in X$ the function

$$p_y(x) = (y|F(x)) = (y|x) + (y|\pi(\xi_1)) = (y|x) + \sum_{k=1}^{\infty} \pi_k(\xi_1)\eta_k.$$

As $\eta_k \to 0$ when $k \to \infty$, this function is continuously Fréchet differentiable also at points of H_0, that is F is weakly continuously Fréchet differentiable.

Step 5. To complete the proof of the theorem, we have to show that the norms of linear operators $A: X \to X$ for which $F - A$ is not regular near zero are greater than $\mathrm{Sur}F(0)$ plus a positive constant.

Clearly the gradient of f_y is defined by

$$\nabla f_y(x) = y + \lambda_{k(\xi_1)}e_1,$$

where

$$\lambda_k(t) = (y|\pi_k'(t)e_k) = \pi_k'(t)\eta_k,$$

so from the properties of π_k' we conclude that $|\lambda_k(t)|$ assumes on Δ_k all values between zero and $|\eta_k|$ and has the same sign as η_k.

Applying Proposition 1 along with the fact that the functions f_y are continuously Fréchet differentiable, we get from Step 2

$$\begin{aligned}\mathrm{Sur}F(0) &= \liminf_{x\to 0} \inf_{\|y\|=1} \|\nabla f_y(x)\| \\ &= \liminf_{x\to 0} \inf\{\|y + \lambda e_1\| : \|y\| = 1,\ 0 \le |\lambda| \le |\eta_k|,\ \lambda\eta_k \ge 0\} = \sqrt{c}.\end{aligned}$$

Suppose now that $A: X \to X$ is a linear bounded operator such that $\mathrm{Sur}(F - A)(0) = 0$. As above, we get that in this case

$$\begin{aligned}0 &= \liminf_{x\to 0} \inf_{\|y=1\|} \|\nabla(y|F - A)(x)\| \\ &= \liminf_{x\to 0} \inf\{\|y - A^*y + \lambda e_1\| : \|y\| = 1,\ 0 \le |\lambda| \le \|\eta_k\|,\ \lambda\eta_k \ge 0\},\end{aligned}$$

so applying the result of Step 3, we conclude that $\|A\| = \|A^*\| \ge \sqrt{c} + \bar{\gamma}$, where $\bar{\gamma}$ is an absolute positive constant not depending on A. This completes the proof.

References

[1] D. Azé, J.-N. Corvellec and R. Lucchetti, Variational pairs and application to stability in nonsmooth analysis, *Nonlinear Analysis TMA*

[2] E. De Giorgi, A. Marino and M. Tosques, Problemi di evoluzione in spazi metrici e curve di massima pendenza, *Atti Acad. Nat. Lincei, Rend. Cl. Sci. Fiz. Mat. Natur.* **68** (1980), 180-187.

[3] A.V. Dmitruk, A.A. Miljutin and N.P. Osmolovskii, The theorem of Ljusternik and theory of extremum, *Uspehi Matem. Nauk'* **35**:6 (1980), 11-46 (in Russian); English translation: *Russian Math. Surveys* **35**:6 (1980), 11-51.

[4] A.L. Dontchev, A.S. Lewis and R.T. Rockafellar, The radius of metric regularity, *Trans. Amer. Math. Soc*, to appear

[5] A.D. Ioffe On the local surjection property, *Nonlinear Analysis TMA*

[6] A.D. Ioffe, Metric regularity and subdifferential calculus, *Uspehi Matem. Nauk* **55**:3 (2000), 103-162 (in Russian); English translation: *Russian Math. Surveys* **55**:3 (2000)

[7] A.D. Ioffe, Towards metric theory of metric regularity, in *Approximation, Optimization and mathematical Economics*, M. Lassonde, ed. Physica Verlag 2001, pp. 165-177.

[8] B.S. Mordukhovich, Coderivative analysis of variational systems, preprint, 2002

Fixed points of increasing operator

Fuyi Li

Department of Mathematics, Shanxi University, Taiyuan 030006 People's Republic of China

Definition 1. Let E be a real Banach space. If P is a nonempty closed convex set in E and satisfies the following two conditions:

(i) $x \in P, \lambda \geq 0 \Rightarrow \lambda x \in P$;

(ii) $x \in P, -x \in P \Rightarrow x = \theta$, where θ denotes the zero element.

Then P is said to be a cone in E.

Given a cone P in E, we define a partial ordering $\leq$ with respect to P by $x \leq y$ iff $y - x \in P$.

Definition 2. Let D be a subset of E and $A : D \to E$ be a operator.

(i) If $Ax \leq Ay$ for any $x, y \in D, x \leq y$, then A is said to be increasing on D;

(ii) If $Ay \leq Ax$ for any $x, y \in D, x \leq y$, then A is said to be decreasing on D.

In general, there are several methods and ways as following to study the existence and uniqueness of fixed point of increasing operators.

1. The way of dismissing the continuity and weakening the compactness

Theorem 1.1. Let the cone P be normal and $[u_0, v_0] = \{x \in E : u_0 \leq x \leq v_0\}$ the order interval in E. Let $A : [u_0, v_0] \to E$ be condensing, in particular, strictly set-contractive, completely continuous operator and increasing operator. Suppose that

$$u_0 \leq Au_0, \quad Av_0 \leq v_0.$$

Then A has the maximal fixed point x^* and the minimal fixed point x_* in $[u_0, v_0]$, and $x^* = \lim_{n\to\infty} v_n, x_* = \lim_{n\to\infty} u_n$, where $v_n = Av_{n-1}, u_n = Au_{n-1}, n = 1, 2, \cdots$, and satisfy

$$u_0 \leq u_1 \leq \cdots \leq u_n \leq \cdots \leq v_n \leq \cdots \leq v_1 \leq v_0. \tag{1}$$

Theorem 1.2. Let the cone P be regular and $A : [u_0, v_0] \to E$ continuous increasing operator. Suppose that $u_0 \leq Au_0, Av_0 \leq v_0$. Then A has the maximal fixed point x^* and the minimal fixed point x_* in $[u_0, v_0]$, and $x^* = \lim_{n\to\infty} v_n, x_* = \lim_{n\to\infty} u_n$, where $v_n = Av_{n-1}, u_n = Au_{n-1}, n = 1, 2, \cdots$, and also satisfy (1).

Theorem 1.3. Let the cone P be strongly minihedral, $A : [u_0, v_0] \to E$ an increasing operator and $u_0 \le Au_0, Av_0 \le v_0$. Then A has the maximal fixed point x^* and the minimal fixed point x_* in $[u_0, v_0]$.

These above theorems of increasing operators are relatively classical. Afterwards, many people would like to get rid of the continuity and weaken the compactness.

Theorem 1.4[14]. Let X be a partial order set, $D = [u_0, v_0]$ an order interval in X and $A : D \to X$ a mapping. Suppose that

(i) there exist a sequence compatible partial order topological space Y and two increasing operators $B : D \to Y$ and $C : [Bu_0, Bv_0] \to X$ such that A can be denoted $A = CB$;

(ii) $u_0 \le Au_0, Av_0 \le v_0$;

(iii) $B(D)$ is a quasiseparable and quasicompact subset of Y.

Then A has the maximal fixed point and the minimal fixed point in D.

Theorem 1.5[14]. Let $D = [u_0, v_0], A : D \to E$ a increasing operator satisfying $u_0 \le Au_0, Av_0 \le v_0$. If one of the following conditions holds, then A has the maximal fixed point and the minimal fixed point in $[u_0, v_0]$.

(i) $A(D)$ is relative compact;

(ii) $A(D)$ is a quasiseparable and quasicompact subset of E;

(iii) P is regular.

Theorem 1.6[25]. Let X be a partial order set, $D = [u_0, v_0]$ an order interval in X and (Y, P) a partial order Banach space. Suppose that the operator $A : D \to X$ satisfies $u_0 \le Au_0, Av_0 \le v_0$. If there exist two increasing operators $B : D \to Y$ and $C : [Bu_0, Bv_0] \to X$ such that $A = CB$ and every chain of $B(D)$ is relative weakly compact subset of Y, then A has the maximal fixed point and the minimal fixed point in D.

Concerning these above work, please read [1,2,5-7,11,13,14,16,17,25,28-31,35,36] for reference.

2. Fixed point of convex and concave operator

Theorem 2.1[10,13]. Let P be a normal cone and $A : P^\circ \to P^\circ$ an α concave increasing operator or $-\alpha$ convex decreasing operator. Then A has an unique fixed point x^* in P° and for any initial value $x_0 \in P^\circ$ the successive approximation $x_n = Ax_{n-1}, n = 1, 2 \cdots$, all converges to x^*, namely $\|x_n - x^*\| \to 0$ as $n \to \infty$ with convergence rate $\|x_n - x^*\| = O(1 - \gamma^{\alpha^n})$, where γ depends on initial value x_0.

Theorem 2.2[23]. Let P be a normal cone and $A : P_h \to P_h$ a φ concave increasing operator. Suppose that there exist $u_0, v_0, w_0 \in P_h$ such that $u_0 \le Au_0 \le Av_0 \le v_0$ and for any $x \in [u_0, v_0]$ and $t \in (0,1), \varphi(t, x) \ge \varphi(t, w_0)$ and $\lim\limits_{s \to t^-} \varphi(t, x) > t$, then A has an unique fixed point $x^* \in P_h$ and for any initial value $x_0 \in P_h$ the successive approximation $x_n = Ax_{n-1}, n = 1, 2 \cdots$, all converges to x^*.

Theorem2.3[23]. Let P be a normal cone, $A : P_h \to P_h$ a φ increasing operator. Suppose that $\varphi(t, x)$ is increasing in x, $\lim\limits_{s \to t^-} \varphi(t, x) > t$ for any $t \in (0,1)$, and there exists an $x_0 \in P_h$ such that $t/\varphi(t, x_0) \to 0$ as $t \to 0^+$, then A has an unique fixed point $x^* \in P_h$ and for any initial value $y_0 \in P_h$, the successive approximation $y_n = Ay_{n-1}, n = 1, 2 \cdots$, all converges to x^*.

We introduce the concepts of φ concave operator and $-\varphi$ convex operator[23], thus unitedly deal with α concave ($-\alpha$ convex) operator[10,13], u_0−concave operator[11,13,18], general concave(convex) operator[7,8] and some decreasing operators[9], obtain better results. For $-\varphi$ convex decreasing operator, we give the similar results, to see [23]. Moreover, [38] gets rid of the condition $\lim\limits_{s \to t^-} \varphi(t, x) > t$. In [20] we discuss the addition case of a concave and a linear operator or a sup-linear operator.

Concerning these above work, please refer to [2-4,7-9,10-13,15,18,20,23,26,27,37].

3. The case that the increasing operators controlled by a linear operator

Theorem 3.1[33]. Let P be a normal cone in E and $D = [u_0, v_0]$ an order interval in E. Suppose that the operator $A : D \times D \to E$ satisfies the following conditions:

(i) $u_0 \le A(u_0, v_0), A(v_0, u_0) \le v_0$;

(ii) for any fixed $x \in D, A(x, y)$ is decreasing in y;

(iii) there exists a constant $M \ge 0$ such that for any fixed $y \in D$ the following holds

$$-M(x_2 - x_1) \le A(x_2, y) - A(x_1, y), u_0 \le x \le y \le v_0;$$

(iv) $A(y, x) - A(x, y) \le L(y - x), u_0 \le x \le y \le v_0$, where $L : E \to E$ is a linear operator and the spectral ridius $r(L) < 1$.

Then the equation $x = A(x, x)$ has an unique solution x^* in D and for any $x_0 \in D$, the successive approximation

$$x_n = \frac{1}{1+M}[A(x_{n-1}, x_{n-1}) + Mx_{n-1}], n = 1, 2 \cdots,$$

all converges to x^* in the norm and for any given $\alpha \in (r(L), 1)$ there exist n_0 such that

$$||x_n - x^*|| \le 2N \left(\frac{\alpha + M}{1 + M}\right)^n ||v_0 - u_0||, \quad n \ge n_0, \tag{2}$$

where N is the normal constant.

Theorem 3.2[33]. Let P be a normal cone in E and $D = [u_0, v_0]$ an order interval in E. Suppose that the operator $A : D \to E$ satisfies the following conditions:

(i) $u_0 \le Au_0, Av_0 \le v_0$;

(ii) there exists a constant $M \ge 0$ such that for any $u_0 \le x \le y \le v_0$ the following holds

$$-M(y - x) \le Ay - Ax \le L(y - x),$$

where $L : E \to E$ is a linear operator and spectral radius $r(L) < 1$. Then the equation $x = Ax$ has an unique solution x^* in D and for any $x_0 \in D$, the successive approximation

$$x_n = \frac{1}{1+M}[Ax_{n-1} + Mx_{n-1}], n = 1, 2 \cdots ,$$

all converges to x^* in the norm and the estimative formula (2) holds.

We use Thompson's metric to unite above fixed point results on φ increasing operators and increasing operator controlled by a linear operator.

Theorem 3.3[24]. Let $(X, d, \leq)$ be a complete metric space and the partial order the normal on the metric. Suppose that $A : [u_0, v_0] \to X$ is an increasing operator and satisfies the following conditions:

(i) $u_0 \leq Au_0, Av_0 \leq v_0$;

(ii) $d(Ay, Ax) \leq \varphi(d(y, x)), x, y \in [u_0, v_0], x \leq y$,

where $\varphi : R_+ \to R_+$ satisfies $\lim_{s \to t+} \varphi(s) > t, \forall t > 0$. Then A has an unique fixed point x^* in $[u_0, v_0]$ and for any $x_0 \in [u_0, v_0]$, the successive approximation $x_n = Ax_{n-1}, n = 1, 2 \cdots$, all converges to x^* in the metric.

When the metric in Theorem 3.3 takes Thompson's metric, we get Theorem 2.3; when the metric in Theorem 3.3 takes the norm metric, we get Theorem 3.2, to see [24].

4. Fixed point theorem of two points expansion type

As above, the fixed point theorems that we discussed are the type of two points contraction, namely we have considered the fixed points of A, which satisfies the condition of super- and sub-solutions in usual order, i.e. $u_0 \leq Au_0, Av_0 \leq v_0, u_0 \leq v_0$. The circumstances generally describe the sub-linear problems, but the sup-linear problems always lead to the circumstances of two points expansion.

Theorem 4.1[31,32]. Let P be a solid cone in E and $A : P \to P$ a completely increasing operator. Suppose that

(i) A satisfies the condition of two point expansion, namely there exists a below solution u_0 and supper solution v_0 of A satisfying $u_0 \leq Au_0, Av_0 \leq v_0, u_0 \not\leq v_0$;

(ii) there exist $u^* \in P$ and a constant $\alpha > 0$ such that $Ax \geq \alpha\|Ax\|u^*, \forall x \in P$;

(iii) $u^* \in P^\circ, v_0 \in P^\circ$.

Then A has lest one nonzero fixed point in P.

For convenience of recounting, we give some conditions beforehand.

(H_0) There exist $u_0, v_0 \in P$ such that $Au_0 \leq u_0, v_0 \leq Av_0, v_0 \not\leq u_0$.

(g) There exist $h \in P \setminus \{\theta\}$ and a functional $g : Q \to R_+$ satisfying $g(x) \to \infty$ as $\|x\| \to \infty$ such that $Ax \geq g(x)h, \forall x \in Q$.

Theorem 4.2[19]. Suppose that $A : P \to P$ is a condensing increasing operator, $A(P) \subset Q$ and satisfies conditions (H_0) and (g). Assume that $A|_Q$ is $e-$continuous,$e \in Q \setminus \{\theta\}$ and

(i) at every point x in Q, for any number $\varepsilon > 0$, there exists a number $\delta > 0$ such that $A(x + \varepsilon e) \geq Ax + \delta e$;

(ii) there exist $x_0 \in P$ and numbers $\lambda, \mu > 0$ such that $Ax_0 \leq u_0 - \lambda e, v_0 \leq \mu h$.

Then A has lest one nonzero fixed point in P.

This above Theorem improves the relative results of [31,32] by dismissing the condition "P be a solid cone" and weakening the condition "$Ax \geq \alpha\|Ax\|u^*$, $\forall x \in P$". Moreover, in the specific problems the condition "$Ax \geq g(x)h, \forall x \in Q$" is easy to test. [34] improves the result of Theorem 4.1 to the extent of strictly set-contractive map. The results of the type of two points expansion is relatively fewer, at present there are only four primary articles which we refer to.

Theorem 4.3[1,2]. Let (E,P) be an OBS whose positive cone has nonempty interior. Suppose that there exist four points $\bar{y}_k, \hat{y}_k \in E, k = 1,2$, with

$$\bar{y}_1 < \hat{y}_1 < \bar{y}_2 < \hat{y}_2$$

and a compact, strongly increasing map $f : [\bar{y}_1, \hat{y}_2] \to E$ such that

$$\bar{y}_1 \leq f(\bar{y}_1),\ f(\hat{y}_1) < \hat{y}_1,\ \bar{y}_2 < f(\bar{y}_2),\ f(\hat{y}_2) \leq \hat{y}_2.$$

Then f has at least three distinct fixed points x_1, x_2, x_3 such that $\bar{y}_1 \leq x_1 \ll \hat{y}_1, \bar{y}_2 \ll x_2 \leq \hat{y}_2$, and $\bar{y}_2 \not\leq x_3 \not\leq \hat{y}_1$.

Theorem 4.4[21]. Let P be a normal cone in E. Suppose that there exist $u_1, v_1, u_2, v_2 \in E$ with $u_1 < v_1 < v_2, u_1 < u_2 < v_2, u_2 \not\leq v_1$ and $A : [u_1, v_2] \to E$ is a strictly set-contractive increasing operator. Assume that there exists $e > \theta$, numbers $\lambda_0, \mu_0 > 0$ and a natural number m such that the operator A is $e-$continuous and

$$u_1 \leq Au_1, A^m v_1 \leq v_1 - \lambda_0 e, u_2 + \mu_o e \leq A^m u_2, Av_2 \leq v_2,$$

Then the operator A has lest three fixed points x_1, x_2, x_3 with $u_1 \leq x_1 < v_1, u_2 < x_2 \leq v_2, u_2 \not\leq x_3 \not\leq v_1$.

This theorem removed the following conditions used by Amann: the cone is solid, the operator A is strongly increasing; We suppose that $A : [u_1, v_2] \to E$ is strictly set-contractive(condensing), but in [1,2] they used the condition $A_e = A \circ j_e|_{[u_1,v_2]}$. Our condition is more natural, more inherent and easier to test.

5. Fixed point theorem of decreasing operator

Theorem 5.1[9]. Suppose that

(i) P is a normal cone, $A : P \to P$ is a condensing operator(especially strictly set-contractive map, completely continuous operator), and A is a decreasing operator;

(ii) $A\theta > \theta, A^2\theta \geq \varepsilon_0 A\theta$, where $\varepsilon_0 > 0$;

(iii) For any given $x > \theta$ and $t \in (0,1)$ there exists $\eta = \eta(x,t) > 0$ such that $A(tx) \leq [t(1+\eta)]^{-1}Ax$.

Then A has an unique fixed point $x^* > \theta$ and uses any $x_0 \in P$ as initial value, all the successive approximation $x_n = Ax_{n-1}, n = 1,2,\cdots$, satisfy $\|x_n - x^*\| \to 0$.

Theorem 5.2[12] Suppose (a) cone P is normal and operator $A : P \to P$ is decreasing, (b) $A\theta > \theta, A^2\theta \geq \varepsilon_0 A\theta$, where $\varepsilon_0 > 0$ and (c) for any $0 < a < b < 1$ there exists $\eta = \eta(a,b) > 0$ such that

$$A(tx) \leq [t(1+\eta)]^{-1}Ax, \quad \forall t \in [a,b], x \in [\theta, A\theta].$$

Then A has exactly one fixed point $x^* \in P$ and $x^* > \theta$. Moreover, constructing successively sequence $x_n = Ax_{n-1}, n = 1, 2, \cdots$, for any initial $x_0 \in P$, we have $\|x_n - x^*\| \to 0,\ n \to \infty$.

Theorem 5.8[22]. Suppose that (1) P is a normal cone, $A : P \to E$ is convex decreasing operator and there exist a number $\varepsilon > 0$ and a natural number m such that

$$\theta \leq A\theta, \quad \varepsilon A\theta \leq A^2\theta, \quad \frac{1}{2}A\theta \leq A^{2m}\theta.$$

Then the operator A has an unique fixed point x^* in P and for any given initial value $x_0 \in [\theta, A\theta]$, all the successive approximation $x_n = Ax_{n-1}, n = 1, 2, \cdots$, satisfy $\|x_n - x^*\| \to 0$.

References

[1] H. Amann, On the number of solutions of nonlinear equations in ordered Banach spaces, *J. Funct. Anal.*, 1972, 11: 346-384

[2] H. Amann, Fixed point equations and nonlinear eigenvalue problems in ordered Banach spaces, *SIAM Review*, 1976, 18(4): 620-709

[3] P. J. Bushell, Hilbert's metric and positive contraction mappings in a Banach space, *Arch. Rational Mech. Anal.*, 1973, 52: 330-338

[4] P. J. Bushell, On a class of Volterra and Fredholm non-linear integral equations, *Math. Proc. Camb. Phil. Soc.*, 1976, 79: 329-335

[5] K. Deimling, *Nonlinear Functional Analysis*. Springer-Verlag Berlin Heidelberg, 1985

[6] Yihong Du, Total order minihedral cones, *J. Sys. Sci. & Math. Scis.*, 1988, 8(1): 19-24 (in Chinese)

[7] Yihong Du, Fixed points of increasing operators in ordered Banach spaces and applications, *Appl. Anal.*, 1990, 38: 1-20

[8] Yihong Du, Fixed points of certain noncompact operators and applications, *Acta Math. Sinica*, 1989,32(5): 618-627 (in Chinese)

[9] Dajun Guo, A fixed point theorem of decreasing operator and its applications, *Kexue Tongbao*, 1984, 29: 189 (in Chinese)

[10] Dajun Guo, Fixed pionts and eigenvectors of some classes of concave and convex operators, *Kexue Tongbao*, 1985, 30(15): 1132-1135 (in Chinese)

[11] Dajun Guo, *Nonolinear Functional Analysis*, Jinan: Shandong Sci. Tech. Publishing,1985 (in Chinese)

[12] Dajun Guo, Positive fixed points and eigenvectors of noncompact decreasing operators with applications to nonlinear integral equations, *Chin. Ann. of Math.*, 1993, 14B(4): 419-426 (in Chinese)

[13] Dajun Guo, V. Lakshmikantham, *Nonlinear Problems In Abstract Cones*, Boston and New York:Academic Prees, Inc, 1988

[14] Dajun Guo, Jingxian Sun, *Ordinary Deffirential Equtions in Abstract Spaces*, Jinan: Shandong Sci. Tech. Publishing,1989 (in Chinese)

[15] Dajun Guo, Jingxian Sun, Zhaoli Liu, *the Functional Methods of Nonliear Ordinary Deffirential equations*, Jinan: Shandong Sci. Tech. Publishing,1995 (in Chinese)

[16] S. Heikkila, On fixed points through a generalized iteration method with applications to differential and integral equations involving discontinuities, *Nonlinear Anal.*, 1990, 14(5): 413-426

[17] S. Heikkila, V. Lakshmikantham, Yong Sun, Fixed point results in ordered normed spaces with applications to abstract and differential equations, *J. Math. Anal. Appl.*, 1992, 163: 422-437

[18] M. A. Krasnoselskii, P. P. Zabreiko, *Geometrical Methods of Nonlinear Analysis*, Springer-Verlag Now York, 1984

[19] Fuyi Li, Two point expansion fixed point theorem and solutions of convex concave operator equations with applications, *Acta Math. Sinica*, 1997, 40(3): 157-164. MR 99b:47080 (in Chinese)

[20] Fuyi Li, Existence and uniqueness of positive solutions of some nonlinear equations ,*Acta Math. Appl. Sinica*,1997, 20(4): 609-615. MR 99e:47085 (in Chinese)

[21] Fuyi Li, Improvement on Amann's three-solution theorem and applications, *J. Sys. Sci. & Math. Scis.*, 1999, 19(4): 403-406 (in Chinese)

[22] Fuyi Li, Jinfeng Feng, Peilong Shen, Fixed point theorems of some decreasing operators and applications, *Acta Math. Sinica*, 1999, 42(2): 193-196. MR 2000c:47106 (in Chinese)

[23] Fuyi Li, Zhandong Liang, Fixed-point theorems of φ-concave($-\varphi$-convex) operators and application, *J. Sys. Sci. & Math. Scis.*, 1994, 14(4): 355-360. MR 96c:47090 (in Chinese)

[24] Li Fuyi, Zhang Qingzheng, Ordering contraction mapping principle and applications, *Acta Math. Scientia*, 1998, 18(4): 457-460

[25] Xiaoying Liu, Congxin Wu, Fixed point of discontinuous weakly compact increasing operators and its application to initial value problem in Banach spaces, *J. Sys. Sci. & Math. Scis.*, 2000, 20(2): 175-180 (in Chinese)

[26] A. J. B. Potter, Applications of Hilbert's projective metric to certain classes of non-homogeneous operators, *Quart. J. Math. Oxford Ser.*, 1977, 28(2): 93-99

[27] Chenglin Qin, Positive fixed points of quasi-homogeneous mapping. *Acta Math. Sinica*, 1984, 27(6): 792-794 (in Chinese)

[28] Jingxian Sun, The fixed points theorems of discontinous increaseing operator and application for nonlinear equations with discontinuous item, *Acta Math. Sinica*, 1988, 31(1): 101-107 (in Chinese)

[29] Jingxian Sun, The fixed points and general fixed points of increasing operator, *Acta Math. Sinica*, 1989, 32(4): 457-463 (in Chinese)

[30] Jingxian Sun, Some new theorem for sequence compactness in Banach space and applications, *Chin. Ann. of Math.*, 1990, 11A(4): 407-412 (in Chinese)

[31] Jingxian Sun, Some new fixed point theorems of increasing operators and applications, *Appl. Anal.*, 1991, 42: 263-273

[32] Jingxian Sun, Fixed point theorems of two-point extension type and applications, *J. Syst. Sci. Math. Sci.*, 1992, 12(3): 284-286 (in Chinese)

[33] Jingxian Sun, Lishan Liu, The iterative solutions of nonlinear operator equations and their applications, *Acta Math. Scientia*, 1993, 13(3): 141-145 (in Chinese)

[34] Jingxian Sun, Zhaoli Liu, A fixed point theorem for k-set-contractions of the type of two-piont extension, *J. Shandong Univ.*, 1993, 28(1): 25-29 (in Chinese)

[35] Jingxian Sun, Yong Sun, Some fixed point theorems of increasing operators, *Appl. Anal.*, 1986, 23: 23-27

[36] Jingxian Sun, Zengqin Zhao, Fixed point theorems of increasing operators and applications to nonlinear integro-differential equations with discontinuous terms, *J. Math. Anal. Appl.*, 1993, 175(1): 33-45

[37] Weisun Wan, Contraction conditions of mapping and fixed piont theorems of Banach type, *Acta Math. Sinica*, 1984, 27: 35-52 (in Chinese)

[38] Zengqin Zhao, Uniqueness and existence of fixed points on some mixed monotone mappings in ordered linear spaces, *J. Sys. Sci. & Math. Scis.* , 1999, 19(2): 217-224 (in Chinese)

Localized critical points in Banach spaces and sign changing solutions of nonlinear p-Laplacian equations

Zhaoli Liu

Department of Mathematics, Shandong University
Shandong 250100, P. R. China

1. Introduction

In this paper, we will report some recent results on the existence and multiplicity of sign changing solutions of the p-Laplacian equation

$$-\Delta_p u = f(x, u), \qquad u \in W_0^{1,p}(\Omega) \tag{1.1}$$

on a bounded domain $\Omega \in \mathbb{R}^N$ with smooth boundary, where $p > 1$. We will also consider the equation

$$-\Delta_p u = \lambda f(x, u), \qquad u \in W_0^{1,p}(\Omega) \qquad (1.1)_\lambda$$

with a positive parameter λ.

In the semilinear case ($p = 2$), these problems have been studied by many authors. For problems with superlinear and subcritical nonlinearities, see [**4, 9, 10, 12, 17, 18, 19, 20, 21, 22, 23, 24**] for existence of one sign changing solution and [**3, 11, 15, 16**] for existence of infinitely many sign changing solutions. Existence of sign changing solutions of problems involve a critical Sobolev exponent was studied in [**2, 13, 14, 25**]. Different approaches have been developed in these papers and they seem not to be applicable to the quasilinear problems (1.1) and $(1.1)_\lambda$ mainly due to two reasons. First of all, the variational setting for the problems (1.1) and $(1.1)_\lambda$ lacks an ordered Hilbert space structure which plays an important role in dealing with semilinear problems. Second, the regularity theory for quasilinear equations is not as strong as the theory for semilinear equations.

Supported by the AvH in Germany and FUKTME in China.

In order to find sign changing solutions of problems (1.1) and $(1.1)_\lambda$, we have to develop a new critical point theory in ordered Banach spaces. In section 2 we give several theorems in which critical points are found inside or outside of given convex subsets of Banach spaces. In section 3 we state results for problem $(1.1)_\lambda$ with f being oscillatory.

Results concerning problem (1.1) with f being subcritical and $f/|t|^{p-2}$ being superlinear will be given in section 4. Proofs of the results can be found in **[5, 6, 7, 8]** and will not be given in this paper.

2. Localized critical points in Banach spaces

Let X, Y, Z be Banach spaces such that Y is densely embedded into X and that Z is compactly embedded into Y. Assume $J \in \mathcal{C}(X,\mathbb{R})$ and $A \in \mathcal{C}(X,X)$ be such that $K = \{u \in X : \ J'(u) = 0\} \subset Y$ and $A(Y) \subset Z$. We assume one of the following hypotheses relating A to J.

(J_1) There exist $1 < p \le 2$, $a_1 > 0$ and $a_2 > 0$ such that

$$\langle J'(u), u - A(u)\rangle_{X^*,X} \ge a_1\|u - A(u)\|_X^2(\|u\|_X + \|A(u)\|_X)^{p-2}, \quad \forall u \in X,$$

$$\|J'(u)\|_{X^*} \le a_2\|u - A(u)\|_X^{p-1}, \quad \forall u \in X.$$

(J_2) There exist $p \ge 2$, $a_1 > 0$ and $a_2 > 0$ such that

$$\langle J'(u), u - A(u)\rangle_{X^*,X} \ge a_1\|u - A(u)\|_X^p, \quad \forall u \in X,$$

$$\|J'(u)\|_{X^*} \le a_2\|u - A(u)\|_X(\|u\|_X + \|A(u)\|_X)^{p-2}, \quad \forall u \in X.$$

Our goal is to find critical points of J inside a given convex set $\mathcal{D}_1$ and outside of other convex sets $\mathcal{D}_i$. For a subset $\mathcal{D} \subset Y$ we set $\widetilde{\mathcal{D}} = \mathrm{clos}_X(\mathcal{D})$. Given a pair of convex sets $\mathcal{D}_2 \subset \mathcal{D}_1 \subset Y$ we need the following hypotheses.

(P_1) $A(\widetilde{\mathcal{D}_1}) \subset \mathcal{D}_1 \cap Z$ is bounded in Z.
(P_2) $A(\widetilde{\mathcal{D}_2}) \subset \mathrm{int}_{\mathcal{D}_1}(\mathcal{D}_2) \cap Z$.
(P_3) $\mathrm{int}_{\mathcal{D}_1}(\mathcal{D}_2)$ is convex;
if $C \subset \mathrm{int}_{\mathcal{D}_1}(\mathcal{D}_2)$ is compact then $\mathrm{clos}_Y \,\mathrm{conv}\, C \subset \mathrm{int}_{\mathcal{D}_1}(\mathcal{D}_2)$;
if $a \in \mathcal{D}_2$ and $b \in \mathrm{int}_{\mathcal{D}_1}(\mathcal{D}_2)$ then $ta + (1-t)b \in \mathrm{int}_{\mathcal{D}_1}(\mathcal{D}_2)$ for $0 \le t < 1$.

In the applications to (1.1) and $(1.1)_\lambda$ we have in mind that $X := W_0^{1,\,p}(\Omega)$, $Y := X \cap \mathcal{C}^1(\overline{\Omega})$, $Z := X \cap \mathcal{C}^{1,\alpha}(\overline{\Omega})$, and $\mathcal{D}_i$ are order intervals in Y between subsolutions and supersolutions. The functional and the operator for problem $(1.1)_\lambda$ (for (1.1), $\lambda = 1$) will be

$$J_\lambda(u) := \frac{1}{p}\int_\Omega |\nabla u|^p - \lambda\int_\Omega F(x,u), \quad u \in X,$$

$$A_\lambda(u) := \lambda(-\Delta_p + \lambda m h_p(\cdot))^{-1}(f(x,u) + m h_p(u)), \quad u \in X,$$

where $F(x,t) = \int_0^t f(x,s)ds$, m is a suitable positive number, and $h_p(t) = |t|^{p-2}t$. It was proved in **[6, 7]** that under general conditions on f the assumptions (J_1) (in the case $1 < p \le 2$), (J_2) (in the case $p \ge 2$) and (P_1)-(P_3) are satisfied.

THEOREM 2.1 ([**5**]). *Assume either* (J_1) *or* (J_2). *Let* $\mathcal{D}_2, \mathcal{D}_3 \subset \mathcal{D}_1 \subset Y$ *be closed convex sets and set* $\mathcal{D}_4 := \mathcal{D}_2 \cap \mathcal{D}_3$. *Suppose the properties* $(P_1) - (P_3)$ *hold for* $\mathcal{D}_i \subset \mathcal{D}_1$, $i = 2, 3, 4$. *If* J *satisfies the Palais-Smale condition and if there exist* $h_1, h_2 \in \mathcal{C}([0,1], \mathcal{D}_1)$ *such that* $h_1(0) = h_2(0) \in \mathcal{D}_2$, $h_1(1) = h_2(1) \in \mathcal{D}_3$, $h_2([0,1]) \subset \mathcal{D}_2 \cup \mathcal{D}_3$, $h_1([0,1]) \cap h_2([0,1]) = \{h_1(0), h_1(1)\}$, *and*

$$-\infty < \inf_{u \in \mathcal{D}_1} J(u) \le \max_{0 \le t \le 1} J(h_1(t)) < \inf_{u \in \mathcal{D}_4} J(u),$$

then the following conclusions hold.

a) *Setting* $\Gamma_1 := \{h \in \mathcal{C}([0,1], \mathcal{D}_1) : h(0) \in \mathcal{D}_2,\ h(1) \in \mathcal{D}_3\}$ *and*

$$c_1 = \inf_{h \in \Gamma_1} \max_{0 \le t \le 1} J(h(t)).$$

there exists $w_1 \in (K \cap \mathcal{D}_1) \setminus (\mathcal{D}_2 \cup \mathcal{D}_3)$ *with* $J(w_1) = c_1$.

b) *Setting* $\Gamma_2 := \{h \in \mathcal{C}([0,1], \mathcal{D}_2 \cup \mathcal{D}_3) : h(0) = h_1(0),\ h(1) = h_1(1)\}$ *and*

$$\gamma_2 = \inf_{h \in \Gamma_2} \max_{0 \le t \le 1} J(h(t))$$

there exists $w_2 \in (K \cap \mathcal{D}_1) \setminus (\mathcal{D}_2 \cup \mathcal{D}_3)$ *with* $c_2 := J(w_2) \ge \gamma_2$.

c) *If* J *has only a finite number of critical points in* $\mathcal{D}_1 \setminus (\mathcal{D}_2 \cup \mathcal{D}_3)$ *and if they lie in* $\operatorname{int}_Y \mathcal{D}_1$ *then*

$$C_1(J, w_1) \ne 0, \quad C_2(J, w_2) \ne 0.$$

THEOREM 2.2 ([**5**]). *Suppose* J *is even,* A *is odd, and the hypotheses of Theorem 2.2 hold with* $\mathcal{D}_1 = -\mathcal{D}_1$, $\mathcal{D}_2 = -\mathcal{D}_3$ *and* $0 \in \mathcal{D}_4$. *If there exists an odd map* $h_0 \in \mathcal{C}(S^{k-1}, \mathcal{D}_1)$, S^{k-1} *being the unit sphere in* $\mathbb{R}^k$, *such that*

$$-\infty < \inf J|_{\mathcal{D}_1} \le \max J \circ h_0 < \inf J|_{\mathcal{D}_4},$$

then J *has at least* $k-1$ *pairs of critical points in* $\mathcal{D}_1 \setminus (\mathcal{D}_2 \cup \mathcal{D}_3)$ *with energy in* $[\inf J|_{\mathcal{D}_1}, \max J \circ h_0]$ *and at least* $k-1$ *pairs of critical points in* $\mathcal{D}_1 \setminus (\mathcal{D}_2 \cup \mathcal{D}_3)$ *with energy in* $[\inf J|_{\mathcal{D}_4}, \max J \circ h]$ *where* $h \in \mathcal{C}(B^k, \mathcal{D}_1)$ *is any odd extension of* h_0 *to the closed unit ball* $B^k \subset \mathbb{R}^k$.

Using A, we can construct a locally Lipschitz continuous operator $B : X \setminus K \to Z$ such that B keeps all the properties of A (see [**5**]). For $u \in X \setminus K$, we use $\varphi^t(u)$ to denote the solution of

$$\dot{\varphi}(t) = -\varphi(t) + B(\varphi(t)), \qquad \varphi(0) = u \tag{2.1}$$

with maximal interval $[0, \tau(u))$ of existence. We also denote $\varphi^t(u) \equiv u$ for $0 \le t < \infty$ if $u \in K$. In Theorem 2.2, the critical values c_j in $[\inf J|_{\mathcal{D}_1}, \max J \circ h_0]$ and d_j in $[\inf J|_{\mathcal{D}_4}, \max J \circ h]$ are defined as, for $j = 1, \dots, k-1$,

$$c_j = \inf_{E \in \Gamma_j} \sup_{u \in E} J(u), \quad d_j = \inf_{E \in \Delta_j} \max_{u \in E} J(u),$$

where

$$\Gamma_j := \{E \subset \mathcal{D}_1 \setminus (C_{\mathcal{D}_1}(\mathcal{O}_2) \cup C_{\mathcal{D}_1}(\mathcal{O}_3)) : E = -E,\ \operatorname{gen}(E) \ge j\},$$

$$\Delta_j := \{E \subset \partial_{\mathcal{D}_1} C_{\mathcal{D}_1}(\mathcal{O}_4) \setminus (C_{\mathcal{D}_1}(\mathcal{O}_2) \cup C_{\mathcal{D}_1}(\mathcal{O}_3)) :\ E = -E,\ \operatorname{gen}(E) \ge j\},$$

gen(E) is the genus of E, and for $i = 2,\ 3,\ 4$, $\mathcal{O}_i := \mathrm{int}_{\mathcal{D}_1}(\mathcal{D}_i)$ and

$$C_{\mathcal{D}_1}(\mathcal{O}_i) := \{u \in \mathcal{D}_1 :\ \text{there exists } t > 0 \text{ such that } \varphi^t(u) \in \mathcal{O}_i\}.$$

In applications to the problems (1.1) and $(1.1)_\lambda$, the sets $\mathcal{D}_i$ ($i = 1, \ldots, 4$) above are order intervals between subsolutions and supersolutions and these sets have nonempty interiors in Y but not in X. So in the proofs of the above theorems the topology of Y plays an important role (cf. **[5, 6]**). By constructing convex sets with nonempty interiors in X other than order intervals we need not to use topologies of other spaces except X. This causes convenience in some cases. Now Let $\mathcal{D}^{\pm}$ be two closed convex subsets of X with nonempty interiors in X. We shall use the following assumptions on $\mathcal{D}^{\pm}$, A and J:

(A_1) $\mathcal{O} := \mathrm{int}(\mathcal{D}^+) \cap \mathrm{int}(\mathcal{D}^-) \neq \emptyset$ and $A(\mathcal{D}^{\pm}) \subset \mathrm{int}(\mathcal{D}^{\pm})$.

(A_2) The map A is compact, that is, A maps bounded subsets of X into pre-compact subsets of X.

(J_3) For any $b \in \mathbb{R}$ there exists a constant $c = c(b) > 0$ such that if $u \in J^b := \{u \in X : J(u) \leq b\}$ then

$$\|u\|_X + \|A(u)\|_X \leq c(1 + \|u - A(u)\|_X).$$

(J_4) There exists a path $h : [0,1] \to X$ such that $h(0) \in \mathrm{int}(\mathcal{D}^+) \setminus \mathcal{D}^-$ and $h(1) \in \mathrm{int}(\mathcal{D}^-) \setminus \mathcal{D}^+$ and

$$\max_{0 \leq t \leq 1} J(h(t)) < \alpha_0 := \inf_{u \in \mathcal{D}^+ \cap \mathcal{D}^-} J(u).$$

(J_5) There exist a number α_1, a sequence $\{X_n\}_{n \in \mathbb{N}}$ of subspaces of X with $\dim X_n = n$ and a sequence $\{R_n\}_{n \in \mathbb{N}}$ of positive numbers such that

$$\sup_{u \in X_n,\, \|u\|_X > R_n} J(u) \leq \alpha_1 < \alpha_0 = \inf_{u \in \mathcal{D}^+ \cap \mathcal{D}^-} J(u).$$

Defining

$$C_X(\mathcal{O}) := \{u \in X :\ \text{there exists } t > 0 \text{ such that } \varphi^t(u) \in \mathcal{O}\}$$

where $\varphi^t(u)$ is defined by (2.1), but now $B : X \to X$ in (2.1) is constructed in accordance with A satisfying (A_1) instead of (P_1)-(P_3) (see **[8]**), we have

THEOREM 2.3 (**[8]**). *Assume (A_1), (A_2), (J_3) and (J_4) in addition to either (J_1) or (J_2). Then J has a critical point in each of the four mutually disjoint sets: $\partial C_X(\mathcal{O}) \setminus (\mathcal{D}^+ \cup \mathcal{D}^-)$, $\partial C_X(\mathcal{O}) \cap \mathrm{int}(\mathcal{D}^+)$, $\partial C_X(\mathcal{O}) \cap \mathrm{int}(\mathcal{D}^-)$, and $\mathcal{O}$. Moreover, the values of J at these critical points are not larger than $\max_{u \in \mathcal{P}} J(u)$, where $\mathcal{P} := \{sh(t) :\ 0 \leq s \leq 1,\ 0 \leq t \leq 1\}$.*

Setting $B_n := \{u \in X_n :\ \|u\|_X \leq R_n\}$, we define for $n \in \mathbb{N}$,

$$G_n := \{h \in C(B_n, X) :\ \text{h is odd and } h(x) = x \text{ for } x \in \partial_{X_n} B_n\}$$

and

$$\Gamma_n := \{h(B_m \setminus B) \setminus (\mathcal{D}^+ \cup \mathcal{D}^-) :\ h \in G_m,\ m \geq n,\ B = -B \subset B_m \text{ open}, \ \mathrm{gen}(B) \leq m - n\}.$$

To state a result under (J_5), we also need

$$K_c^* := \{u \in K \setminus (\mathcal{D}^+ \cup \mathcal{D}^-) : J(u) = c\},$$

The next theorem is taken from [8] and can be proved with the same argument as in [**5**].

THEOREM 2.4. *Assume (A_1), (A_2), (J_3) and (J_5) in addition to either (J_1) or (J_2). Assume also that A is odd, J is even and $\mathcal{D}^+ = -\mathcal{D}^-$. Define*

$$d_n := \inf_{A \in \Gamma_n} \sup_{u \in A} J(u) \quad \text{for } n \geq 2.$$

Then $d_n \to +\infty$ as $n \to \infty$ and $K_{d_n}^* \neq \emptyset$ for $n \geq 2$.

3. Results for problem $(1.1)_\lambda$

Results in this section can be derived from Theorems 2.1 and 2.2. Recalling $h_p(t) = |t|^{p-2}t$, we assume that:

(H_0) $f \in \mathcal{C}(\overline{\Omega} \times \mathbb{R})$, $f(x,0) = 0$, $\liminf_{t \to 0} \dfrac{f(x,t)}{h_p(t)} > -\infty$.

(H_1) There exists $t^- < 0$ with $\min f|_{\overline{\Omega} \times \{t^-\}} > 0$, or $\lim_{t \to 0^-} \dfrac{f(x,t)}{|t|^{p-1}} = 0 =: t^-$ uniformly in x.

There exists $t^+ > 0$ with $\max f|_{\overline{\Omega} \times \{t^+\}} < 0$, or $\lim_{t \to 0^+} \dfrac{f(x,t)}{|t|^{p-1}} = 0 =: t^+$ uniformly in x.

(H_2) There exist $s^- < t^-$ and $s^+ > t^+$ such that

$$\min F|_{\overline{\Omega} \times \{s^-, s^+\}} > \max F|_{\overline{\Omega} \times [t^-, t^+]}.$$

(H_3) $\lim_{|t| \to \infty} f(x,t)/|t|^{p-1} = 0$ uniformly in x.

THEOREM 3.1 ([**7**]). *Assume $(H_0) - (H_3)$. Then there exists $\Lambda > 0$ such that $(1.1)_\lambda$ with $\lambda \geq \Lambda$ has two positive solutions u_1, u_2, two negative solutions v_1, v_2 and two sign changing solutions w_1, w_2. These solutions lie in Y, outside of the order interval $[t^-, t^+] := \{u \in Y : t^- \leq u \leq t^+\}$. More precisely, for $i = 1,\ 2$,*

$$\min_{\Omega} v_i,\ \min_{\Omega} w_i < t^-, \quad \max_{\Omega} u_i,\ \max_{\Omega} w_i > t^+.$$

THEOREM 3.2 ([**7**]). *Assume $(H_0) - (H_3)$ and that $f(x,t)$ is odd in t. Then for any $k \in \mathbb{N}$, there exists $\Lambda_k > 0$ such that $(1.1)_\lambda$ with $\lambda \geq \Lambda_k$ has k pairs of sign changing solutions with energy less than α_λ, and k pairs of sign changing solutions with energy at least α_λ, where $\alpha_\lambda = \inf J_\lambda|_{[t^-, t^+]}$.*

(H_4) There exist real numbers $t_n^- < \cdots < t_1^- < 0 < t_1^+ < \cdots < t_n^+$ such that

$$\min f|_{\overline{\Omega} \times \{t_i^-\}} > 0 > \max f|_{\overline{\Omega} \times \{t_i^+\}} \quad \text{for } i = 1, \ldots, n,$$

$$\min F|_{\overline{\Omega} \times \{t_i^-, t_i^+\}} > \max F|_{\overline{\Omega} \times [t_{i-1}^-, t_{i-1}^+]} \quad \text{for } i = 2, \ldots, n.$$

THEOREM 3.3 ([**7**]). *Assume* (H_0) *and* (H_4). *Then there exists* $\Lambda > 0$ *such that* $(1.1)_\lambda$ *with* $\lambda \geq \Lambda$ *has* $2(n-1)$ *positive solutions* u_i, $\hat{u}_i$, $2(n-1)$ *negative solutions* v_i, $\hat{v}_i$, *and* $2(n-1)$ *sign changing solutions* w_i, $\hat{w}_i$, $i = 1, \cdots, n-1$. *The six solutions with index* i *lie in the order interval* $[t_{i+1}^-, t_{i+1}^+] \subset Y$ *and outside of* $[t_i^-, t_i^+]$, $i = 1, \cdots, n-1$. *More precisely, for* $i = 1, \ldots, n-1$,

$$t_{i+1}^- \leq \min_\Omega v_i, \ \min_\Omega \hat{v}_i, \ \min_\Omega w_i, \ \min_\Omega \hat{w}_i < t_i^-,$$

$$t_i^+ < \max_\Omega u_i, \ \max_\Omega \hat{u}_i, \ \max_\Omega w_i, \ \max_\Omega \hat{w}_i \leq t_{i+1}^+.$$

If (H_0) and (H_4) are assumed with the number n of t_i^- and of t_i^+ being replaced with ∞ then the number of positive solutions, the number of negative solutions, and the number of sign changing solutions go to ∞ as $\lambda \to +\infty$. We allow $\{t_i^-\}_1^\infty$ to be decreasing and $\{t_i^+\}_1^\infty$ to be increasing, or vice versa.

Structure of positive solutions of $(1.1)_\lambda$ under similar conditions to (H_4) has been studied by many people. The existence of sign changing solutions under these conditions has not been studied yet, to the best knowledge of the author.

4. Results on problem (1.1)

For formulating the conditions, we set $p^* := Np/(N-p)$ for $p < N$ and $p^* := \infty$ for $p \geq N$. Let λ_1 be the first eigenvalue of: $-\Delta_p u = \lambda h_p(u)$, $u \in X = W_0^{1,p}(\Omega)$. Then we assume:

(F_0) $f \in \mathcal{C}(\overline{\Omega} \times \mathbb{R}, \mathbb{R})$ and $f(x, 0) = 0$.

(F_1) $\lim_{|t|\to\infty} f(x,t)/|t|^{p^*-1} = 0$ uniformly in x if $1 < p < N$, $\lim_{|t|\to\infty} \ln|f(x,t)|/t^{N/(N-1)} = 0$ uniformly in x if $p = N$, and no such condition if $p > N$.

(F_2) There exist $\mu > p$ and $M > 0$ such that

$$0 < F(x,t) := \int_0^t f(x,s)ds \leq \frac{1}{\mu} t f(x,t) \qquad \text{for } |t| \geq M.$$

(F_3) $\limsup_{t\to 0} |f(x,t)|/|t|^{p-1} < \lambda_1$ uniformly in x.

(F_3') There exist $t_- < 0 < t_+$ such that $f(x, t_-) \geq 0 \geq f(x, t_+)$.

Using Theorems 2.3 and 2.4, we can prove

THEOREM 4.1 ([**8**]). *Assume* (F_0)-(F_2) *in addition to either* (F_3) *or* (F_3'). *Then problem* (1.1) *has a sign-changing solution.*

THEOREM 4.2 ([**8**]). *Assume* (F_0)-(F_2) *and that* f *is odd in* t. *Then problem* (1.1) *has infinitely many sign changing solutions.*

Under the assumptions of Theorem 4.1, the problem (1.1) has also a positive solution and a negative solution. This can be proved by the mountain pass theorem of Ambrosetti and Rabinowitz [**1**].

With an additional technical assumption: $N < p + p^2/(p-1)$ if $1 < p < N$, Theorem 4.1 was first proved in [**6**]. This additional condition was used in [**6**] only to deduce the Y-convergence of Palais-Smale sequences from their X-convergence.

In the semilinear case: $p = 2$, the third solution was first proved to exist by Wang **[26]**. That the third solution is a sign changing solution was proved by many authors. In the quasilinear case: $p > 1$, there seems no results except those in **[6]** concerning the existence of the third solution.

References

[1] Ambrosetti, A. and Rabinowitz, P.H.: Dual variational methods in critical point theory and applications. J. Funct. Anal., **14**, 349-381 (1973).

[2] Atkinson, F.V., Brezis, H., and Peletier, L.A.: Nodal solutions of elliptic equations with critical Sobolev exponents. J. Diff. Equ. **85**, 151-170 (1990).

[3] Bartsch, T.: Critical point theory on partially ordered Hilbert spaces. J. Funct. Anal. **186**, 117–152 (2001).

[4] Bartsch, T., Chang, K.C., and Wang, Z.-Q.: On the Morse indices of sign changing solutions of nonlinear elliptic problems. Math. Z. **233**, 655–677 (2000).

[5] Bartsch, T. and Liu, Z.L.: Location and critical groups of critical points in Banach spaces with an application to nonlinear eigenvalue problems. Preprint.

[6] Bartsch, T. and Liu, Z.L.: On a superlinear elliptic p-Laplacian equation. Preprint.

[7] Bartsch, T. and Liu, Z.L.: Multiple sign changing solutions of a quasilinear elliptic eigenvalue problem involving the p-Laplacian. Preprint.

[8] Bartsch, T., Liu, Z.L., and Weth, T.: Sign changing solutions of a p-Laplacian equation. Preprint.

[9] Bartsch, T. and Wang, Z.-Q.: On the existence of sign changing solutions for semilinear Dirichlet problems. Top. Meth. Nonlin. Anal. **7**, 115–131 (1996).

[10] Bartsch, T. and Wang, Z.-Q.: Sign changing solutions of nonlinear Schrödinger equationss. Top. Meth. Nonlin. Anal. **13**, 191–198 (1999).

[11] Bartsch, T. and Willem, M.: Infinitely many radial solutions of a semilinear elliptic problem on $\mathbb{R}^N$. Arch. Rational Mech. Anal., **124**, 261-276 (1993).

[12] Castro, A., Cossio, J., and Neuberger, J.M.: A sign changing solution for a superlinear Dirichlet problem. Rocky Mountain J. Math. **27**, 1041–1053 (1997).

[13] Cerami, G., Solimini, S., and Struwe, M.: Some existence results for superlinear elliptic boundary value problems involving critical exponents. J. Funct. Anal., **69**, 289-306 (1986).

[14] Comte, M. and Tarantello, G.: A Neumann problem with critical Soblev exponent. Houston J. Math., **18**, 279-294 (1992).

[15] Conti, M., Merizzi, L., and Terracini, S.: Radial solutions of superlinear equations on $\mathbb{R}^N$, Part I: a global variational approach. Arch Rational Mech. Anal., **153**, 291-316 (2000).

[16] Conti, M. and Terracini, S.: Radial solutions of superlinear equations on $\mathbb{R}^N$, Part II: the forced case. Arch Rational Mech. Anal., **153**, 317-339 (2000).

[17] Dancer, E.N. and Du, Y.: Existence of changing sign solutions for some semilinear problems with jumping nonlinearities at zero. Proc. Royal Soc. Edinburgh **124A**, 1165-1176 (1994)

[18] Dancer, E.N. and Du, Y.: On sign-changing solutions of certain semilinear elliptic problems. Appl. Anal. **56**, 193-206 (1995).

[19] Dancer, E.N. and Du, Y.: Multiple solutions of some semilinear elliptic equations via the generalized Conley index. J. Math. Anal. Appl. **189**, 848–871 (1995).

[20] Li, S.J. and Wang, Z.-Q.: Mountain pass theorems in order intervals and multiple solutions for semilinear elliptic Dirichlet problems. J. Analyse Math. **81**, 373–396 (2000).

[21] Li, S.J. and Wang, Z.-Q.: Ljusternik-Schnirelman theory in partially ordered Hilbert spaces. Tran. Amer. Math. Soc. **354**, 3207–3227 (2002).

[22] Li, S.J. and Zhang, Z.T.: Sign-changing solutions and multiple solutions theorems for semilinear elliptic boundary value problems. Acta Math. Sinica, New Series, **16**, 113-122 (2000).

[23] Liu, Z.L. and Sun, J.X.: Invariant sets of descending flow in critical point theory with applications to nonlinear differential equations. J. Diff. Equ. **172**, 257–299 (2001).

[24] Liu, Z.L. and Sun, J.X.: Four versus two solutions of semilinear elliptic boundary value problems. Calc. Var. **14**, 319-327 (2002).
[25] Tarantello, G.: Nodal solutions of semilinear elliptic equations with critical exponent. Diff. Int. Equ., **5**, 25-42 (1992).
[26] Wang, Z.-Q.: On a semilinear elliptic equation. Ann. Inst. H. Poincaré. Anal. Non Linéaire **8**, 43–57 (1991).

Collinear central configurations in celestial mechanics

Yiming Long[1] and Shanzhong Sun[2]

[1] *Nankai Institute of Mathematics, Nankai University Tianjin 300071, The People's Republic of China*

[2] *LMAM, School of Mathematics, Peking University Beijing 100871, The People's Republic of China*

Let $\mathbf{N}$ and $\mathbf{R}$ denote the sets of natural integers and real numbers respectively. Given any mass vector $m = (m_1, \ldots, m_n) \in (\mathbf{R}^+)^n$ with $\mathbf{R}^+ = \{r \in \mathbf{R} \,|\, r > 0\}$, the potential of the n-body problem on the motion of n particles $q = (q_1, \cdots, q_n) \in (\mathbf{R}^3)^n$ with the mass m_i at q_i under the gravitation is defined by

$$U(q) = \sum_{i<j} \frac{m_i m_j}{r_{ij}}, \tag{1}$$

where $r_{ij} = |q_i - q_j|$. When one studies homographic solutions of the n-body problem, the motion at any fixed time must satisfy the following nonlinear algebraic equation system:

$$\lambda m_i q_i + U'_{q_i}(q) = 0, \qquad \text{for } 1 \le i \le n, \tag{2}$$

with a parameter $\lambda \in \mathbf{R}$, where U'_{q_i} is the derivative of U with respect to q_i. By the homogeneity of $U(q)$ of degree -1, we obtain

$$\lambda = \frac{U(q)}{Mq \cdot q} > 0, \tag{3}$$

where $M = \mathrm{diag}(m_1, m_1, m_1, m_2, m_2, m_2, \ldots, m_n, m_n, m_n)$. Let

$$\Delta = \cup\{q = (q_1, \cdots, q_n) \in (\mathbf{R}^3)^n \,|\, q_i = q_j \text{ for some } i \ne j\}$$

be the collision set. As usual, we have

The first author was partially supported by the 973 Program of STM, Yangzi River Professorship, MCME, RFDP, PMC Key Lab of EM of China, S. S. Chern Foundation, CEC of Tianjin, and Nankai University.

e-mail: `longym@nankai.edu.cn` `sz_sun@math.pku.edu.cn`.

Definition 1. A configuration $q = (q_1, \ldots, q_n) \in (\mathbf{R}^3)^n \setminus \Delta$ is a **central configuration** (CC for short) for the mass vector $m = (m_1, \ldots, m_n) \in (\mathbf{R}^+)^n$, if q is a solution of the system (2) for some constant $\lambda \in \mathbf{R}$. q is **collinear**, if all the q_is lie on a straight line in $\mathbf{R}^3$. By this definition, $aAq = (aAq_1, \ldots, aAq_n)$ is also a CC for any $A \in \mathrm{SO}(3)$ and $a > 0$, if q is a CC. If p and q are configurations, we write $p \sim q$ if there exist some $A \in \mathrm{SO}(3)$ and $a > 0$ such that $p = aAq$, otherwise we write $p \not\sim q$. This defines an equivalent relation in central configurations, and as usual we denote the equivalent class of central configurations of q still by q, and also call these equivalent classes the central configurations. For any given $m \in (\mathbf{R}^+)^n$, we denote by $L(n,m)$ the set of all n-body collinear central configurations for m, and use ${}^{\#}B$ to denote the number of elements in a set B.

Note that in another words, an n-point geometric configuration

$$q = (q_1, \ldots, q_n) \in (\mathbf{R}^3)^n$$

is a central configuration, if there exists a solution $q(t)$ of the classical n-body problem such that $q(t)$ is similar to q for all time t.

It is still an open problem to understand the structure of central configurations, and even that there is a long standing conjecture on whether the number of CCs is finite for any given mass vector $m \in (\mathbf{R}^+)^n$ (cf. §360 of [**Win1**] and [**Sma2**]).

Collinear central configuration is one of a few families of central configurations with given masses which are well-known to have been completely understood. Nevertheless, it is still strange enough that historically there exist two contradictory results on collinear CCs.

(i) F. R. Moulton in [**Mou1**] of 1910 gave a proof for the claim

$${}^{\#}L(n,m) = \frac{n!}{2}. \tag{4}$$

This claim was reconfirmed by S. Smale in Part II of [**Sma1**] of 1970 together with a new variational proof which is different from that of Moulton. Moulton's claim (4) and a slightly modified proof of Smale were described as Theorem 10.4.10 in [**AbM1**] of R. Abraham and J. Marsden in 1978. The claim (4) was also quoted in p.65 of [**AKN1**] of V. I. Arnold, V. V. Kozlov, and A. I. Neishtadt in 1997.

(ii) In §358 of [**Win1**] of 1941 and 1952, A. Wintner claimed the following result for $m = (m_1, m_2, m_3) \in (\mathbf{R}^+)^3$: ${}^{\#}L(3,m) = 3$ if m_1, m_2, and m_3 are mutually distinct, ${}^{\#}L(3,m) = 2$ if two of m_1, m_2, and m_3 are equal but not the third, ${}^{\#}L(3,m) = 1$ if $m_1 = m_2 = m_3$.

In the recent paper [**LoS1**], we systematically studied collinear CCs. Unfortunately, we found that both of the above two contradictory claims of Moulton and Wintner are incomplete for 3-body collinear central configurations as shown in our Theorem 3 below. We also generalized the result to the n-body case. We refer to [**LoS1**] for all the details of results in this report. We start from the 3-body case. Here we need to introduce certain singular surfaces in the 3-dimensional mass half space first.

Definition 2. We denote 3 planes and a straight line in the **mass half space** $\mathbf{R}^+ \times \mathbf{R}^2$ by

$$\Pi(y) = \{(x,y,z) \in \mathbf{R}^+ \times \mathbf{R}^2 \mid y = 0\}, \quad (5)$$
$$\Pi(z) = \{(x,y,z) \in \mathbf{R}^+ \times \mathbf{R}^2 \mid z = 0\}, \quad (6)$$
$$\Pi(y+z) = \{(x,y,z) \in \mathbf{R}^+ \times \mathbf{R}^2 \mid y + z = 0\}, \quad (7)$$
$$A(x) = \{(x,y,z) \in \mathbf{R}^+ \times \mathbf{R}^2 \mid y = z = 0\}. \quad (8)$$

Note that these three planes $\Pi(y)$, $\Pi(z)$, and $\Pi(y+z)$ divide the mass half space into 6 open connected components. On the three pairs of opposite components, we define three homogeneous polynomials $f_1(x,y,z)$, $f_2(x,y,z)$, and $f_3(x,y,z)$ of degree 6 in 3 variables $(x,y,z) \in \mathbf{R}^+ \times \mathbf{R}^2 \setminus (\Pi(y) \cup \Pi(z) \cup \Pi(y+z))$ by

$$\begin{aligned} f_1(x,y,z) &= z^6 + (4y+2x)z^5 + (6y+5x)yz^4 + (3y+4x)y^2z^3 \\ &\quad -(3y+4x)y^3z^2 - (3y+5x)y^4z - (y+2x)y^5, \quad \text{if } yz > 0, \end{aligned} \quad (9)$$
$$\begin{aligned} f_2(x,y,z) &= z^6 + (4y+2x)z^5 + (6y+5x)yz^4 + (5y+4x)y^2z^3 \\ &\quad +(5y+6x)y^3z^2 + (3y+5x)y^4z + (y+2x)y^5, \quad \text{if } y(y+z) < 0, \end{aligned} \quad (10)$$
$$\begin{aligned} f_3(x,y,z) &= z^6 + (4y+2x)z^5 + (6y+5x)yz^4 + (5y+6x)y^2z^3 \\ &\quad +(3y+4x)y^3z^2 + (3y+5x)y^4z + (y+2x)y^5, \quad \text{if } z(y+z) < 0. \end{aligned} \quad (11)$$

Define surfaces F_i for $1 \le i \le 3$ in $\mathbf{R}^+ \times \mathbf{R}^2$ by

$$F_i = \{(x,y,z) \in \mathrm{dom}(f_i) \subset \mathbf{R}^+ \times \mathbf{R}^2 \mid f_i(x,y,z) = 0\}, \quad (12)$$

and

$$H_3 = F_1 \cup F_2 \cup F_3. \quad (13)$$

For $1 \le i \le 3$ we call f_i, F_i, and H_3 a **singular polynomial**, a **singular surface**, and the **total singular surface for 3-body collinear central configurations** in the mass half space respectively. We define the total singular set Ψ_3 in the mass half space by

$$\Psi_3 = H_3 \cup \Pi(y) \cup \Pi(z) \cup \Pi(y+z). \quad (14)$$

Note that

$$A(x) = \overline{H_3} \cap \Pi(y) \cap \Pi(z) \cap \Pi(y+z), \quad (15)$$

where $\overline{H_3}$ is the closure of H_3 in $\mathbf{R}^+ \times \mathbf{R}^2$.

To understand the singular surface F_1, note that by Descartes' rule of signs (cf. p.300 of [**Jac1**]), for any given $(x_0, y_0) \in (\mathbf{R}^+)^2$, the polynomial $f_1(x_0, y_0, z)$ possesses one and only one positive root $z = z_0$. Geometrically this means that the vertical line passing through the point $(x_0, y_0, 0)$ intersects F_1 at precisely one point (x_0, y_0, z_0). Thus F_1 in $(\mathbf{R}^+)^3$ is diffeomorphic to $(\mathbf{R}^+)^2$ through these vertical lines.

Theorem 3. *Let H_3 be the non-empty total singular surface in $\mathbf{R}^+ \times \mathbf{R}^2$ defined by (13). Then for any mass vector $m = (m_1, m_2, m_3) \in (\mathbf{R}^+)^3$ one and only one of the following four cases must happen:*

(i) $\#L(3,m) = 3$, if m_1, m_2, and m_3 are mutually distinct, and $(m_1, m_2 - m_1, m_3 - m_2) \notin H_3$;

(ii) $^{\#}L(3,m) = 2$, *if* m_1, m_2, *and* m_3 *are mutually distinct, and* $(m_1, m_2 - m_1, m_3 - m_2) \in H_3$;

(iii) $^{\#}L(3,m) = 2$, *if two of* m_1, m_2, *and* m_3 *are equal to each other but not the third;*

(iv) $^{\#}L(3,m) = 1$, *if* $m_1 = m_2 = m_3$.

Note that Ψ_3 is the singular set in the mass half space which decreases the number of collinear central configurations.

The main difference between our Theorem 3 and the claim of Wintner in §358 of [**Win1**] comes from the existence of the non-empty singular surface H_3 and the above case (ii), which Wintner overlooked. Moulton in fact overlooked all the cases (ii)-(iv). The following concrete Example 4 further explains the case (ii).

Here for any $n \in \mathbf{N}$ we denote by $P(n)$ the set of all permutations of $\{1, \ldots, n\}$. For any element $\tau \in P(n)$, we use $\tau = (\tau(1), \tau(2), \ldots, \tau(n))$ to denote the permutation τ in stead of the notation $\tau = \tau(1)\tau(2)\cdots\tau(n)$ to avoid possible confusions. We also denote a permutation of $\{m_1, m_2, \ldots, m_n\}$ by $m(\tau) = (m_{\tau(1)}, m_{\tau(2)}, \ldots, m_{\tau(n)})$ for $\tau \in P(n)$ in stead of $m_{\tau(1)}m_{\tau(2)}\cdots m_{\tau(n)}$. We define the converse permutation $c(\tau)$ of τ by

$$c(\tau) = (\tau(n), \ldots, \tau(1)). \tag{16}$$

Denote by $[a] = \max\{k \in \mathbf{Z} \mid k \le a\}$ for any $a \in \mathbf{R}$, and by $C(m,k)$ the number of k-combinations in a set of m elements.

Example 4. Let $m_1 = 1$, $m_2 = 2$, and $m_3 - 2$ be the unique positive solution of the equation

$$x^6 + 6x^5 + 11x^4 + 7x^3 - 7x^2 - 8x - 3 = 0. \tag{17}$$

Then there holds $m_1 = 1 < m_2 = 2 < m_3 < 3$, and for this mass vector $m = (m_1, m_2, m_3)$ there exist precisely two distinct equivalent classes of central configurations which are correspond to the two permutations (m_1, m_2, m_3) and (m_1, m_3, m_2) of the given masses on the line. This example shows that Moulton and Wintner's claims for the 3-body collinear central configurations are incomplete for the mutually distinct 3 positive masses.

To generalize our Theorem 3 to the n-body collinear central configurations with $n \ge 4$, we need the following definitions and notations to represent central configurations corresponding to masses with possible repetitions. When $q = (q_1, \ldots, q_n) \in (\mathbf{R}^3)^n$ is collinear, without loss of generality, we suppose that all of q_is are points located on the real line with coordinate $(q_i, 0, 0) \in \mathbf{R}^3$ and $q_1 < q_2 < \cdots < q_n$.

Definition 5. For any $n \in \mathbf{N}$ and vector $m = (m_1, \ldots, m_n) \in (\mathbf{R}^+)^n$, we call $(n, m; k, p)$ a *multi-set* of masses with repetition (k, p), if there exist $k \in \mathbf{N}$, $p = (p_1, \ldots, p_k) \in \mathbf{N}^k$, and $1 \le j_1 < \cdots < j_k \le n$ such that $n = \sum_{i=1}^k p_i$, the masses m_{j_1}, …, m_{j_k} are mutually distinct, and

$$^{\#}\{j \in \{1, \ldots, n\} \mid m_j = m_{j_i}\} = p_i, \quad \text{for} \quad 1 \le i \le k. \tag{18}$$

We call $q = (q_1, \ldots, q_n) \in \mathbf{R}^n$ with $q_1 < q_2 < \cdots < q_n$ a **collinear configuration corresponding to the multi-set** $(n, m; k, p)$, if the mass on q_j is $m_{\tau(j)}$ for $1 \le j \le n$ and some $\tau \in P(n)$. We denote the set of all equivalent classes of collinear central configurations corresponding to $(n, m; k, p)$ by $L(n, m; k, p)$. Note that now we allow the masses to be repeated on different points.

Definition 6. Let $m = (m_1, \ldots, m_n) \in (\mathbf{R}^+)^n$. Given a permutation $m(\alpha) \in P(m)$ with $\alpha \in P(n)$, it is well known that there exists a unique collinear central configuration $q(\alpha)$ for the mass $m(\alpha)$. Let $P(m) = \{m(\alpha) \,|\, \alpha \in P(n)\}$, the set of all the permutations of m. We call two elements $m(\alpha)$ and $m(\beta)$ in $P(m)$ **substantially equal**, if $m(\alpha) = m(\beta)$ or $m(\alpha) = m(c(\beta))$, otherwise we call $m(\alpha)$ and $m(\beta)$ **substantially distinct**. The concept of substantially equal defines an equivalent relation on $P(m)$. Let $\hat{P}(m)$ be the set of equivalent classes of $P(m)$ under this equivalent relation. Let $\{A_1(m), \ldots, A_{d(m)}(m)\}$ be a partition of $\hat{P}(m)$ for some positive integer $d(m)$ such that

$$\hat{P}(m) = \bigcup_{j=1}^{d(m)} A_j(m), \quad A_i(m) \cap A_j(m) = \emptyset, \quad \text{if } i \neq j, \tag{19}$$

and for $m(\alpha) \in A_i(m)$ and $m(\beta) \in A_j(m)$ there hold

$$q(\alpha) \sim q(\beta), \qquad \text{if } i = j, \tag{20}$$

$$q(\alpha) \not\sim q(\beta), \qquad \text{if } i \neq j. \tag{21}$$

The partition $\{A_1(m), \ldots, A_{d(m)}(m)\}$ of $\hat{P}(m)$ is uniquely determined by m and our above definitions. There holds $1 \leq d(m) \leq n!/2$. For $m(\alpha) \in A_i(m)$ with $1 \leq i \leq d(m)$, we define the **singular dimension** of $m(\alpha)$ and $A_i(m)$ by

$$\dim_s(m(\alpha)) = \dim_s(A_i(m)) = {}^{\#}A_i(m) - 1. \tag{22}$$

We define the **singular dimension** of m by

$$\dim_s(m) = \sum_{i=1}^{d(m)} \dim_s(A_i(m)). \tag{23}$$

By definition, we have

$$0 \leq \dim_s(m) \leq \frac{n!}{2}, \qquad \forall m \in (\mathbf{R}^+)^n. \tag{24}$$

Note that $\dim_s(m) > 0$ implies the existence of singularity, which decreases the number of collinear CCs corresponding to m, generated not by equal masses.

For any given multi-set $(n, m; k, p)$ with repetition (k, p), we define two integers by

$$T(n, k, p) = \frac{n!}{p_1! \cdots p_k! 2}, \qquad S(n, k, p) = \frac{[\frac{n}{2}]!}{[\frac{p_1}{2}]! \cdots [\frac{p_k}{2}]! 2}. \tag{25}$$

Theorem 7. *For any given integer $n \geq 3$, there exists a non-empty singular set H_n besides the hyperplanes yielded by equal masses in the mass half space $\mathbf{R}^+ \times \mathbf{R}^{n-1}$ such that for any $m = (m_1, \ldots, m_n) \in (\mathbf{R}^+)^n$ with repetition (k, p), if we denote the corresponding multi-set by $(n, m; k, p)$, then*

$$\dim_s(m) > 0, \qquad \textit{if } (m_1, m_2 - m_1, \ldots, m_n - m_{n-1}) \in H_n, \tag{26}$$

and the following hold:

(i) If either n is odd and there is one and only one $i \in \{1,\dots,k\}$ such that p_i is odd, or n and p_i are even for all $i = 1,\dots,k$, then there hold

$$^{\#}L(n,m;k,p) = T(n,k,p) + S(n,k,p), \qquad \text{if } (m_1, m_2 - m_1, \dots, m_n - m_{n-1}) \notin H_n, \tag{27}$$

and

$$^{\#}L(n,m;k,p) = T(n,k,p) + S(n,k,p) - \dim_s(m), \quad \text{if } (m_1, m_2 - m_1, \dots, m_n - m_{n-1}) \in H_n. \tag{28}$$

(ii) For all the other cases, there hold

$$^{\#}L(n,m;k,p) = T(n,k,p), \qquad \text{if } (m_1, m_2 - m_1, \dots, m_n - m_{n-1}) \notin H_n, \tag{29}$$

and

$$^{\#}L(n,m;k,p) = T(n,k,p) - \dim_s(m), \qquad \text{if } (m_1, m_2 - m_1, \dots, m_n - m_{n-1}) \in H_n. \tag{30}$$

Corollary 8. *Let $(n,m;k,p)$ be any given multi-set.*

(i) If $k = 1$ and $p_1 = n$, i.e., $m_1 = \cdots = m_n$, there holds

$$^{\#}L(n,m;1,n) = 1. \tag{31}$$

(ii) If $k = n$ and $p_i = 1$ for $1 \le i \le n$, i.e., $m_1, \dots, m_n$ are mutually distinct, there hold

$$^{\#}L(n,m;n,(1,\dots,1)) = \frac{n!}{2}, \quad \text{if } (m_1, m_2 - m_1, \dots, m_n - m_{n-1}) \notin H_n, \tag{32}$$

$$^{\#}L(n,m;n,(1,\dots,1)) = \frac{n!}{2} - \dim_s(m), \quad \text{if } (m_1, m_2 - m_1, \dots, m_n - m_{n-1}) \in H_n. \tag{33}$$

Remark 9. That a configuration $q = (q_1, \dots, q_n) \in \mathbf{R}^n \setminus \Delta$ is a CC is a concept related to the masses on points q_is of q. But that two configurations q and $\hat{q}$ are equivalent to each other is a concept which only related to the geometric property of q and $\hat{q}$ on whether there is an $E \in \mathrm{SO}(3)$ and a constant $r > 0$ such that $q = rE\hat{q}$.

Note that proofs given in [**Mou1**], [**Sma1**], and [**AbM1**] only yield an upper bound $n!/2$ for $^{\#}L(n,m;k,p)$. They did not consider possible repetitions of masses, did not try to prove that the CCs found by their methods represent distinct equivalent classes of CCs, and then missed the existences of the singular set H_n in the mass space and the hyperplanes generated by equal masses which decrease the number of equivalent classes of central configurations. Furthermore, because there are two natural group actions on the configuration space generated by equal masses and by the rotation of angle π which is a $\mathbf{Z}_2$ action, and because their actions overlap to each other on some elements in the configuration space in certain cases (the two cases of (i) of Theorem 1.9), the $\mathbf{Z}_2$-action in proofs of [**Sma1**] and [**AbM1**] has fixed points and thus is not free in these cases. So the corresponding quotient space is not $\mathbf{RP}^{n-2}$ in these cases. The proof given in §358 of [**Win1**] for the 3-body case also missed the existence of singular surface H_3 in the mass space.

Remark 10. The following problems on collinear central configurations may be further considered.

(i) It is interesting to understand the precise structure of the singular set H_n for $n \geq 4$. We tend to believe that H_n is also a finite union of algebraic hypersurfaces when $n \geq 4$ similar to the case of $n = 3$ in our Theorem 3.

(ii) Viewing the system (2) for collinear central configurations as a linear system in unknown variable masses $m_1, \ldots, m_n$ for a fixed $q = (q_1, \ldots, q_n)$, it is easy to see that the dimension of the solution space allowing m possessing zero or negative components is at most 2. Because of this reason, in the sense of Definition 6 we suspect

$$\dim_s(m(\alpha)) \leq 1, \qquad \forall \alpha \in P(n). \tag{34}$$

Note that (34) holds when $n = 3$ by our Theorem 3. It is interesting to find good estimate on and to further understand $\dim_s(m)$ for any given m.

(iii) We suspect that such a phenomenon of singular set in the mass space which decreases the number of central configurations should also exist for planar non-collinear as well as non-flat central configurations, i.e., there exists a non-collinear central configuration $q = (q_1, \ldots, q_n) \in (\mathbf{R}^3)^n$ for some mass $m = (m_1, \ldots, m_n) \in (\mathbf{R}^+)^n$ such that q is also a central configuration for $m(\tau)$ with some $\tau \in P(n)$ and that $m(\tau)$ is substantially distinct from m.

Acknowledgements. The authors thank the hospitality of the Department of Mathematics, Shanxi University at Taiyuan during their visit on August 14-17, 2002.

References

[AbM1] R. Abraham and J. E. Marsden, Foundations of Mechanics. 2nd ed. Benjamin/Cummings Publ. Com. London. 1978.

[AKN1] V. I. Arnold, V. V. Kozlov, and A. I. Neishtadt, Mathematical Aspects of Classical and Celestial Mechanics. 2nd Ed. Springer. Berlin. 1993, 2nd printing 1997.

[Bru1] R. Brualdi, Introductory Combinatorics. North Holland, New York. 1977.

[Eul1] L. Euler, De motu rectilineo trium corporum se mutuo attahentium, Novi Comm. Acad. Sci. Imp. Petrop. 11 (1767) 144-151.

[Lag1] J. L. Lagrange, Essai sur le problème des trois corps. Euvres, vol. 6, Gauthier-Villars, Paris, 1772, pp.272-292

[Jac1] N. Jacobson, Basic Algebra I. W. H. Freeman and Com. 1974.

[LoS1] Y. Long and S. Sun, Collinear central configurations and singular hypersurfaces in the mass space. Preprint, Nankai Inst. of Math. Nankai Univ. (2002). Submitted.

[Mou1] F. R. Moulton, The straight line solutions of the n-body problem. *Ann. of Math.* II Ser. 12 (1910) 1-17.

[Sma1] S. Smale, Topology and mechanics.(I) *Invent. Math.* 10 (1970) 303-331, (II) *Invent. Math.* 11 (1970) 45-64.

[Sma2] S. Smale, Mathematical problems for the next century, *Math. Intelligencer.* 20 (1998), 7-15.

[Win1] A. Wintner, The Analytical Foundations of Celestial Mechanics. Princeton Math. Series 5, 215. Princeton Univ. Press, Princeton, NJ. 1941. 3rd printing 1952.

A maximum principle for bounded solutions of the telegraph equation and the upper and lower solution method

Jean Mawhin[1], Rafael Ortega[2,*] and Aureliano M. Robles-Pérez[3]

[1]*Département de Mathématique Université Catholique de Louvain*
1348-Louvain-la-Neuve, Belgium

[2,3]*Departamento de Matemática Aplicada Facultad de Ciencias Universidad de Granada*
18071-Granada, Spain

ABSTRACT. A maximum principle is proved for the weak solutions $u \in L^\infty(\mathbb{R} \times \mathbb{T}^3)$ of the telegraph equation $u_{tt} - \Delta_x u + cu_t + \lambda u = f(t,x)$, in space dimension three, when $c > 0$, $\lambda \in (0, c^2/4]$ and $f \in L^\infty(\mathbb{R} \times \mathbb{T}^3)$ (Theorem 1). The result is extended to a solution and a forcing belonging to a suitable space of bounded measures (Theorem 2). Those results provide a method of upper and lower solutions for the semilinear equation $u_{tt} - \Delta_x u + cu_t = F(t,x,u)$. Applications are given to the semilinear telegraph equation with Duffing and with sine-Gordon nonlinearities.

1. Introduction

It has been proved in [**5**] that a maximum principle holds for the solutions of one-dimensional telegraph equation

$$u_{tt} - u_{xx} + cu_t + \lambda u = f(t,x), \quad (c > 0), \tag{LT}$$

which are T-periodic in t and 2π-periodic in x if and only if $\lambda \in (0, \nu(c)]$, where $\nu(c)$ is some number contained in the interval $\left(\frac{c^2}{4}, \frac{c^2}{4} + \left(\frac{\pi}{T}\right)^2\right]$. This maximum principle

*Supported by DGES PB98-1294 (Spain).

[1]e-mail: `mawhin@math.ucl.ac.be`.

[2]e-mail: `rortega@ugr.es`.

[3]e-mail: `arobles@ugr.es`.

on a torus has been used in [**5**] to develope a method of upper and lower solutions for the corresponding solutions of nonlinear telegraph equations of the form

$$u_{tt} - u_{xx} + cu_t = F(t, x, u), \tag{NT}$$

when the function $u \mapsto F(t, x, u) + \nu(c)u$ is nondecreasing.

A similar maximum principle has been obtained in [**2**] for the solutions $u(t, x)$ of equation (LT) which are 2π-periodic with respect to x and bounded over $\mathbb{R}$ with respect to t (namely $u \in L^\infty(\mathbb{R} \times \mathbb{T}^3)$), if $\lambda \in \left(0, \frac{c^2}{4}\right]$ and $f \in L^\infty(\mathbb{R} \times \mathbb{T}^3)$. Furthermore, the constant $\frac{c^2}{4}$ is optimal and the maximum principle is not strong. In the same paper, using an approximation argument, the maximum principle has been generalized to the case where f is replaced by an element in a suitable class of measures. Using this tool, a method of (weak) upper and lower solutions for the (weak) solutions which are 2π-periodic in x and bounded in t. has been developed for equation (NT), when the function $u \mapsto F(t, x, u) + \frac{c^2}{4}u$ is nondecreasing.

This method has been applied to the forced dissipative sine-Gordon equation

$$u_{tt} - u_{xx} + cu_t + a \sin u = p(t, x), \tag{SG}$$

when $0 < a \leq \frac{c^2}{4}$ and $p \in L^\infty(\mathbb{R} \times \mathbb{T}^3)$. Various conditions upon p provide existence and/or uniqueness of a solution $u \in W^{1,\infty}(\mathbb{R} \times \mathbb{T}^3)$ of equation (SG). In particular, when p is continuous and almost-periodic on $\mathbb{R} \times \mathbb{T}^3$, and $\|p\|_{L^\infty} < a$, the existence-uniqueness result above together with an argument of Amerio's type [**1**] implies the existence of a unique almost periodic solution for equation (SG).

Some extensions to the telegraph equation with space dimension three have been recently obtained in [**3**]. We describe those results and sketch the corresponding proofs, refering to the above mentioned paper for more details.

2. A maximum principle in $L^\infty(\mathbb{R} \times \mathbb{T}^3)$

DEFINITION 1. *Let $c > 0$ and $f \in L^\infty(\mathbb{R} \times \mathbb{T}^3)$. A* bounded solution *of the problem*

$$\mathcal{L}u + \lambda u := u_{tt} - \Delta_x u + cu_t + \lambda u = f(t, x) \quad in \quad \mathbb{R} \times \mathbb{R}^3, \tag{2.1}$$

$$u(t, x_1 + 2\pi, x_2, x_3) = u(t, x_1, x_2 + 2\pi, x_3) = u(t, x_1, x_2, x_3 + 2\pi) = u(t, x)$$

is a function $u \in L^\infty(\mathbb{R} \times \mathbb{T}^3)$ satisfying

$$\int_{\mathbb{R} \times \mathbb{T}^3} (\mathcal{L}^* \phi + \lambda \phi) u = \int_{\mathbb{R} \times \mathbb{T}^3} f \phi, \quad for\ all \quad \phi \in \mathcal{D}(\mathbb{R} \times \mathbb{T}^3),$$

where $\mathcal{L}^* \phi = \phi_{tt} - \Delta \phi - c\phi_t$, i.e.

$$\mathcal{L}u + \lambda u = f \quad in \quad \mathcal{D}'(\mathbb{R} \times \mathbb{T}^3), \quad u \in L^\infty(\mathbb{R} \times \mathbb{T}^3). \tag{2.2}$$

THEOREM 1. *For each $\lambda \in (0, c^2/4]$ and $f \in L^\infty(\mathbb{R} \times \mathbb{T}^3)$, problem (2.2) has a unique solution u. Moreover, if $f \geq 0$ a.e. in $\mathbb{R} \times \mathbb{T}^3$, then $u \geq 0$ a.e. in $\mathbb{R} \times \mathbb{T}^3$.*

Before sketching the proof of Theorem 1, we make a few remarks about its statement.

REMARK 1. Remarks 1 and 2 in [**2**] prove that $\frac{c^2}{4}$ *is optimal and the maximum principle is not strong.*

REMARK 2. By comparison with the case where f is constant, one deduces that *the bounded solution of equation (2.1) satisfies*

$$\|u\|_{L^\infty} \leq \frac{1}{\lambda}\|f\|_{L^\infty}.$$

Sketch of the proof of Theorem 1. *Uniqueness.* By an indirect argument, assume that u is a nontrivial bounded solution of (2.2) for $f = 0$. Convoluting u with an appropriate function $S \in \mathcal{D}(\mathbb{R} \times \mathbb{T}^3)$ provides another nontrivial solution $v = u * S \in L^\infty(\mathbb{R} \times \mathbb{T}^3) \cap C^\infty(\mathbb{R} \times \mathbb{T}^3)$. The Fourier coefficient of v associated to a suitable eigenfunction $\gamma(x) = e^{i(\nu \cdot x)}$ of the Laplace-Beltrami operator on $\mathbb{T}^3$ gives a bounded function $w(t)$ solution of

$$\ddot{v} + c\dot{v} + \eta v = 0$$

for some positive constant η depending upon λ and the eigenvalue associated to γ, a contradiction.

Existence for $\lambda = \frac{c^2}{4}$. The function

$$\chi(x) = \frac{1}{4\pi|x|}e^{-\frac{c|x|}{2}}, \quad x \in \mathbb{R}^3 \setminus \{0\},$$

belongs to $L^1(\mathbb{R}^3)$ with $\|\chi\|_{L^1(\mathbb{R}^3)} = 4/c^2$. The action

$$\langle \mathcal{U}_3, \phi\rangle := \frac{1}{4\pi}\int_{\mathbb{R}^3} e^{-\frac{c|x|}{2}}\frac{\phi(|x|, x)}{|x|}\,dx,$$

on functions $\phi \in C_0(\mathbb{R} \times \mathbb{R}^3)$ ($C_0(\mathbb{R} \times \mathbb{R}^3)$ the space of continuous functions on $\mathbb{R} \times \mathbb{R}^3$ with compact support) defines a positive measure $\mathcal{U}_3$ on $\mathbb{R} \times \mathbb{R}^3$ such that $\|\mathcal{U}_3\|_{M(\mathbb{R}\times\mathbb{R}^3)} = 4/c^2$. The change of unknown $\psi(t, x) := e^{-\frac{ct}{2}}\phi(t, x)$ and Kirchhoff's formula ([**6**], p. 177)

$$\phi(0,0) = \frac{1}{4\pi}\int_{\mathbb{R}^3}\frac{(\Box\phi)(|x|, x)}{|x|}\,dx$$

imply that $\mathcal{U}_3$ is a solution of

$$\mathcal{L}u + \frac{c^2}{4}u = \delta \quad \text{in} \quad \mathcal{D}'(\mathbb{R} \times \mathbb{R}^3),$$

with δ the Dirac measure, i.e. a fundamental solution of $\mathcal{L} + \frac{c^2}{4}$. Hence, for each $f \in L^\infty(\mathbb{R} \times \mathbb{R}^3)$,

$$u(t, x) := (\mathcal{U}_3 * f)(t, x) = \int_{\mathbb{R}\times\mathbb{R}^3} f(t - \tau, x - \xi)d\mathcal{U}_3(\tau, \xi), \quad \text{a.e. in} \quad \mathbb{R} \times \mathbb{R}^3$$

is a $L^\infty(\mathbb{R} \times \mathbb{R}^3)$-function satisfying

$$\|u\|_{L^\infty} \leq \|\mathcal{U}_3\|_M\|f\|_{L^\infty} = \frac{4}{c^2}\|f\|_{L^\infty}, \quad \mathcal{L}u + \frac{c^2}{4}u = f \quad \text{in} \quad \mathcal{D}'(\mathbb{R} \times \mathbb{R}^3).$$

It is easy to show that $u \in L^\infty(\mathbb{R} \times \mathbb{T}^3)$ if $f \in L^\infty(\mathbb{R} \times \mathbb{T}^3)$. To go from the case of test functions in $\mathcal{D}(\mathbb{R} \times \mathbb{R}^3)$ to test functions in $\mathcal{D}(\mathbb{R} \times \mathbb{T}^3)$ one uses, like in Lemma 5.1 of [5], a partition of unity periodic in x.

Existence for $\lambda \in \left(0, \frac{c^2}{4}\right)$. The linear operator

$$\mathcal{R} : L^\infty(\mathbb{R} \times \mathbb{T}^3) \to L^\infty(\mathbb{R} \times \mathbb{T}^3), \quad \mathcal{R}(f) = \mathcal{U}_3 * f$$

is bounded and has norm $\frac{4}{c^2}$. Problem (2.2) is equivalent to

$$\left[I - \left(\frac{c^2}{4} - \lambda\right) \mathcal{R}\right] u = \mathcal{R} f, \quad u \in L^\infty(\mathbb{R} \times \mathbb{T}^3)$$

with $\left\| \left(\frac{c^2}{4} - \lambda\right) \mathcal{R} \right\| < 1$, so that

$$u = \left[I - \left(\frac{c^2}{4} - \lambda\right) \mathcal{R}\right]^{-1} \mathcal{R} f = \sum_{n=0}^{\infty} \left(\frac{c^2}{4} - \lambda\right)^n \mathcal{R}^{n+1} f. \tag{2.3}$$

This leads to the existence and positivity principles.

3. A maximum principle for a class of bounded measures

We define in the space of measures $M(\mathbb{R} \times \mathbb{T}^3)$, for each $h \in \mathbb{R}$ and $K_h = [h - \pi, h + \pi] \times \mathbb{T}^3$, the seminorm

$$||\mu||_h = \sup\{\langle \mu, \phi \rangle : \phi \in C_0(\mathbb{R} \times \mathbb{T}^3), \ ||\phi||_{L^\infty} = 1, \ \operatorname{supp}(\phi) \subset K_h\},$$

and denote by $\mathcal{E}$ the class of measures for which this family of seminorms is bounded independently of h, a Banach space with the norm

$$||\mu||_{\mathcal{E}} = \sup_{h \in \mathbb{R}} ||\mu||_h.$$

Defining a norm in $L^\infty(\mathbb{R} \times \mathbb{T}^3)$ by

$$|||f||| = \sup_{h \in \mathbb{R}} \int_{K_h} |f(t,x)| \, dt \, dx = ||f \, dt \, dx||_{\mathcal{E}},$$

one proves that, given $\mu \in \mathcal{E}$ a sequence (f_n) in $L^\infty(\mathbb{R} \times \mathbb{T}^3)$ exists such that $f_n \, dt \, dx \to \mu$ vaguely, $|||f_n||| \leq 2||\mu||_{\mathcal{E}}$ and $f_n \geq 0$ a.e. in $\mathbb{R} \times \mathbb{T}^3$ if $\mu \geq 0$.

LEMMA 1. *The solution* u *of problem (2.2) satisfies the inequality*

$$|||u||| \leq \frac{1}{\lambda} |||f|||.$$

Sketch of the proof. Case $\lambda = \frac{c^2}{4}$. From $u = \mathcal{U}_3 * f$ we deduce that

$$|||u||| = \sup_{h \in \mathbb{R}} \left| \int_{K_h} \int_{\mathbb{R}^3} \frac{e^{-\frac{c}{2}|\xi|}}{4\pi|\xi|} f(t - |\xi|, x - \xi) \, d\xi \, dt \, dx \right|$$

$$= \sup_{h \in \mathbb{R}} \left| \int_{\mathbb{R}^3} \frac{e^{-\frac{c}{2}|\xi|}}{4\pi|\xi|} \int_{K_h} |f(t - |\xi|, x - \xi)| \, dt \, dx \right| \leq \frac{4}{c^4} |||f|||.$$

Case $\lambda \in \left(0, \frac{c^2}{4}\right)$. We employ the formula (2.3) where $\mathcal{R}f = \mathcal{U}_3 * f$ (convolution in $\mathbb{R} \times \mathbb{R}^3$). Then

$$|||u||| \leq \left[\sum_{n=0}^{\infty} \left(\frac{c^2}{4} - \lambda\right)^n \left(\frac{4}{c^2}\right)^{n+1}\right] |||f||| = \frac{1}{\lambda}|||f|||.$$

Given $\mu \in \mathcal{E}$ we consider the problem

$$\mathcal{L}\eta + \lambda\eta = \mu \quad \text{in} \quad \mathcal{D}'(\mathbb{R} \times \mathbb{T}^3), \quad \eta \in \mathcal{E}. \tag{3.1}$$

THEOREM 2. *Problem (3.1) has a unique solution η, and $\eta \geq 0$ if $\mu \geq 0$.*

Sketch of the proof. Uniqueness. Let $\eta \in \mathcal{E}$ be a solution of (3.1) with $\mu = 0$. For any $S \in \mathcal{D}(\mathbb{R} \times \mathbb{T}^3)$, $v = \eta * S$ (convolution in the topological group $\mathbb{R} \times \mathbb{T}^3$) is bounded and continuous and, given $\phi \in \mathcal{D}(\mathbb{R} \times \mathbb{T}^3)$, one shows that

$$\int_{\mathbb{R}\times\mathbb{T}^3} (\mathcal{L}^*\phi)v = -\lambda \int_{\mathbb{R}\times\mathbb{T}^3} \phi v.$$

Thus $v = 0$. Since S is arbitrary one has also $\eta = 0$.

Existence and positivity. We approximate $\mu \in \mathcal{E}$ (in the vague sense) by a sequence $f_n \in L^\infty(\mathbb{R} \times \mathbb{T}^3)$ with $|||f_n||| \leq 2||\mu||_\mathcal{E}$. The solution $u_n \in L^\infty(\mathbb{R} \times \mathbb{T}^3)$ of $\mathcal{L}u_n + \lambda u_n = f_n$ satisfies

$$|||u_n||| \leq \lambda^{-1}|||f_n||| \leq 2\lambda^{-1}||\mu||_\mathcal{E}.$$

Hence $(\eta_n) = (u_n \, dt \, dx)$ contains a subsequence (η_{n_k}) vaguely convergent to some $\eta \in \mathcal{E}$, which is a solution of (3.1), such that

$$||\eta||_\mathcal{E} \leq \sup ||\eta_n||_\mathcal{E} \leq 2\lambda^{-1}||\mu||_\mathcal{E}.$$

As positivity is preserved by vague limits we get the positivity principle.

REMARK 3. A consequence of the uniqueness of bounded solutions shows that the results extend to doubly periodic solutions : *if* $f \in L^\infty(\mathbb{T}\times\mathbb{T}^3)$, $u \in L^\infty(\mathbb{T}\times\mathbb{T}^3)$, *extending the results of* [5] *to the three dimensional case.*

4. Weak upper and lower solutions

Consider the nonlinear equation

$$\mathfrak{L}u \equiv u_{tt} - \Delta u + cu_t = F(t, x, u) \quad \text{in } \mathfrak{D}'(\mathbb{R} \times \mathbb{T}^3) \tag{4.1}$$

where $F : \mathbb{R} \times \mathbb{T}^3 \times \mathbb{R} \to \mathbb{R}$ satisfies L^∞-Carathéodory conditions.

DEFINITION 2. *A* lower solution *of (4.1) is a function* $u_* \in L^\infty(\mathbb{R} \times \mathbb{T}^3)$ *satisfying*

$$\mathfrak{L}u_* \leq F(t, x, u_*) \quad \text{in } \mathfrak{D}'(\mathbb{R} \times \mathbb{T}^3),$$

i.e.

$$\int_{\mathbb{R}\times\mathbb{T}^3} u_*\mathfrak{L}^*\phi \leq \int_{\mathbb{R}\times\mathbb{T}^3} F(t, x, u_*)\phi, \quad \forall\phi \in \mathfrak{D}(\mathbb{R} \times \mathbb{T}^3), \;\; \phi \geq 0.$$

An upper solution *is a function satisfying the reversed inequality.*

THEOREM 3. *Let u_* and u^* be lower and upper solutions of (4.1) with*

$$u_* \le u^* \quad a.e.\ \mathbb{R} \times \mathbb{T}^3.$$

In addition, assume that

$$\frac{F(t,x,u_1) - F(t,x,u_2)}{u_1 - u_2} \ge -\frac{c^2}{4}$$

if $u_*(t,x) \le u_2 < u_1 \le u^*(t,x)$, a.e. $(t,x) \in \mathbb{R}^2$. Then (4.1) has a solution u in $L^\infty(\mathbb{R} \times \mathbb{T}^3)$ satisfying

$$u_* \le u \le u^*.$$

Sketch of the proof. The proof of the result is based upon the classical iterative scheme

$$(u^{n+1})_{tt} - \Delta u^{n+1} + c u_t^{n+1} + \frac{c^2}{4} u^{n+1} = F(t,x,u^n) + \frac{c^2}{4} u^n$$

with $u^0 = u_*$ and $u^0 = u^*$ respectively. In going to the limit we do not need the compactness of the sequences of iterates. These sequences are bounded and monotone and one can use dominated convergence. As only weak upper and lower solutions are used, the monotonicity properties of the sequences of iterates require the following auxiliary result.

LEMMA 2. *Let $u \in L^\infty(\mathbb{R} \times \mathbb{T}^3)$ be a function such that*

$$\mathfrak{L}u + \lambda u \ge 0 \quad in\ \mathfrak{D}'(\mathbb{R} \times \mathbb{T}^3),$$

where $\lambda \in (0, \frac{c^2}{4}]$. Then $u \ge 0$.

Sketch of the proof. The distribution on the cylinder

$$\mu(\phi) = \int_{\mathbb{R} \times \mathbb{T}^3} \{u\mathfrak{L}^*\phi + \lambda u \phi\}, \quad \forall \phi \in \mathfrak{D}(\mathbb{R} \times \mathbb{T}^3)$$

can be extended to a positive measure μ in $\mathfrak{E}$. One then applies Theorem 2 to finish the proof.

5. Examples

As a first easy example showing how to apply Theorem 3, we consider the equation

$$u_{tt} - \Delta u + c u_t + bu + au^3 = p(t,x) \tag{5.1}$$

where $c > 0$, $b \ge 0$, $a > 0$ and $p \in L^\infty(\mathbb{R} \times \mathbb{T}^3)$. It is immediate to check that the unique positive root $R^* = R^*(a, b, ||p||_{L^\infty})$ of equation $aR^3 + bR = ||p||_{L^\infty}$ is an upper solution of (5.1) and its negative a lower solution. The monotonicity condition in Theorem 3 holds if

$$3aR^{*2} + b \le \frac{c^2}{4}, \tag{5.2}$$

Hence it follows that

COROLLARY 1. *If inequality (5.2) holds, equation (5.1) has at least one (weak) solution* $u \in L^\infty(\mathbb{R} \times \mathbb{T}^3)$ *such that* $\|u\|_{L^\infty} \le R^*$.

We now consider the forced dissipative sine-Gordon equation

$$u_{tt} - \Delta u + cu_t + a \sin u = p(t,x), \tag{5.3}$$

where $a > 0$ and $p \in L^\infty(\mathbb{R} \times \mathbb{T}^3)$. The application of Theorem 3 requires that

$$a \le \frac{c^2}{4}. \tag{5.4}$$

COROLLARY 2. *Assume that (5.4) holds and* p *admits a decomposition* $p = p^* + p^{**}$ *with*

$$p^* = \mathfrak{L}U \quad in\ \mathfrak{D}'(\mathbb{R} \times \mathbb{T}^3) \quad and \quad |p^{**}(t,x)| \le a \cos U(t,x) \quad a.e.\ (t,x),$$

for some $U \in L^\infty(\mathbb{R} \times \mathbb{T}^3)$. *Then (5.3) has a solution* $u \in L^\infty(\mathbb{R} \times \mathbb{T}^3)$ *satisfying*

$$\|u - U\|_{L^\infty} \le \frac{\pi}{2}.$$

Proof. Define $u_* = -\frac{\pi}{2} + U$ and $u^* = \frac{\pi}{2} + U$ and apply Theorem 3.

The simplest choice of U is $U \equiv 0$, giving a solution in $[-\frac{\pi}{2}, \frac{\pi}{2}]$.

6. Final remarks

In one dimension we could prove $u \in W^{1,\infty}(\mathbb{R}\times\mathbb{T})$, and this was used to obtain some existence results about the almost periodic solutions of the one-dimensional sine-Gordon equation in [**2**]. For this the compactness inherited by the regularity $u \in W^{1,\infty}(\mathbb{R}\times\mathbb{T})$ was important. Other techniques provide a similar result in space dimension three. The details will be given in [**4**].

As observed by Dancer in a private communication, the results for dimension three imply that similar ones hold for space dimension two. One uses essentially a method of descent. In this case, a regularity result is obtained for the solution.

THEOREM 4. *For each* $\lambda \in (0, c^2/4]$ *and* $f \in L^\infty(\mathbb{R} \times \mathbb{T}^2)$, *problem (2.2) has a unique solution* u. *Moreover,* u *is continuous, and if* $f \ge 0$ *a.e. in* $\mathbb{R} \times \mathbb{T}^2$, *then* $u \ge 0$ *in* $\mathbb{R} \times \mathbb{T}^2$.

The proof will be given in [**4**].

References

[1] Fink, A.M., *Almost periodic differential equations,* Lecture Notes Math. vol. **377**, Springer-Verlag, Berlin, 1974.

[2] Mawhin, J., Ortega, R., Robles-Pérez, A.M., A maximum principle for bounded solutions of the telegraph equations and applications to nonlinear forcings, *J. Math. Anal. Appl.* **251** (2000), 695–709.

[3] Mawhin, J., Ortega, R., Robles-Pérez, A.M., A maximum principle for bounded solutions of the telegraph equation in space dimension three, *C.R. Acad. Sci. Paris* Ser. I **334** (2002), 1089-1094

[4] Mawhin, J., Ortega, R., Robles-Pérez, A.M., Maximum principles for bounded solutions of the telegraph equation in space-dimensions two and three and applications, *in preparation*

[5] Ortega, R., Robles-Pérez, A.M., A maximum principle for periodic solutions of the telegraph equation, *J. Math. Anal. Appl.* **221** (1998), 625–651.

[6] Vladimirov,V.S., *Equations of Mathematical Physics*, Marcel Dekker, New York, 1971.

Periodic solutions of nonlinear problems with holomorphic type of the nonlinearity

Wacław Marzantowicz

Faculty of Mathematics and Computer Science Adam Mickiewicz University of Poznań ul. Umultowska 87 61-614 Poznań, Poland

ABSTRACT. We study nonlinear ODE problems in the complex plane, with the right hand side $f(u, t)$ being complex analytic function of the space variable u with the coefficients being nonconstant periodic functions in the time variable t and having non-zero Fourier coefficients for non-negative indices only. Assuming first that $f(u, t) = h\big(u, z_0 + r \exp(\frac{2\pi}{T} \imath t)\big)$, where h is holomorphic in the two variables, we drop out the local a priori estimate for the existence of T-periodic solutions of previous works. We prove that there exists $r_0 > 0$, not depending of T, such that for every $r < r_0$ the equation has periodic solution. Next we show that the same still holds for the mentioned above class of nonlinear problems. This improves previous author results, also joint with A. Borisovich.

1. Introduction.

In this paper we study a non-autonomous system of ordinary differential equations in the plane of the form

$$\dot{u}(t) = f(u(t), t), \tag{1.1}$$

where $f(u, t)$ is given by a holomorphic function $h(u, z)$ of two complex variables $(u, z) \in \mathbb{C} \times \mathbb{C}$ in the following way:

1991 *Mathematics Subject Classification.* 34C25, 34Cxx,46J15.

Key words and phrases. periodic solution, Fourier expansions, averaging, Leray-Schauder degree, holomorphic function, singularities.

Research supported by KBN grant nr 2 PO3A 0 04522.

e-mail: `marzan@math.amu.edu.pl`.

Let $h(u, z)$ be a holomorphic function of two variables. Let next fix a point $z_0 \in \mathcal{U}$ and $r > 0$. We define

$$f(u, t) = f_{z_0, r}(u, t) := h\left(u, z_0 + r \exp\left(\frac{2\pi}{T} \imath t\right)\right). \tag{1.2}$$

Note that f is T-periodic with respect to the time variable t.

DEFINITION 1.1. *The space of all functions $f : \mathbb{C} \times T \to \mathbb{C}$ satisfying (1.2) is called the class $\mathcal{A}$.*

We shall study the existence of T-periodic solutions of the equation (1). The special form (1.2) of the right hand side allows to make a use also of the analytic functions theory to get an existence theorem. Moreover there are some hopes that the theory of differential equations in the complex domain let us also to show the non-existence of periodic solutions for some equations of this form.

Remind that equations with the right hand side belonged to some classes containing $\mathcal{A}$ were studied in [**1**] and [**12**]. In [**1**] we assumed that $f \in \mathcal{P}(u, C_+(T))$, where $\mathcal{P}$ denote the class of the functions of the form

$$\mathcal{P}(u, C_+(T)) := \{ \sum_{j=0}^{n} a_j(t) u^j \}$$

where $a_j(t) \in C_+(T)$ the space of continuous T-periodic complex valued functions with non-zero Fourier coefficients at the non-negative indices only. Equivalently, the later are the restrictions to the boundary of functions from the disc algebra (cf. [**1, 12**] for details). Next in [**12**] we supposed that $f \in \mathcal{H}(\mathcal{U}, C_+(T))$, where for an open $\mathcal{U} \subset \mathbb{C}$, $\mathcal{H}(\mathcal{U}, C_+(T))$ denotes the space of all holomorphic with respect to $u \in \mathcal{U}$ functions with the coefficients in $C_+(T)$ (cf. [**12**]). Observe that $f \in \mathcal{A}$ is a polynomial then $f \in \mathcal{P} \subset \mathcal{H}(\mathbb{C}, C_+(T))$, because for a fixed u the restriction function $t \mapsto h\left(u, z_0 + r \exp\left(\frac{2\pi}{T} \imath t\right)\right)$, h-holomorphic belongs to $C_+(T)$.

REMARK 1.2. *Let us remark that well-known and intensively studied by many authors ([**3, 4, 10, 16, 19, 20**] complex Riccati*

$$\dot{u} = u^2 + p(t) + q(t),$$

where $p(t), q(t)$ are complex valued, continuous T-periodic functions is of the form (1)provided the coefficients are of the form $p(t), q(t)$ are of the form $g(z_0 + r \exp(\frac{2\pi}{T} \imath t))$ g a holomorphic function.

Also the complex Bernoulli equation

$$\dot{u} = q(t) u^n + p(t) u,$$

where $p(t), q(t)$ are complex valued, continuous T-periodic functions, $p(t), q_n(t) \neq 0$, is also of this form as p, q are as above.

In order to get an existence result one can study only solutions that belong to $C_+(T) \subset C(T)$. Note that $C_+(T)$ is a subalgebra of $C(T)$. It was observed in

[1] (also [12] for more details) that averaging (1) with respect the time we get a complex analytic equation

$$0 = \mathbf{f}(\mathbf{u}), \tag{1.3}$$

where $\mathbf{f}(u) := \sum_0^\infty \mathbf{c}_j(u-u_0)^j$, with $\mathbf{c}_j := \frac{1}{T}\int_0^T c_j(t)dt$, and $\mathbf{u} := \frac{1}{T}\int_0^T u(t)dt$ for a $f(u,t) \in \mathcal{H}(u, C_+(T))$. Then it was showed that $u(t)$ is a periodic solutions (in the discussed algebra) only if $\mathbf{u}$ is a root of $\mathbf{f}$ i.e. the T-periodic solutions of 1 are only into the fibres $\{F_l\}$, over (isolated !) zeros $\{u_l,\}$ of $\mathbf{f}$ ([1, 12]).

To give a sufficient condition for the existence of T-periodic solution of (1) over a zero of $\mathbf{f}$ the function f was expressed as

$$f(u,t) = (u-u_0)\tilde{f}(u,t) + \rho(t) \tag{1.4}$$

([12]). The main result of [12] says that if for $\tilde{\mu} > 0$ the multiple of T by the sum of C^0 norms of $\sup_{|u-u_0|\le\tilde{\mu}} \|\tilde{f}(u,t)\|$ and $\|\rho\|$ is smaller than 2 then there exists a T-periodic solution $u(t) \in C^0_+(T)$ of (1) with the mean value equal to u_0, and which is non-constant if $\rho \neq 0$. The proof of main theorem is essentially the same as corresponding theorem of [1] i.e. it is a kind of continuation method of Krasnosielski & el. and Mawhin & el. [14]. Indeed our hypothesis is equivalent to an assumption that equation (1) is either small perturbation (but effectively estimated!) of equation (1.3) or the frequency is large enough, which resembles previous approaches to the problem ([13], [11], [19]).

Let us first remark that for a function $f = \sum_{i,j\ge 0} \bar{c}_{i,j} u^j z^i \sum_{i,j\ge 0} c_{i,j}(u-u_0)^j z^i \in \mathcal{A}$ the expansion (1.4) at (u_0, z_0) has the form

$$\begin{aligned} f(u,t) &= \sum_{i\ge 0, j\ge 1} c_{i,j}(u-u_0)^j \left(z_0 + r\exp\left(\tfrac{2\pi}{T}\imath t\right)\right)^i + \sum_{i\ge 0, j=0} c_{i,0}\left(z_0 + r\exp\left(\tfrac{2\pi}{T}\imath t\right)\right)^i \\ &= \tilde{f}(u,t)(u-u_0) + \rho(t). \end{aligned} \tag{1.5}$$

Next the averaging corresponding to the substitution $r = 0$ to the formula (1.2) or to the above expansion. Consequently the necessary condition for the existence of a T-periodic solution $u \in C_+(T)$ for (1) with $f \in \mathcal{A}$ reads as follows.

PROPOSITION 1.3. *Let $h(u, z)$ be a holomorphic function of two variables, $z_0 \in \mathbb{C}$, $r > 0$, and $f = f_{z_0,r}$ be the function defined in (1.2). Then $u(t) \in C(T)$ is a solution of (1) only if $u_0 = \mathbf{u} = \frac{1}{T}\int_T u(t)\,dt$ satisfies the equation*

$$\mathbf{f}(\mathbf{u}) = h(\mathbf{u}, z_0) = 0,$$

where $\mathbf{f} := \frac{1}{T}\int_T f(u,t)\,dt$. This is equivalent to $\frac{1}{T}\int_T r(t)\,dt = \sum_{i\ge 0} c_{0,i} z_0^i = 0$.

Note that the radius $r \in [0, \infty)$ can be considered as a parameter. Assume that $\mathbf{f}(\mathbf{u}) = h(u_0, z_0) = 0$. Furthermore $\|\rho(t)\|$ and $\|\tilde{f}(u,t)\|$ are continuous function of ρ, and u. Moreover for $r = 0$ we have
$r(t) \equiv 0$, $|\tilde{f}(u, z_0)| \le M|(u-u_0)|$ if $|(u-u_0)|$ is small enough. Consequently

for a fixed period T the mentioned above sufficient condition of **[12]** guarantees the existence of a T-periodic solution $u \in C_+(T)$ provided r is sufficiently small. For this equation we have $\mathbf{u} = u_0$ and can be regarded as bifurcating from the stationary solution $u(t) \equiv u_0$.

The aim of this work is to show that for the equation (1) the necessary condition (cf. Prop. 1.3) is also sufficient provided $r < r_0$, where r_0 is the radius of convergence of a power series defined by h, u_0, and z_0 (Theorem 2.1 and Corollary 2.6). Thus we get rid of the condition on the period T, or in other word the local bifurcation of the T-periodic solutions from $u(t) \equiv u_0$ extends onto the whole interval $[0, r_0)$. In our approach we find a use for the fact that the right hand side of our equation (1) is determined by a holomorphic function h. We associate with (1) a differential equation of the complex variable (2.4) but with the right side having a polar. Remarkable is that (2.4) does not depend on T. Despite the singularity this equation has a holomorphic solution $u(z)$ near (u_0, z_0), because the coefficients of its formal power expansion are the same as the Fourier coefficients of a solution $u \in C_+(T)$ and we can use the previous result of **[12]** to get its local convergence. Next, extending the series of complex solution $u(z)$ of (2.4) till its radius of convergence we note that the restriction $u(t) := u\big(z_0 + r\exp\big(\frac{2\pi}{T}\imath t\big)\big)$ is a T-periodic solution of (1).

Finally we show that the thesis of Theorem 2.1 still holds if the function $f(u, t)$ of (1) belongs to $|\mathcal{H}(\mathbb{C}; C_+(T))$ (Theorem 2.7) which essentially improves the main results of **[1]** and **[12]**.

2. The existence of solutions.

In this section we show our main theorem.

THEOREM 2.1. *Let us consider the equation*

$$\dot{u}(t) = f(u(t), t),$$

with $f \in \mathcal{A}$ *being of the form* (1.2)*:*

$$f(u, t) = f_{z_0, r}(u, t) := h\Big(u, z_0 + r\exp\Big(\frac{2\pi}{T}\imath t\Big)\Big), \qquad h\text{-holomorphic in } \mathbb{C} \times \mathbb{C}.$$

Assume that (u_0, z_0) *satisfies the necessary condition for the existence of a periodic solution* $u \in C_+(T)$ *with* $\mathbf{u} = u_0$.

Then there exists a $r_0 > 0$, *depending on* h, u_0 *and* z_0, *but not on* T, *such that for every* $r < r_0$ *this equation has unique solution in* $C_+(T)$ *satisfying* $\mathbf{u} = u_0$.

Let (u_0, z_0) be such that $h(u_0, z_0) = 0$, i.e the necessary condition of Proposition (1.3) is satisfied. To simplify we introduce a new variable putting $\xi := z - z_0 = r\exp\big(\frac{2\pi}{T}\imath t\big)$ if $z = z_0 + r\exp\big(\frac{2\pi}{T}\imath t\big)$. Then $h(u_0, 0) = 0$. Considering ξ as a function of t and differentiating we get

$$d\xi = dz = r\left(\frac{2\pi}{T}\imath\right)\exp\Big(\frac{2\pi}{T}\imath t\Big)\,dt \tag{2.1}$$

Note that (1) can be written as

$$\frac{du}{dt} = h(u(t), \xi(t)). \tag{2.2}$$

Supposing next that our unknown function $u(t)$ is of the form $u(t) = u(\xi(t))$ we get

$$\frac{du}{dt} = \frac{du}{d\xi}\frac{d\xi}{dt} = \frac{du}{d\xi} r \left(\frac{2\pi}{T}\imath\right) \exp\left(\frac{2\pi}{T}\imath t\right) = \left(\frac{2\pi}{T}\imath\right)\xi. \tag{2.3}$$

Substituting (2.3 into (2.2) we obtain a differential equation of the complex variable:

$$\frac{du}{d\xi} = \frac{-\imath T}{2\pi}\frac{1}{\xi} h(u, \xi). \tag{2.4}$$

As a direct consequence of the above we get the following.

PROPOSITION 2.2. *Suppose that complex differential equation (2.4) has a solution $u(\xi)$, with the initial solution $u(0) = u_0$, which is holomorphic in some neighborhood $\mathcal{U}$ of 0.*

Then for every $r > 0$ such that $D_r(0) \subset \mathcal{U}$ the function

$$u(t) := u\left(r \exp\left(\frac{2\pi}{T}\imath t\right)\right)$$

is a T-periodic solution of (1).

□

We are left with the task to prove local solvability of the equation (2.4). It is worth of emphasize that if the right hand side of it would be holomorphic at $(u_0, 0)$ then the local solvability (and that a solution is analytic) follows from the Cauchy-Kowalewsky theorem. However there is a singularity (formal) at this point.

Assume that $u(t) \in C_+(T)$ is a T-periodic solution of (1). Then it corresponds to a function $\tilde{u}$ of the disc algebra $\tilde{A}(D_\mu)$, where D_μ is the disc of radius $\mu = \frac{T}{2\pi}$ (cf. [**12**] Th. 2.8). Consequently, for $r < \mu$ it can be represented as $u(t) = \tilde{u}\left(r \exp\left(\frac{2\pi}{T}\imath t\right)\right)$. This implies that for a fixed $r < \mu$ the formal Fourier expansion of u, is of the form

$$u(t) \sim u_0 + \sum_{k=1}^{\infty} \alpha_k(T) r^k \exp\left(\frac{2\pi}{T} k \imath t\right) \qquad \text{where} \quad \alpha_k(T) \in \mathbb{C}. \tag{2.5}$$

REMARK 2.3. *Since $\tilde{u} \in \tilde{A}(D_\mu)$, it is holomorphic in the interior of D_μ. The term $r^k \exp\left(\frac{2\pi}{T} k \imath t\right)$ corresponds to the power ξ^k of the power expansion of $\tilde{u}$ at 0.*

REMARK 2.4. *Assuming that a T-periodic function $u(t)$ with the Fourier coefficients for non-negative indices only is a solution of (1) we can derive (uniquely) these coefficients substituting the expansion to the both sides of (1) and comparing* *the coefficients at consecutive* $\exp(\frac{2\pi}{T} k\imath t)$. *It is possible because the right hand side of (1) is given as the power series of* $u - u_0$, *or* $\exp(\frac{2\pi}{T}\imath t)$ *respectively (cf. (1.5)).*

Let us next take the formal Taylor expansion of a solution $u(\xi)$ of the equation (2.4)

$$u(\xi) \sim u_0 + \sum_{k=1}^{\infty} \beta_k \xi^k . \tag{2.6}$$

The last means that we suppose that $u(\xi)$ has a power expansion, substitute it to the both sides of (2.4) and derive the coefficients β_k by a recursive procedure comparing the coefficients at the same powers of ξ. This procedure gives the coefficients β_k uniquely but does not ensure whether the series (2.6) is convergent at all (cf. [**8**]). Anyway purely algebraic computation let us to show the following fact.

PROPOSITION 2.5. *Let* $\{\alpha_k(T)\}$ *be the coefficients of the formal Fourier expansion (2.5) of a solution* $u(t) \in C_+(T)$, *and* $\{\beta_k\}$ *be the coefficients of the formal Taylor expansion (2.6) of a formal solution* $u(\xi)$ *of (2.4).*

Then $\forall \quad k \geq 0 \qquad \alpha_k(T) \;=\; \beta_k$.

In particular $\alpha_k(T)$ *does not depend on* T, , *i.e.* $\alpha_k(T) = \alpha_k(T') \;\; \forall \;\; T, T'$.

Proof. The algebraic procedure of finding coefficients of a differential equation with the right hand side being analytic is by expanding it first in the (formal) power series and next finding, in a unique way, the (formal) solution is classical (cf. [**8**]). Note also that for the expressions $\{r^j \exp(\frac{2\pi}{T}\imath\, j\, t)\}$ we have

$$r^j \exp\Big(\frac{2\pi}{T}\imath\, j\, t\Big)\, r^k \exp\Big(\frac{2\pi}{T}\imath\, k\, t\Big) \;=\; r^{j+k} \exp\Big(\frac{2\pi}{T}\imath\,(j+k)\, t\Big) \tag{2.7}$$

thus they form a multiplicative set as the powers of z. Observe next that the equation (2.4) could be written as

$$\Big(\frac{2\pi}{T}\imath\Big)\, \xi\, \frac{du}{d\xi} \;=\; h(u(\xi), \xi) . \tag{2.8}$$

Let $u(\xi) \sim u_0 + \sum_{k=1}^{\infty} \beta_k \xi^k$ and $u(\xi) \sim u_0 + \sum_{k=1}^{\infty} r^k \alpha_k(T) \exp(\frac{2\pi}{T}\imath k\, t)$ be the formal Taylor and Fourier expansions. Recall that

$$h(u, \xi) \;=\; \sum_{i\geq 0, j\geq 1} c_{i,j} (u - u_0)^j (z_0 + r\,\xi)^i + \sum_{i\geq 0, j=0} c_{i,0} (z_0 + r\,\xi)^i$$

that gives

$$f(u,t) \;=\; \sum_{i\geq 0, j\geq 1} c_{i,j} (u-u_0)^j \Big(z_0 + r \exp\Big(\tfrac{2\pi}{T}\imath t\Big)\Big)^i + \sum_{i\geq 0, j=0} c_{i,0} \Big(z_0 + r \exp\Big(\tfrac{2\pi}{T}\imath t\Big)\Big)^i . \tag{2.9}$$

Substituting

$$u - u_0 = \sum_{k=1}^{\infty} \beta_k \, \xi^k$$

to the first equation of (2.9) and

$$\sum_{k=1}^{\infty} r^k \alpha_k(T) \, \exp\left(\frac{2\pi}{T} \imath k t\right)$$

to the second, next multiplying, adding and ordering with respect the powers of ξ and $(r \, \exp\left(\frac{2\pi}{T} \imath t\right)$, we see that at β_k and $\alpha_k(T)$ are the same expressions of $c_{i,j}$ and β_k, and $c_{i,j}$ and $\alpha_k(T)$, respectively, at the right hand side of the equalities (2.8) and (1).

We are with task to compare the left hand sides of them. If $u(\xi) = u_0 + \sum_{k=1}^{\infty} \beta_k \, \xi^k$ then

$$\left(\frac{2\pi}{T} \imath\right) \xi \, \frac{du}{d\xi} = \left(\frac{2\pi}{T} \imath\right) \xi \sum_{k=1}^{\infty} \beta_k \, k \, \xi^{k-1} = \left(\frac{2\pi}{T} \imath\right) \sum_{k=1}^{\infty} \beta_k \, k \, \xi^k .$$

On the other hand if $u(t) = u_0 + \sum_{k=1}^{\infty} r^k \alpha_k(T) \, \exp\left(\frac{2\pi}{T} \imath t^k t\right)$ then

$$\frac{du}{dt} = \sum_{k=1}^{\infty} \left(\frac{2\pi \, k}{T} \imath\right) r^k \alpha_k(T) \, \exp\left(\frac{2\pi}{T} \imath k t\right) = \left(\frac{2\pi}{T} \imath\right) \sum_{k=1}^{\infty} k \, r^k \alpha_k(T) \, \exp\left(\frac{2\pi}{T} \imath k \, t\right).$$

Consequently we have the same expression $\left(\frac{2\pi \, k}{T} \imath\right) k \, \beta_k$, or $\left(\frac{2\pi \, k}{T} \imath\right) k \, \alpha_k(T)$ at ξ^k, or $\exp\left(\frac{2\pi}{T} \imath k \, t\right)$ respectively. This shows the proposition 2.5. □

Having derived the coefficients β_k we have to show that the series (2.6) is convergent, which is the usual second step of such procedure.

Let $r_0(T)$ denotes the radius of convergence of a power series with the coefficients $\{\alpha_k(T)\}$ and r_0 the radius of convergence of a series with coefficients $\{\beta_k\}$. As direct consequence of Proposition 2.5 we have the following.

COROLLARY 2.6. *For every $T > 0$ we have $r_0(T) = r_0$.*

□

We turn back to the prove of Theorem 2.1. Note that the case $r_0(T) = r_0 = 0$ is not excluded appriori in Corollary 2.6. However, from Theorem 4.6 of **[12]** it follows that that for any $f(u, t) \in \mathcal{H}(\mathbb{C}, C_+(T))$, thus for any $f \in \mathcal{A}$, there exists $T_0 > 0$ such that $r_0(T) > 0$ provided $T < T_0$. This proves the existence part statement of Theorem 2.1 with respect to Corollary 2.6.

From Corollary 2.6 follows also that the series (2.5) is uniformly convergent if $r < r_0$. This gives the constant r_0 of the statement of Theorem 2.1.

Finally, the uniqueness of solutions follows from the fact that it is given by a convergent series which coefficients are given unique. The proof of Theorem 2.1 is complete. □

Next we claim that the statement of Theorem 2.1 still holds if $f \in \mathcal{H}(\mathbb{C}; C_+(T))$, in particular if $f \in \mathcal{P}$, which improves the results of [**12**] and [**1**] correspondingly.

THEOREM 2.7. *Let us consider the equation (1)*

$$\dot{u}(t) = f(u(t), t),$$

with $f \in \mathcal{H}(\mathbb{C}; C_+(T))$. Assume that $(u_0, z_0 = 0)$ satisfies the necessary condition $\mathbf{f}(u_0) = 0$ for the existence of a periodic solution $u \in C_+(T)$ with $\mathbf{u} = u_0$.

Then there exists a holomorphic at 0 function $u(\xi)$ and $r_0 > 0$, depending on f, u_0, but not on T, such that for every $r < r_0$ the equation (1) has a unique solution u of the form $u(t) := u\big(r \exp\big(\frac{2\pi}{T} \imath t\big)\big)$ satisfying $\mathbf{u} = u_0 = u(\xi)_{|\xi=0}$.

Proof. We give a proof for the case when $f(u, t) \in \mathcal{P}$. A proof of the general case is similar. Let

$$f(u, t) = p(u, t) = u^n + c_1(t)u^{n-1} + c_2(t)u^{n-2} + \cdots + c_{n-1}(t)u^1 + c_n(t), \; c_j(t) \in C_+(T)$$

be an polynomial in $\mathcal{P}$. Consequently there exist functions $\tilde{c}_j(t) \in \mathcal{A}(D_\mu)$ in the disc algebra of the disc of radius $\mu = \frac{T}{2\pi}$, which means that $\tilde{c}_j$ is a holomorphic function in the interior $\overset{o}{D}_\mu$ which has a continuous extension on the closed disc with $c_j(t) = \tilde{c}_j(t)_{|\partial D_\mu}$. From it follows that we can form a function

$$p(u, \xi) := u^n + \tilde{c}_1(\xi)u^{n-1} + \tilde{c}_2(\xi)u^{n-2} + \cdots + \tilde{c}_{n-1}(\xi)u^1 + \tilde{c}_n(\xi),$$

holomorphic in $\mathbb{C} \times \overset{o}{D}_\mu$. Moreover the Fourier coefficients of $\tilde{c}_j(t)$ are equal to the Taylor coefficients of $\tilde{c}_j$ at 0 ([**12**] Th. 2.2). Note also that the necessary condition (Prop. 1.3) at $(u_0, 0)$ is satisfied. Consequently we apply the recursive method of deriving the coefficients of formal solution $u(\xi)$ for the equation (2.4) with $h(u, \xi) = p(u, \xi)$. As in the Theorem 2.1 this series defines a holomorphic function $u(\xi()$ in the disc D_{r_0} of its convergence, and for this function $u(t) := u\big(r \exp\big(\frac{2\pi}{T}\imath t\big)\big)$ is a T-periodic solution of (1) provided $r < r_0$. The existence $r_0 > 0$ and its estimate follow by the same argument as in Theorem 2.1. □

There is many natural questions connected with a description of of the set of periodic solutions of an equation of the form (1).

REMARK 2.8. *Remark that the estimate $r < r_0$ is not the sharp in general for the existence of periodic solutions, even for the existence of solutions in $C_+(T)$, i.e. the interval $(0, \tilde{r})$ such that for every $r \in (0, \tilde{r})$ there is a a periodic solution of (1.2), is larger than $(0, r)$. This follows form the fact that the solution function $u(\xi)$ could be holomorphic in a region larger than the disc of convergence of its Taylor series at (u_0, z_0).*

But it is impossible if there is a polar of $u(\xi)$ on the circle $\{\xi : |\xi| = r_0\}$. It is not difficult to show that there is not a solution $u(t) \in C_+(T)$ then. In [**16**] *Miklaszewski showed that the Riccati equation*

$$\dot{u} = u^2 + r \exp(\imath t) \tag{2.10}$$

has not a 2π-periodic solution if $r = r_0$. Next in [**20**] *Żołądek filled up a gap in the proof left by Miklaszewski describing additionally the set $\{r_i\}$ of all radii for which the equation (2.10) has not periodic solutions. One can show that $r_0 \sim 1.445798...$ and the point $\{-r_0\}$ is a polar then (cf.* [**16**]*).*

To emphasize the difference between the result we got and previous results of [**1**] *and* [**12**] *we remind that a direct application to the above Riccati equation of those theorems requires*

$$T(1+r) < 2$$

as a sufficient condition for the existence of T-periodic solutions. In particular $T = 2\pi$ is never in its range. To show for this particular equation the existence of 2π-periodic solutions of it, for any $r < r_0$, Miklaszewski used the same Fourier series we have discussed, and its convergence followed from the definition of r_0 then.

In view of the above one can conjecture that the equation (1.2) has not T-periodic solutions for $r = r_0$ if there is singularity of the solution $u(\xi)$ of (2.4) on the convergence circle $\{\xi : |\xi| = r_0\}$.

We are able to show that for $r < r_0$ all T-periodic solutions of (1.2) belong to $C_+(T)$. Moreover, under the above assumption this equation for $r = r_0$ this equation has not T-periodic solutions satisfying some geometric condition.

References

[1] A. Borisovich, W. Marzantowicz, Multiplicity of periodic solutions for the planar polynomial equation, Nonlinear Analysis, 43 (2001), 217-231.

[2] Th. Bröcker, L. Lander, Differentialable Germs and Catastrophes, Cambridge University Press, (1975).

[3] J. Campos, Möbius transformation and periodic solutions of complex Riccati equations, *Bull. London Math. Soc.*, **9**, (1997), pp. 205–213.

[4] J. Campos, R. Ortega, Nonexistence of periodic solutions of a complex Riccati equation, *Differential and Integral Equations*, **9 No. 2**, (1996), pp. 247–249.

[5] A. Capietto, J. Mawhin, F. Zanolin, Continuation theorems for periodic perturbations of autonomous systems, *Trans. of AMS*, **329 No. 1**, (1992), pp. 41–71.

[6] J. B. Conway, Functions of One Complex Variable II, Graduate Texts in Mathematics 159, Springer Varlag, (1995).

[7] J. B. Garnett, Bounded Analytic Functions, Academic Press, (1981).

[8] V. V. Golubiev, Lectures on the Analytic Theory of Differential Equations, Gosud. Izdat. Tehn. - Teor. Lit., Moscow-Leningrad, (1950) (in Russian).

[9] V. V. Golubiev, Vorlesungen über Differentialgleichungen im kompleksen, Hochschulbücher für Mathematik, Bd. 43, VEB Deutscher Verlag der Wissenschaften, Berlin, (1958).

[10] N. Lloyd, The number of periodic solutions of the equations $\dot{z} = z^N + p_1(t)z^{N-1} + \ldots + p_N(t)$, *Proc. London Math. Soc.*, **27**, (1973), pp. 667–700.

[11] R. Mansevich, J. Mawhin, F.Zanolin, Periodic solutions of complex-valued differential equations and systems with periodic coefficients, *J. Differential Equations*, **126 no. 2**, (1996), pp. 355–373.

[12] W. Marzantowicz, Periodic solutions of nonlinear problems with positive oriented periodic coefficients, *Variational and Topological Methods in the Study of Nonlinear Phenomena*, Editors - Benci, Cerami, Degiovanni, Fortunato, Giannoni, and Micheletti. Birkhauser Series - PNLDE, Vol. 49, pp. 43-64.

[13] J. Mawhin, Periodic solutions of some planar nonautonomous polynomial differential equations, *Diff. and Int. Equat.*, **7**, (1994), pp. 1055–1061.

[14] J. Mawhin, Continuation theorems and periodic solutions of ordinary differential equations, *Proc. Conf. on Topological Methods in Diff. Equations and Inclusions Universite de Montreal*, NATO Sci. Ass. (1994).

[15] J. Mawhin, M. Willem, Critical Point Theory and Hamiltonian Systems, *Applied Mathematical Sciences 74*, Springer-Verlag, (1989).

[16] D. Miklaszewski, An equation $\dot{z} = z^2 + p(t)$ with no 2π-periodic solutions, *Bull. Belg. Math. Soc.*, **3** (1996), pp. 239–242.

[17] L. Nirenberg, Topics in Nonlinear Functional Analysis, Courant Institute, New York, (1974).

[18] L. Nirenberg, Functional Analysis, Courant Institute of Mathematics, New York University, (1972) .

[19] R. Srzednicki, On periodic solutions of planar polynomial differential equations with periodic coefficients, *J. Diff. Eq.*, **114**, (1994), pp. 77–100.

[20] H. Zoladek, The method of holomorphic foliations in planar periodic systems: the case of Riccati equations, *J. Diff. Eq.*, **165 no. 1**, (2000), 143–173.

Blow-up of quasilinear parabolic equations in R^n –A unified approach

Y.W. Qi

Department of Mathematics, UCSB, Santa Barbara, CA 93106

1. Introduction

The study of blow-up, caused by growth of solutions rather than lack of smoothness, on nonlinear partial differentail equations (PDEs) probably originates from the simple yet revealing works of Fujita [4, 5] in 1966. He considered the following Cauchy problem

$$u_t = \Delta u + u^p, \qquad x \in \mathbf{R}^n,\ t > 0, \tag{1.1}$$
$$u(x,0) = u_0(x) \geq 0, \qquad x \in \mathbf{R}^n,$$

where the exponent $p > 1$. What he showed is the now famous Fujita type of result:

(a) If $1 < p < 1 + 2/n$, then every solution of (1.1) blows up in finite time provided u_0 is not identically zero;

(b) If $p > 1 + 2/n$ and $u_0(x) \leq \delta e^{-|x|^2}$ for some $0 < \delta \ll 1$, then u is a global solution.

For the critical exponent case of $p = 1+2/n$, it was shown first by Hayakawa [9] for dimensions $n = 1, 2$, and then by Kobayashi, Siaro and Tanaka [11], and later Aronson and Weinberger [1] that for all $n \geq 1$ every non-trivial solution blows up in finite time. That is

$$||u(\cdot\,,t)||_\infty \to \infty \qquad \text{when} \qquad t \to T > 0.$$

The above formula now serves as the standard definition of blow-up for parabolic PDEs with strong nonlinear term.

These elegant works revealed a new phenomenon of nonlinear PDEs and stimulated the study of similar features for various nonlinear evolution equations. Especially, the following Cauchy problem of porous medium type of equations with

---

e-mail: maqi@math.ucsb.edu.

complex reaction term

$$u_t = \Delta u^m + t^s|x|^\sigma u^p, \qquad x \in \mathbf{R}^n, \ t > 0, \tag{1.2}$$
$$u(x,0) = u_0(x) \geq 0, \qquad x \in \mathbf{R}^n,$$

were studied, for various special cases, by many authors [2, 3, 6, 12, 13, 14], where $m > (n-2)_+/n$, $s \geq 0$, and $\sigma > -1$ if $n = 1$ or $\sigma > -2$ if $n \geq 2$ and $p > \max\{m, 1\}$.

A very interesting observation can be made is that although study of different cases shows a remarkable uniformity in results, a unified approach did not appear until late 1990's [16]. In addition, the kind of ad hoc approach used by various authors fails to demonstrate the cause and effect of competition between diffusion and reaction, which is the underlying reason for this kind of interesting behaviour. For instance, for the case of

$$u_t = \Delta u^m + u^p, \qquad x \in \mathbf{R}^n, \ t > 0, \tag{1.3}$$
$$u(x,0) = u_0(x) \geq 0, \qquad x \in \mathbf{R}^n,$$

the study of $m > 1$ [6] and $m < 1$ [13] both demonstrate the following result which is a close resemblance of the original work of Fujita:

(a) If $1 < p < m + 2/n$, then every solution of (1.3) blows up in finite time provided u_0 is not identically zero;
(b) If $p > m + 2/n$ and $u_0(x)$ is small, then u is a global solution.

Yet, the two works use very different approaches. Another difficulty of the old approaches is how to deal with the critical exponent case which is more subtle than the generic cases [7], [14].

Furthermore, the reaction term, $t^s|x|^\sigma u^p$, though has a neat scaling proves to be tougher to tackle using the old approaches. Luckily, a very simple argument is found by the author which gives not only a complete answer to all cases represented by (1.2), but also the following $m-$ *Laplace* equation:

$$u_t = \operatorname{div}(|\nabla u|^{m-1}\nabla u) + t^s|x|^\sigma u^p, \qquad x \in \mathbf{R}^n, \ t > 0, \tag{1.4}$$
$$u(x,0) = u_0(x) \geq 0, \qquad x \in \mathbf{R}^n,$$

where $(n-1)_+/n < m < 1$, $s \geq 0$, and $\sigma > -1$ if $n = 1$ or $\sigma > -2$ if $n \geq 2$ and $p > \max\{m, 1\}$.

I now state our main results on (1.2) and (1.4). For simplicity, I assume the initial value u_0 is continuous.

THEOREM 1. *([16]) Assume that* $m + (m-1)s + (2 + \sigma + 2s)/n > 1$, *then* $p_c = m + (m-1)s + (2 + \sigma + 2s)/n$ *is the critical exponent of (1.2); i.e. when* $1 < p \leq p_c$ *every non-trivial solution of (1.2) blows up in finite time, and when* $p > p_c$, *(1.2) admits small positive global solution.*

THEOREM 2. *([17]) Assume that* $(n-1)/(n+1) < m < 1$, $s \geq 0$, $p > 1$ *and* $\sigma > n(1-m) - (1 + m + 2s)$. *Then* $p_c = m + (1 + m + 2s + \sigma)/n > 1$ *is the critical exponent of (1.4). That is, if* $1 < p \le p_c$ *then every non-trivial solution of (1.4) blows up in finite time; if* $p > p_c$ *then (1.4) has non-trivial global solutions.*

Remark. The result of Theorem 2 is obtained in a joint paper [17] with Prof. Mingxin Wang of Southeast University of China. His effort and contribution to that work is acknowledged by the author.

The plan of this paper is that the blow-up result will be proved for the simple case of $s = 0$ for (1.2) in section 2. The purpose of so doing is it not only makes the presentation more transparent by demonstrating the essential aspect of my approach, but also it saves some space. In section 3, I shall show the existence of global solution by constructing ground state solutions of an elliptic equation which is closely related to the self-similar solutions of (1.2).

Due to the restriction on length, it is not possible to present the complete proof of Theorem 1 and Theorem 2 here. Interested reader is referred to [16] and [17] for a more detailed demonstration.

2. The Blow-up Case

In this section, I prove that if $s = 0$ in (1.2), then every solution blows up in finite time provided $p \le p_c = m + (2+\sigma)/n$.

For simplicity of demonstration, in particular, to avoid the singularity of reaction term at $x = 0$, I shall consider the following initial-boundary value problem in an exterior domain:

$$(I)\begin{cases} u_t = \Delta u^m + |x|^\sigma u^p, & x \in \Omega, \quad t > 0, \\ u(x,0) = u_0(x) \ge 0, & x \in \Omega, \\ u(x,t) = 0, & x \in \partial\Omega, \quad t > 0, \end{cases}$$

where $\Omega = R^n \setminus B_R$ with B_R being the ball with radius R and centered at the origin. One implicit assumption is that u_0 is positive somewhere in Ω. It is apparent that a solution u satisfies $\partial u/\partial n \ge 0$ on $\partial\Omega$. It is also clear, by a simple application of maximum principle, that if any non-trivial solution u of (I) blows up in finite time, then so does every solution of the Cauchy problem.

Let $\psi(x)$ be a smooth, radially symmetric and non-increasing function which satisfies

$$0 \le \psi(x) \le 1, \ \psi(x) \equiv 1 \ \text{ for } x \le 1 \ \text{ and } \ \psi(x) \equiv 0 \ \text{ for } \ x \ge 2.$$

Set $\psi_l(x) = \psi(x/l)$. It follows that for $l \ge 1$, $\psi_l(x)$ is smooth, radially symmetric and non-increasing function which satisfies

$$\begin{aligned} &0 \le \psi(x) \le 1, \ \psi(x) \equiv 1 \ \text{ for } x \le l \ \text{ and } \ \psi(x) \equiv 0 \ \text{ for } \ x \ge 2l, \\ &|\nabla\psi_l| \le C/l, \qquad |\Delta \psi_l| \le C/l^2. \end{aligned} \tag{2.1}$$

Let $w_l(t) = \int_\Omega u\phi_l \, dx$. Then simple calculation shows

$$\frac{dw_l(t)}{dt} \ge \int_\Omega u^m \Delta \psi_l \, dx + \int_\Omega |x|^\sigma u^p \psi_l \, dx \tag{2.2}$$

and

$$\int_\Omega u^m |\triangle \psi_l|\, dx \;\le\; \left(\int_\Omega |x|^\sigma u^{mk/(k-1)} \psi_l\, dx\right)^{(k-1)/k} \times \tag{2.3}$$
$$\left(\int_\Omega |x|^{-(k-1)\sigma} |\triangle \psi_l|^k \psi_l^{-(k-1)}\, dx\right)^{1/k}$$

for any $k > 1$. Since $p > m$, by choosing $k = p/(p-m)$, we have

$$\int_\Omega |x|^\sigma u^{mk/(k-1)} \psi_l\, dx = \int_\Omega |x|^\sigma u^p \psi_l\, dx.$$

Consequently we derive from (2.2) and (2.3) that

$$\frac{dw_l(t)}{dt} \ge -C(\psi_l, l)\left(\int_\Omega |x|^\sigma u^p \psi_l\, dx\right)^{m/p} + \int_\Omega |x|^\sigma u^p \psi_l\, dx.$$

It is easy to verify that

$$\begin{aligned} C(\psi_l, l) &= \left(\int_{R^n} |x|^{-m\sigma/(p-m)} |\triangle \psi_l|^{p/(p-m)} \psi_l^{-m/(p-m)}\, dx\right)^{(p-m)/p} \\ &= l^{n-2-(n+\sigma)m/p} C_1, \end{aligned}$$

where $C_1 = C(\psi, 1)$ is a constant independent of l. Thus,

$$\frac{dw_l(t)}{dt} \ge \left(\int_\Omega |x|^\sigma u^p \psi_l\, dx\right)^{m/p} \left\{ -C_1 l^{n-2-(n+\sigma)m/p} + \left(\int_\Omega |x|^\sigma u^p \psi_l\, dx\right)^{(p-m)/p} \right\}. \tag{2.4}$$

By Hölder's inequality we get

$$\int_\Omega |x|^\sigma u^p \psi_l\, dx \ge \left(\int_\Omega |x|^\sigma u \psi_l\, dx\right)^p \left(\int_\Omega |x|^{-\sigma/(p-1)} \psi_l\, dx\right)^{-(p-1)}.$$

Hence,

$$\int_\Omega |x|^\sigma u^p \psi_l\, dx \ge \begin{cases} C_2(n,\sigma,p) w_l^p l^{-n(p-1)+\sigma} & \text{if } \sigma < n(p-1), \\ C_3(n,\sigma,p) w_l^p (\log l)^{-(p-1)} & \text{if } \sigma = n(p-1), \\ C_4(n,\sigma,p) w_l^p & \text{if } \sigma > n(p-1). \end{cases}$$

The following simple lemma is what I need to continue my proof.

LEMMA 1. *Let w be a positive continuous function which satisfies the following inequality $dw(t)/dt \ge Cw^q$ in the distributional sense, where $C > 0$ is a constant and $q > 1$. Then w is an increasing function, and there exists a finite $T > 0$ such that $w(t) \to \infty$ as $t \to T$.*

I start with the case $\sigma < n(p-1)$. It follows from (2.4) and (2.5) that

$$\frac{dw_l(t)}{dt} \geq \left(\int_\Omega |x|^\sigma u^p \psi_l \, dx\right)^{m/p} \Big\{ -C_1 l^{n-2-(n+\sigma)m/p} + C_2 w_l^{p-m} l^{[\sigma - n(p-1)](p-m)/p} \Big\}. \tag{2.6}$$

First I consider the subcritical case.

(i) $p < m + (2+\sigma)/n$. Under this assumption, the exponent in l of the first term in (2.6) is smaller than the second term, that is

$$n - 2 - (n+\sigma)m/p < [\sigma - n(p-1)](p-m)/p$$

and consequently

$$\frac{l^{n-2-(n+\sigma)m/p}}{l^{[\sigma-n(p-1)](p-m)/p}} \to \infty \qquad \text{as } l \to \infty. \tag{2.7}$$

Using this fact and the property that w_l is an increasing function of l, we deduce from (2.6) and (2.7) that there exist $\epsilon > 0$, $l \gg 1$ such that

$$\frac{dw_l(t)}{dt} \geq \epsilon w_l^p l^{\sigma - n(p-1)} \tag{2.8}$$

for all $t > 0$. Thus, w_l, and in consequence, u blows up in finite time by Lemma 1.

Now I consider the more subtle critical case.

(ii) $p = m + (2+\sigma)/n$. Under this assumption, the exponent in l of the first term in (2.6) is equal to the second term, that is

$$n - 2 - (n+\sigma)m/p = [\sigma - n(p-1)](p-m)/p < 0.$$

If the limiting function $\lim_{l\to\infty} w_l(t)$ is an unbounded function of t, I am done. Otherwise, $u(\cdot,t) \in L^1(\Omega)$ for all $t > 0$ and furthermore there exists an $M > 0$ such that

$$||u(\cdot,t)||_{1,\Omega} < M \qquad \text{for all } t > 0. \tag{2.9}$$

I prove this is impossible. Suppose the contrary. It is clear from (2.9) that

$$\int_\Omega u(x,t) |\triangle \psi_l| \, dx \to 0 \qquad \text{as } l \to \infty.$$

Hence, $w(t) = ||u(\cdot,t)||_{1,\Omega}$ satisfies

$$w(t) - w(0) \geq \int_0^t \Xi(s) \, ds. \tag{2.10}$$

where $\Xi(t) = \min(1, \int_\Omega u^p(x,t)|x|^\sigma \, dx)$.

Now I need to treat different values of m separately.

If $m > 1$, we know $u(x,t) \geq E(x,t;r_0)$ in Ω for some $r_0 > 0$, where $E(x,t;r_0)$ is the Barenblatt-Pattle solution of $v_t = \Delta v^m$. Using this lower bound, I get

$$w(t) - w(0) \geq C(1+t)^{-1} \tag{2.11}$$

for all $t \gg 1$. Thus, $w(t) \to \infty$ as $t \to \infty$. But this contradicts (2.9). Thus, u blows up in finite time.

If $m = 1$, then $u(x,t) \geq \epsilon G(x,t)$, where G is the Gaussian and ϵ is a small positive number. Again, we get (2.11) and u must blow up in finite time.

If $(n-2)_+/n < m < 1$, thanks to the Herrero and Pierre [9], we have an inequality of the form

$$u(x,t) \geq \epsilon(1+t)^{-1/\gamma}(1 + C|x|^2/(1+t)^{2/\nu})^{-1/(1-m)} \tag{2.12}$$

for $|x| \geq 1$, where ϵ and C are positive constants. It is an easy matter using this to show (2.11) is valid and then u must blow up in finite time. This completes the proof for $\sigma < n(p-1)$.

The situation of $\sigma \geq n(p-1)$ is much simpler as we shall demonstrate. Observing that $\sigma \geq n(p-1)$ and $m > (n-2)_+/n$ imply that the exponent in l of the first term in (2.4) is strictly less than zero.

When $\sigma = n(p-1)$, instead of (2.6), we find

$$\frac{dw_l(t)}{dt} \geq \left(\int_\Omega |x|^\sigma u^p \psi_l \, dx\right)^{m/p} \Big\{ - C_1 l^{n-2-(n+\sigma)m/p} + C_2 w_l^{p-m} (\log l)^{-(p-1)(p-m)/p} \Big\}. \tag{2.13}$$

Then, by repeating the argument of $\sigma < n(p-1)$ with p subcritical, one can show u blows up in finite time.

The case of $\sigma > n(p-1)$ can be handled similarly using the third inequality in (2.5). This completes the proof of blow-up case. Next we turn to global existence.

3. Global existence for $p > p_c$

The global existence of solutions can be dealt with using similarity solutions when p is supercritical. They take the form

$$u(x,t) = t^{-\alpha} v(|x|/t^\beta), \quad \alpha = \frac{2+\sigma+2s}{2(p-1)+(m-1)\sigma}, \quad \beta = \frac{p-m-s(m-1)}{2(p-1)+(m-1)\sigma}.$$

The resulting ODE for v, after an appropriate scaling, with $r = |x|/t^\beta$, is

$$\begin{aligned} &(v^m)'' + \frac{n-1}{r}(v^m)' + \frac{r}{2}v' + kv + r^\sigma v^p = 0, \\ &v(0) = \eta > 0, \quad (v^m)'(r) \to -r^{\sigma+1} v^p(0)/(n+\sigma), \quad \text{as } r \to 0, \end{aligned} \tag{3.1}$$

where $k = \alpha/2\beta$. The case of $p > p_c$ corresponds to $k < n/2$. The case of $\sigma = 0$ is due to Haraux and Weissler [8] and we shall assume throughout that $\sigma \neq 0$.

We observe that a function $U(x,t) = U(|x|/t^\beta)$ is a super-solution iff it satisfies the inequality

$$(U^m)'' + \frac{n-1}{r}(U^m)' + \frac{r}{2}U' + kU + r^\sigma U^p \leq 0. \tag{3.2}$$

The approach we adopt is to demonstrate the existence of a ground state to (3.1) or a positive function satisfies (3.2) when $k < n/2$. For simplicity, we shall provide detailed demonstration only for $m = 1$.

LEMMA 2. *Let $\sigma > 0$. Then $U = \eta e^{-Ar^2/4}$ with $0 < \eta \ll 1$ and $k/2n < A < 1/4$ satisfies inequality (3.2).*

Proof: It is elementary to verify the claim and we omit the detail.

Q.E.D.

LEMMA 3. *Let $n \geq 2$ and $-2 < \sigma < 0$ or $n = 1$ and $-1 < \sigma < 0$. Then there exists a ground state solution to (3.1).*

Proof: Let v be a solution of (3.1). Then $r^{n-1}v'(r)|_{r=0} = 0$. It is easy to see that v satisfies the following integral equation

$$v'r^{n-1} + \frac{r^n}{2}v = \int_0^r (\frac{n}{2} - k)vs^{n-1}\, ds - \int_0^r v^p s^{n-1+\sigma}\, ds. \tag{3.3}$$

We shall look for a ground state when $\eta \ll 1$. First we note that v is a decreasing function of r before it reaches zero. Since the first term dominates the second term when $r \geq 1$ as $\sigma < 0$, the effort is to show v as well as $v'r^{n-1} + vr^n/2$ stay positive in $[0, M]$ for some $M > 1$. In this connection we consider

$$w'' + \frac{n-1}{r}w' + \frac{r}{2}w' + kw + r^\sigma \eta^p = 0 \tag{3.4}$$

with the same initial values as v. It is obvious that v as well as $v'r^{n-1} + vr^n/2$ stay positive as long as the corresponding terms of w do. An elementary calculation gives that w and $w'r^{n-1} + wr^n/2$ are both positive in $[0, R]$, where $R^{\sigma+2} = (\sigma + 2)(\sigma + n)\eta^{-1/(p-1)}$. From (3.3), we get

$$v'r^{n-1} + \frac{r^n}{2}v = v'(r_0)r_0^{n-1} + \frac{r_0^n}{2}v(r_0) + \int_{r_0}^r (\frac{n}{2} - k)vs^{n-1}\, ds - \int_{r_0}^r v^p s^{n-1+\sigma}\, ds, \tag{3.5}$$

where $r_0^{|\sigma|} = 2/(n-2k)$. Thus, v is a ground state if $\eta \ll 1$.

This completes the proof of global existence.

References

[1] D.G. Aronson and H.F. Weinberger, Multidimensional nonlinear diffusion arising in population genetics, *Adv. in Math.* **30** (1978), 33 - 76.

[2] C. Bandle and H.A. Levine, On the existence and non-existence of global solutions of reaction-diffusion equations in sectorial domains, *Trans. Amer. Math. Soc.* **316** (1986), 595 - 622.

[3] A. Friedman and J. B. McLeod, Blow-up of positive solutions of semilinear heat equation, *Indiana Univ. Math. J.*, **34** (1985), 425 -447.

[4] H. Fujita, On the blowing up of solutions of the Cauchy problem for $u_t = \triangle u + u^{1+\alpha}$, *J. Fac. Sci. Univ. Tokyo* Sect 1, **13** (1966), 109 - 124.

[5] H. Fujita, On some nonexistence and nonuniqueness theorems for nonlinear parabolic equations, *Proc. Sympos. Pure Math.* **18** (1969), 105 - 113.

[6] V.A. Galaktionov, S.P. Kurdyumov, A. P. Mikhailov and A. A. Samarskii, Unbounded solutions of the Cauchy problem for the parabolic equation $u_t = \nabla(u^\sigma \nabla u) + u^\beta$, *Sov. Phys. Dokl.,* **25** (1980), 458-459.

[7] V.A. Galaktionov, Blow-up for quasilinear heat equations with critical Fujita's exponents, *Proc. Roy. Soc. Edinburgh, Sec. A,* **124**(1994), 517 - 525.

[8] A. Haraux and F. Weissler, Nonuniqueness for a semilinear initial value problem, *Indiana Univ. Math. J.*, **31** (1982), 167-189.

[9] K. Hayakawa, On nonexistence of global solutions of some semilinear parabolic equations, *Proc. Japan Acad.,* **49** (1973), 503 - 525.

[10] M.A. Herrero and M. Pierre, The Cauchy problem for $u_t = \triangle u^m$ when $0 < m < 1$, *Trans. Amer. Math. Soc.,* **291(1)** (1985), 145 - 158.

[11] K. Kobayashi, T. Siaro and H. Tanaka, On the blowing up problem for semilinear heat equations, *J. Math. Soc. Japan,* **29** (1977), 407 - 424.

[12] K. Mochizuki and K. Mukai, Existence and nonexistence of global solutions to fast diffusions with source, *Math. Appl. Anal.,* **2** (1994), 92 - 102.

[13] Y.W. Qi, On the equation $u_t = \triangle u^\alpha + u^\beta$, *Proc. Roy. Soc. Edinburgh,* **123:A** (1993), 373 - 390.

[14] Y.W. Qi, Critical exponents of degenerate parabolic equations, *Science in China,* **38:A** (1995), 1153 - 1162.

[15] Y.W. Qi, The global existence and nonuniqueness of a nonlinear degenerate equation, *Nonlinear Analysis, TMA,* **31** (1998), 117 - 136.

[16] Y.W. Qi, The critical exponents of parabolic equations and blow-up in $\mathbf{R}^n$, *Proc. Roy. Soc. Edinburgh, Sec. A,* **128(1)** (1998), 123 - 136.

[17] Y.W. Qi and M.X. Wang, Critical Exponents of Quasilinear Parabolic Equations , *J. Math. Anal. Appl.*, **267** (2002), 264-280.

[18] F.B. Weissler, Existence and nonexistence of global solutions for a semilinear heat equation, *Israel J. Math.,* **38** (1981), 29 - 40.

Remarks on a priori estimates for superlinear elliptic problems

Miguel Ramos

CMAF and Faculty of Sciences, University of Lisbon Av. Prof. Gama Pinto, 2 1649-003 Lisboa, Portugal

ABSTRACT. We make some remarks on the method of obtaining a priori estimates for the solutions of superlinear elliptic equations having prescribed Morse index. In particular, we show that sometimes these estimates can be obtained without performing blow-up procedures.

1. Remarks on superlinear elliptic problems

In this note we discuss compactness properties of the solutions of elliptic equations in $\mathbb{R}^N$ with superlinear and subcritical nonlinear terms. For definiteness, we take $N \geqslant 3$ and consider first the following boundary value problem:

$$-\Delta u = m(x)u + g(x,u), \qquad u \in H_0^1(\Omega), \tag{P}$$

where Ω is a C^2 bounded domain, $m(x) \in C^1(\overline{\Omega})$ and $g \in C^1(\overline{\Omega} \times \mathbb{R}; \mathbb{R})$ satisfies the following superlinear and subcritical growth conditions (we denote $2^\star := 2N/(N-2)$): *there exist* $A, B, R > 0$, $\mu > 1$ *and* $2 < p < 2^\star$ *such that*

$$0 < g(x,s)\, s \leqslant p\, G(x,s), \qquad \forall x \in \overline{\Omega},\ \forall |s| > R, \tag{H1}$$

$$\frac{\partial g(x,s)}{\partial s}\, s^2 \geqslant \mu\, g(x,s)\, s, \qquad \forall x \in \overline{\Omega},\ \forall |s| > R, \tag{H2}$$

and

$$\left| \frac{\partial G(x,s)}{\partial x} \right| \leqslant A\, |G(x,s)| + B, \qquad \forall x \in \overline{\Omega},\ \forall s \in \mathbb{R}, \tag{H3}$$

1991 *Mathematics Subject Classification.* 35J20, 35J35, 35J50, 58E05.

Key words and phrases. Critical point theory, superlinear elliptic problems, biharmonic operator, elliptic systems.

Work partially supported by FCT.

e-mail: `mramos@ptmat.fc.ul.pt`.

where $G(x,s) := \int_0^s g(x,\xi)\,d\xi$. Observe that (H1) and (H2) imply that

$$C_1\,|s|^{\mu+1} \leqslant g(x,s)\,s \leqslant C_2\,|s|^{p}, \qquad \forall x \in \overline{\Omega},\ \forall |s| > R, \tag{1.1}$$

and that

$$C_1\,g(x,s)\,s \leqslant G(x,s) \leqslant C_2\,g(x,s)\,s, \qquad \forall x \in \overline{\Omega},\ \forall |s| > R, \tag{1.2}$$

for some positive constants C_1 and C_2. Clearly, any solution u of problem (P) actually lies in $C^2(\Omega) \cap C^1(\overline{\Omega})$. Moreover, solutions of (P) correspond to critical points of the energy functional

$$I(u) := \frac{1}{2}\int_\Omega |\nabla u|^2 - \frac{1}{2}\int_\Omega m(x)u^2 - \int_\Omega G(x,u), \qquad u \in H_0^1(\Omega).$$

For every critical point u of I, we denote by $i(u)$ its Morse index, that is, the supremum of the dimensions of the linear subspaces of $H_0^1(\Omega)$ over which the quadratic form $I''(u)$ is negative definite.

Theorem 1. *Under assumptions* (H1), (H2) *and* (H3), *for any* $k \in \mathbb{N}$ *there exists* $C = C(\Omega,k)$ *such that any solution* u *of* (P) *having Morse index* $i(u) \leqslant k$ *satisfies*

$$\int_\Omega |\nabla u|^2 + \int_\Omega g(x,u)u \leqslant C\left(1 + \int_\Omega u^2\right). \tag{1.3}$$

In particular, $||u||_{H_0^1(\Omega)} \leqslant \widetilde{C}(\Omega,k)$.

The proof of Thorem 1 is postponed to Section 2. Some remarks are in order. A result similar to Theorem 1 goes back to the paper [BL] for the case where, roughly,

$$\lim_{|s|\to\infty} \frac{\partial g(x,s)}{\partial s}\,\frac{1}{|s|^{p-2}} = a(x) > 0 \tag{1.4}$$

uniformly in $x \in \overline{\Omega}$ ($2 < p < 2^\star$). More recently, other results on a priori bounds for finite Morse index solutions were obtained by several authors. For example, the case where $g(s) = A(s^+)^{p-1} - B(s^-)^{q-1}$ with $A,B > 0$, $2 < p,q < 2^\star$ is considered in [HRS1,2] (we have denoted $s^\pm := \max\{\pm s, 0\}$), while [DR,RTT] deal with $g(x,s) = a(x)|s|^{p-2}s$ where $a(x)$ is no longer a positive function. The papers [RR] and [AV] concern the biharmonic operator and systems of elliptic equations, respectively. Another example goes as follows. Let (u_j) satisfy

$$-\varepsilon_j^2\,\Delta u_j + u_j = g(u_j), \qquad u_j \in H_0^1(\Omega), \qquad i(u_j) \leqslant 1,$$

for some sequence $\varepsilon_j \to 0$, where g satisfies (1.4) with, say, $a(x) \equiv p-1$. Then one can show that whenever $||u_j||_{L^\infty}$ is attained at points x_j which are bounded away from $\partial\Omega$ there exists a non zero function $v \in H^1(\mathbb{R}^N)$ such that $-\Delta v + v = |v|^{p-2}v$ and

$$\frac{1}{\varepsilon_j^N}\,I(u_j) - I_\infty(v) \to 0 \qquad \text{as } j \to \infty,$$

where we have denoted $I_\infty(v) = \frac{1}{2}\int(|\nabla v|^2 + v^2) - \frac{1}{p}\int |v|^p$. A refinement of this property is used in [RY] in the discussion of the shape of the positive solutions of a system of the form

$$-\varepsilon^2\Delta u + u = v^{q-1}, \quad -\varepsilon^2\Delta v + v = u^{p-1} \text{ in } \Omega, \qquad \frac{\partial u}{\partial n} = \frac{\partial v}{\partial n} = 0 \text{ on } \partial\Omega,$$

when $\varepsilon \to 0$.

All the quoted authors establish a priori bounds in the basis of an argument by contradiction followed by a blow-up procedure. In the present paper we purpose a direct approach in the proof of Theorem 1. We do not rely on rescaling arguments and thus asymptotic conditions such as (1.4) are not needed.

This approach is useful in at least two situations. The first situation concerns the case when the "limit problem" in $\mathbb{R}^N$ which comes out from the blow-up procedure admits non zero solutions with finite Morse index, so that the mentioned argument by contradiction becomes more delicate. This happens for example in [HRS2], where the authors made use of Harnack inequalities and of the construction of barrier functions to deal with the case where, essentially, $g(s) = A(s^+)^{p-1} - B(s^-)^{q-1}$ with $A, B > 0$ and $2 < p, q < 2^*$; as a constraint of their method, in [HRS2] the authors assume that $3 < p, q < 5$ in case $N = 3$, while Theorem 1 above shows that in fact we may allow $2 < p, q < 5$ if $N = 3$.

The second situation we have in mind concerns elliptic problems in unbounded domains in which the L^∞ norms of the solutions are attained at points of $\mathbb{R}^N$ having arbitrarily large norms, thus making rescaling arguments more difficult or even impossible to apply. For a concrete example, we mention a fourth order equation such as:

$$\Delta^2 u = \lambda h(x)u + a(x)g(u), \qquad u \in \mathcal{D}^{2,2}(\mathbb{R}^N), \tag{Q}$$

where $\lambda \in \mathbb{R}$ and $N \geqslant 5$. The notation $\mathcal{D}^{2,2}(\mathbb{R}^N)$ stands for the closure of $\mathcal{D}(\mathbb{R}^N)$ with respect to the norm $\int(\Delta u)^2$. Our next theorem complements previous work in [Ra]. We assume the following:

(H) $a \in C^3(\mathbb{R}^N;\mathbb{R})$ is such that $a(0) > 0$, $\limsup_{|x|\to\infty} a(x) \in [-\infty, 0[$ and $a(x)$ has only nondegenerate zeros. Moreover, $0 < h \in L^\infty(\mathbb{R}^N) \cap L^{N/4}(\mathbb{R}^N) \cap C^1(\mathbb{R}^N)$. As for g, we assume that $g \in C^1(\mathbb{R};\mathbb{R})$, $g(s)s > 0$ for all $s \neq 0$,

$$g(0) = 0 = g'(0) \text{ and } \lim_{|s|\to\infty} \frac{g'(s)}{(p-1)|s|^{p-2}} = \ell > 0 \quad (2 < p < \frac{2N}{N-4})$$

and

$$\exists\delta > 0\ \exists C > 0 : \ G(s) \leqslant C\, g(s)\, s, \qquad \forall |s| \leqslant \delta,$$

where $G(s) := \int_0^s g(\xi)\, d\xi$.

A discussion on relevant examples is provided in [Ra, Section 4] (in [Ra], the asymptotic property of g stated in (H) is replaced by a similar thought different assumption, namely: $g(s)s \leqslant pG(s)$ for every s and $pG(s)/sg(s) \to 1$ as $|s| \to \infty$). We point out that, similarly to [Ra], we do not assume that $a(x)$ is bounded; on the other hand, we do assume that a^+ has compact support.

Now, under our assumptions on $h(x)$, the map $u \mapsto \int h\, u^2$ is compact in $\mathcal{D}^{2,2}(\mathbb{R}^N)$. We denote by $(\lambda_i(h))_{i\in\mathbb{N}}$ the sequence of eigenvalues of the linear problem

$\Delta^2 u = \mu h(x)u$, $u \in \mathcal{D}^{2,2}(\mathbb{R}^N)$. Under the present framework and for any $\lambda \in \mathbb{R}$, Theorem 4.1 in [Ra] establishes the existence of infinitely many solutions for (Q) in case g is odd (i.e. $g(-s) = -g(s)$). In case g is no longer odd, the existence of a non zero solution is proved in Theorem 4.10 of [Ra], provided that $\lambda < \lambda_1(h)$. Here we prove the following generalization of the latter result.

Theorem 2. *Under assumptions* (H), *if* $\lambda \neq \lambda_i(h)$ *for every* $i \in \mathbb{N}$ *then problem* (Q) *has a non zero solution.*

The proof of Theorem 2 is given in Section 3. The non zero solution is obtained by taking the limit of a sequence of solutions $u_R(x)$ in $H_0^2(B_R(0))$ with $R \to \infty$. The main difficulty consists in proving that this sequence is bounded independently of R. In [Ra], the assumption that $\lambda < \lambda_1(h)$ was used as a way to bound the corresponding energies; here we follow the procedure used in [CGR], which consisted in recovering compactness thanks to the information on the Morse index of the solutions. However, and in contrast to the second order problem in [CGR], here it may happen that $||u_R||_{L^\infty} = |u_R(x_R)|$ with $|x_R| \to \infty$. This phenomenon will be bypassed through a variant of Theorem 1.

Acknowledgment. We warmly thank Armando Machado for a technical useful suggestion in the proof of Lemma 2.1.

2. Proof of Theorem 1

The proof of Theorem 1 is based on the following lemma. Unless otherwise stated, integrations take place in Ω.

Lemma 2.1. *Under assumptions* (H1), (H3) *and* (1.2), *there exist* $C > 0$, $r > 0$ *and a finite number of points* $x_1, \dots, x_m \in \overline{\Omega}$ *such that* $\overline{\Omega} \subset \cup_{i=1}^m B_r(x_i)$ *and*

$$\int (|\nabla u|^2 + g(x,u)u)\varphi^2 \leqslant C\left(1 + \int (|\nabla u|^2 + g(x,u)u)\,|\nabla\varphi^2| + \int u^2(|\nabla\varphi|^2 + \varphi^2)\right)$$

for any solution u *of* (P) *and any function* $\varphi \in \mathcal{D}(B_{2r}(x_i))$.

Proof. Fix $x_0 \in \partial\Omega$ and let $\alpha : \mathbb{R}^{N-1} \to \partial\Omega$ be a C^2 diffeomorphism onto a neighborhood of x_0 in $\partial\Omega$, such that $\alpha(0) = x_0$. Let

$$\Psi(y,t) := \alpha(y) + t\,\nu_0,$$

where ν_0 denotes the unit outward normal of Ω at the point x_0 and $(y,t) \in \mathbb{R}^{N-1} \times \mathbb{R}$. Since $\frac{\partial\Psi}{\partial t}(0) = \nu_0$ and $\nabla_y\Psi(0) = \nabla\alpha(0)$, Ψ admits a C^2 local inverse

$$\Psi^{-1} : B_r(x_0) \subset \mathbb{R}^N \to \mathbb{R}^N$$

for some $r > 0$, such that $\Psi^{-1}(x_0) = 0$. Consider the C^1 vector field

$$V(x) := D\Psi_{\Psi^{-1}(x)}(\Psi^{-1}(x)),$$

where $D\Psi_z$ denotes the derivative of Ψ evaluated at the point z. Thus,

$$V(x) = D\alpha_y(y) + t\,\nu_0, \qquad \forall x = \alpha(y) + t\,\nu_0 \in B_r(x_0),$$

so that $V(x_0) = 0$, $DV_{x_0} = Id$ and, by construction,

$$\langle V(x), \nu_x \rangle = 0, \qquad \forall x \in B_r(x_0) \cap \partial\Omega. \tag{2.1}$$

In particular, given $\varepsilon > 0$ we can choose $r > 0$ sufficiently small so that

$$||V - (Id - x_0)||_{C^1(B_r(x_0))} \leqslant \varepsilon. \tag{2.2}$$

Now, let u be a solution of (P) and consider the Pohožaev-type vector field

$$W := \varphi^2 \langle \nabla u, V \rangle \nabla u - \varphi^2 \frac{1}{2} |\nabla u|^2 V + \varphi^2 G(x,u) V,$$

where φ has its support in $B_r(x_0)$. It follows from (2.1) and the boundary condition $u|_{\partial\Omega} = 0$ that $\int_\Omega \operatorname{div} W = 0$. By computing $\operatorname{div} W$ and using (2.2) we get that

$$\int (2^\star + \mathrm{o}(1)) G(x,u) \varphi^2 \leqslant \int (1 + \mathrm{o}(1)) |\nabla u|^2 \varphi^2 + C\left(1 + \gamma(u)\right), \tag{2.3}$$

where $\mathrm{o}(1) \to 0$ as $r \to 0$ uniformly in x, $C > 0$ is a generic constant independent of u and φ, and we have denoted

$$\gamma(u) := \int (|\nabla u|^2 + |G(x,u)|) |\nabla \varphi^2| + \int u^2 \varphi^2.$$

We observe that the integral term $\int \langle \nabla_x G(x,u), V \rangle \varphi^2$ can be absorbed into the left-hand member of (2.3) if r is sufficiently small, thanks to (2.2) and (H3).

On the other hand, multiplying the equation in (P) by $u\varphi^2$ and integrating yields, thanks also to (H1), that

$$\int |\nabla u|^2 \varphi^2 \leqslant p \int G(x,u) \varphi^2 + C\left(1 + \int u^2 |\nabla \varphi|^2 + \int u^2 \varphi^2\right). \tag{2.4}$$

Since $p < 2^\star$, it follows from (2.3) and (2.4) that

$$\int G(x,u) \varphi^2 \leqslant C\left(1 + \int u^2 |\nabla \varphi|^2 + \gamma(u)\right). \tag{2.5}$$

Using (1.2), (2.4) and (2.5) we conclude that

$$\int |\nabla u|^2 \varphi^2 + \int g(x,u) u \varphi^2 \leqslant C\left(1 + \int u^2 |\nabla \varphi|^2 + \gamma(u)\right),$$

for any $\varphi \in \mathcal{D}(B_r(x_0))$, where C depends only on x_0, Ω, $||m||_{L^\infty}$ and on the constants appearing in (H1), (H3) and (1.2). A compactness argument yields the conclusion of the lemma (of course, we let $V(x) = x - x_0$ in case we fix $x_0 \in \Omega$). ∎

Proof of Theorem 1 completed. Let $r > 0$ and $x_i \in \overline{\Omega}$ be given by Lemma 2.1. For simplicity of notations, assume that $x_i = 0$ and, for any numbers $r < s_0 < s_1 < 2r$, denote

$$\mathcal{B}(s_0, s_1) := \{x : s_0 < |x| < s_1\} \subset B_{2r}(0) \setminus B_r(0).$$

Let $\psi \in \mathcal{D}(\mathbb{R}^N)$ have support in a neighborhood of $\mathcal{B}(s_0, s_1)$, with $\psi = 1$ in $\mathcal{B}(s_0, s_1)$. Since $i(u) \leqslant k$, we can choose $r < s_0 < s_1 < 2r$ such that $I''(u)u\psi, u\psi \geqslant 0$. Using the equation in (P) and assumption (H2), the latter inequality implies that

$$\int |\nabla u|^2 \psi^2 + \int g(x,u) u \psi^2 \leqslant C\left(1 + \int u^2 |\nabla \psi|^2\right),$$

and so

$$\int_{\mathcal{B}(s_0,s_1)} |\nabla u|^2 + \int_{\mathcal{B}(s_0,s_1)} g(x,u)u \leqslant C\left(1 + \int_\Omega u^2\right). \tag{2.6}$$

Now, let $\varphi \in \mathcal{D}(B_{s_1}(0))$ be such that $\varphi = 1$ in $B_{s_0}(0)$. Observing that $\nabla\varphi = 0$ except in $\mathcal{B}(s_0, s_1)$, it follows from Lemma 2.1 and from (2.6) that

$$\int_{B_{s_0}(0)} |\nabla u|^2 + \int_{B_{s_0}(0)} g(x,u)u \leqslant C\left(1 + \int_\Omega u^2\right).$$

This shows that for every $x_1, \ldots, x_m$ it holds:

$$\int_{B_r(x_i)} |\nabla u|^2 + \int_{B_r(x_i)} g(x,u)u \leqslant C\left(1 + \int_\Omega u^2\right).$$

To conclude, it remains to recall that $\overline{\Omega} \subset \cup_{i=1}^m B_r(x_i)$. The final statement in the theorem is a consequence of the first inequality in (1.1). ∎

3. Proof of Theorem 2

In this section we assume that (H) holds and that $\lambda_k(h) < \lambda < \lambda_{k+1}(h)$ for some $k \in \mathbb{N}$. The energy functional associated to (Q) is formally given by

$$I(u) = \frac{1}{2}\int (\Delta u)^2 - \frac{\lambda}{2}\int h(x)\, u^2 - \int a(x)\, G(u).$$

In order to avoid duplications, we ask the reader to check in [CGR] that Theorem 2 can be deduced as a consequence of the following claim, that we prove below:

Claim. If $R_n \to \infty$, any sequence of solutions of

$$\Delta^2 u_n = \lambda h(x)u_n + a(x)g(u_n), \qquad u_n \in H_0^2(B_{R_n}(0)), \tag{3.1}$$

such that $\limsup i(u_n) < \infty$ *is bounded in* $H_0^2(\mathbb{R}^N)$. (See also the Remark at the end of the present section.)

The rest of the section is devoted to the proof of this claim. In view of a contradiction, we assume that (u_n) satisfies (3.1), $\limsup i(u_n) < \infty$ and that $t_n := \|\Delta u_n\|_2 \to \infty$. As in [CGR, eq. (38)] or [Ra, Lemma 4.5], the latter property implies that, up to a subsequence,

$$\frac{u_n}{t_n} \rightharpoonup 0 \quad \text{weakly in } H_0^2(\mathbb{R}^N). \tag{3.2}$$

Since the $\lambda h(x)$ term in (3.1) is compact, we see from (3.1) and (3.2) that

$$t_n^2 + \int a^- g(u_n)u_n = \int a^+ g(u_n)u_n + \mathrm{o}(t_n^2), \qquad \text{as } n \to \infty. \tag{3.3}$$

Except when otherwise is indicated, all integrals take place in $B_{R_n}(0)$. For $\varepsilon > 0$ we denote by $\mathcal{A}_\varepsilon$ the compact set:

$$\mathcal{A}_\varepsilon := \{x \in \mathbb{R}^N : a(x) \geqslant \varepsilon\}.$$

Lemma 3.1. *For every ε,* $\displaystyle\int_{\mathcal{A}_\varepsilon} a^+ g(u_n)u_n = \mathrm{o}(t_n^2)$.

Proof. The function $a(x)g(s)$ satisfies conditions (H1)-(H3) on $\mathcal{A}_\varepsilon \times \mathbb{R}$ (with $2^\star$ replaced by $2N/(N-4)$). The conclusion of Lemma 3.1 follows then from an argument similar to the one in the proof of Theorem 1. We observe that here the situation is somehow simpler, since $\mathcal{A}_\varepsilon \cap \partial B_R(0) = \emptyset$ for every large $R > 0$. As for the required Pohožaev-type estimate similar to (2.3), it can be checked in [RR, pp. 5-6]. ∎

Lemma 3.2. *There exists $C > 0$ such that, for every $n \in \mathbb{N}$,*

$$\int_{\{a\geqslant 0\}} g(u_n)u_n \leqslant C\,t_n^2 + C\int a^+ g(u_n)u_n.$$

Proof. Since $a(x)$ has only nondegenerate zeros, we can fix a small $\delta > 0$ in such a way that $|\nabla a|$ is bounded away from zero in a neighborhood of the compact set $\{|a| \leqslant \delta\}$. Thanks to Lemma 3.1, to prove Lemma 3.2 it is enough to show that

$$\int_{\{|a|\leqslant\delta\}} g(u_n)u_n \leqslant C\,t_n^2 + C\int a^+ g(u_n)u_n,$$

for some constant C (which may depend on δ). Thanks to (3.3) and to our assumptions on g (cf. (H)), it is sufficient to prove that

$$\int_{\{|a|\leqslant\delta\}} |\nabla a|^2 G(u_n) \leqslant C\,t_n^2 + C\int_{\{|a|\leqslant 2\delta\}} |a|\,G(u_n).$$

This, in turn, follows from a Pohožaev-type estimate based on the computation of $\operatorname{div}(\psi a G(u_n)\nabla a)$, where ψ is any smooth cut-off function which separates the sets $\{|a| \leqslant \delta\}$ and $\{|a| \geqslant 2\delta\}$ (see [Ra, Lemma 4.7] for details). ∎

Lemma 3.3. *We have* $\int a^+ g(u_n)u_n = \mathrm{o}(t_n^2)$.

Proof. Given $\varepsilon > 0$, we deduce from Lemmas 3.1 and 3.2 that

$$\begin{aligned}\int a^+ g(u_n)u_n &= \int_{\{0\leqslant a\leqslant\varepsilon\}} a^+ g(u_n)u_n + \mathrm{o}(t_n^2)\\ &\leqslant \varepsilon\int_{\{a\geqslant 0\}} g(u_n)u_n + \mathrm{o}(t_n^2)\\ &\leqslant \varepsilon\,C\,t_n^2 + \varepsilon\,C\int a^+ g(u_n)u_n + \mathrm{o}(t_n^2),\end{aligned}$$

and this proves Lemma 3.3. ∎

Now, Lemma 3.3 and (3.3) show that we have a contradiction, and this proves the claim stated at the beginning of the section. As we mentioned before, Theorem 2 follows from this claim.

Remark. The fact that (3.1) has indeed a non zero solution u with bounded Morse index (namely, $i(u) \leqslant k+1$, where k was defined at the beginning of Section 3) in every space $H_0^2(B_R(0))$ with $R > 0$ large but *fixed* is essentially proved in [RR]. There the authors work in the space $H^2 \cap H_0^1(B_R(0))$ but in the present case this is irrelevant, since $\partial B_R(0)$ does not intersect the zero set of $a(x)$, if R

is sufficiently large. In fact, the precedent estimates, together with the truncation argument in [RR], can be used to prove directly that (3.1) has a non zero solution u with $i(u) \leqslant k+1$, for every $R = R_n$ sufficiently large.

References

[AV] S.B. ANGENENT, R.C. VAN DER VORST, A priori bounds and renormalized Morse indices of solutions of an elliptic system, *Ann. Inst. H. Poincaré Anal. Non Linéaire* 17 (2000), 277–306.

[BL] A. BAHRI, P.L. LIONS, Solutions of superlinear elliptic equations and their Morse indices, *Comm. Pure Appl. Math.* 45 (1992), 1205–1215.

[DR] A.R. DOMINGOS, M. RAMOS, Solutions of semilinear elliptic equations with superlinear sign changing nonlinearities, *Nonlinear Analysis* 50 (2002), 149–161.

[CGR] D.G. COSTA, Y. GUO, M. RAMOS, Existence and multiplicity results for nonlinear elliptic problems in $\mathbb{R}^N$ with an indefinite functional, *Electron. J. Diff. Eqns.*, Vol. 2002(2002), No. 25, pp. 1–15.

[HRS1] A. HARRABI, S. REBHI, A. SELMI, Solutions of superlinear elliptic equations and their Morse indices, I, *Duke Math. J.* 94 (1998), 141–157.

[HRS2] A. HARRABI, S. REBHI, A. SELMI, Solutions of superlinear elliptic equations and their Morse indices, II, *Duke Math. J.* 94 (1998), 159–179.

[Ra] M. RAMOS, Uniform Estimates for the Biharmonic Operator in $\mathbb{R}^N$ and Applications, submitted.

[RR] M. RAMOS, P. RODRIGUES, On a fourth order superlinear elliptic problem, in: *USA-Chile Workshop on Nonlinear Analysis, Electron. J. Diff. Eqns.*, Conf. 06, 2001, pp. 243–255.

[RTT] M. RAMOS, S. TERRACINI, C. TROESTLER, Superlinear indefinite elliptic problems and Pohožaev type identities, *J. Funct. Anal.* 159 (1998), 596–628.

[RY] M. RAMOS, Jianfu YANG, in preparation.

Miguel Ramos
CMAF and Faculty of Sciences, University of Lisbon
Av. Prof. Gama Pinto, 2
1649-003 Lisboa, Portugal
mramos@ptmat.fc.ul.pt

Asymptotics for some nonlinear elliptic systems in Maxwell-Chern-Simons vortex theory

Tonia Ricciardi

Dipartimento di Matematica e Applicazioni Università di Napoli Federico II
Via Cintia 80126 Naples, Italy

ABSTRACT. We provide a unified proof of the asymptotics of the self-dual Maxwell-Chern-Simons vortices, as the Maxwell term is neglected, in both the $U(1)$ and $CP(1)$ case. This result is achieved by identifying and analyzing a suitable class of nonlinear elliptic systems with exponential type nonlinearities.

1. Introduction and main result

The vortex solutions for the $U(1)$ Maxwell-Chern-Simons model introduced in [**9**], correspond to (distributional) solutions $(\widetilde{u}, v)$ for the system:

$$-\Delta\widetilde{u} = \varepsilon^{-1}(v - e^{\widetilde{u}}) - 4\pi\sum_{j=1}^{n}\delta_{p_j} \qquad \text{on } \Sigma \tag{1.1}$$

$$-\Delta v = \varepsilon^{-1}\left\{e^{\widetilde{u}}(1-v) - \varepsilon^{-1}(v - e^{\widetilde{u}})\right\} \qquad \text{on } \Sigma, \tag{1.2}$$

where Σ is a compact Riemannian 2-manifold without boundary, $n \geq 0$ is an integer, $p_j \in \Sigma$ for $j = 1, \ldots, n$, Δ denotes the Laplace-Beltrami operator and $\varepsilon > 0$ a constant. We shall be interested in the asymptotic behavior of solutions when $\varepsilon \to 0$.

Physically, $e^{\widetilde{u}}$ represents a *density* of particles; it vanishes exactly at the points p_j, $j = 1, \ldots, n$ (the *vortex points*). The function v is a neutral scalar field and

---

1991 *Mathematics Subject Classification.* 35J60.

Key words and phrases. nonlinear elliptic system, Chern-Simons vortex theory.

Partially supported by PRIN 2000 "Variational Methods and Nonlinear Differential Equations".

e-mail: `tonia.ricciardi@unina.it` fax: +39 081 675665.

$\varepsilon > 0$ is the coupling constant for the Maxwell term. In particular, letting $\varepsilon \to 0$ corresponds to dropping the Maxwell term in the Lagrangian.

The limit $\varepsilon \to 0$ is meaningful in view of the following

THEOREM 1.1 ([**13**]). *If $\varepsilon 4\pi n/|\Sigma|$ is sufficiently small, then there exist at least two solutions for* (1.1)–(1.2).

The proof of Theorem 1.1 is variational. The two solutions are obtained as a local minimum and a mountain pass for a suitable functional. We refer to [**13**] for the detailed proof.

By a formal analysis of (1.1)–(1.2), we expect that as $\varepsilon \to 0$, e^u should converge to a solution u_∞ for the equation

$$-\Delta u_\infty = e^{u_\infty}(1 - e^{u_\infty}) - 4\pi \sum_{j=1}^{n} \delta_{p_j} \qquad \text{on } \Sigma. \tag{1.3}$$

We observe that solutions for (1.3) correspond to vortex solutions for the Chern-Simons model introduced in [**7**] and [**6**]. In [**11**] we provided a rigorous proof of this formal argument, in any relevant norm. Namely, we showed

THEOREM 1.2 ([**11**]). *Suppose (u, v) are (distributional) solutions for* (1.1)–(1.2) *with $\varepsilon \to 0$. Then there exists a solution u_∞ for the equation* (1.3) *such that, up to subsequences, $(e^{\widetilde{u}}, v) \to (e^{u_\infty}, e^{u_\infty})$ in $C^h(\Sigma) \times C^h(\Sigma)$, for any $h \geq 0$.*

Note that $e^{\widetilde{u}}$, e^{u_∞} are smooth. Theorem 1.2 completed our previous convergence result obtained with Tarantello [**13**], where the asymptotics for v was established in the L^2-sense only. See also Chae and Kim [**2**].

At this point it is natural to seek a more general class of systems which exhibit an asymptotic behavior as in Theorem 1.2. A further motivation to this question is provided by the analysis of the $CP(1)$ Maxwell-Chern-Simons model in [**3**]. In [**3**] the authors analyze an elliptic system, whose solutions correspond to vortex solutions for the self-dual $CP(1)$ Maxwell-Chern-Simons model introduced in [**4**]. Their system (in a special case) is given by:

$$\Delta U = 2Q\Big(-V + S - \frac{1-e^U}{1+e^U}\Big) + 4\pi \sum_{j=1}^{n} \delta_{p_j} \qquad \text{on } \Sigma \tag{1.4}$$

$$\Delta V = -4Q^2\Big(-V + S - \frac{1-e^U}{1+e^U}\Big) + Q\frac{4e^U}{(1+e^U)^2}V \qquad \text{on } \Sigma \tag{1.5}$$

where Σ and $p_1, \ldots, p_n$ are as in (1.1)–(1.2), U, V are the unknown functions and $S \in \mathbb{R}$, $Q > 0$ are given constants. They prove the existence of at least one solution for (1.4)–(1.5); furthermore, they derive an asymptotic behavior as $Q \to +\infty$ *analogous* to that of system (1.1)–(1.2).

With this motivation, we consider (distributional) solutions $(\widetilde{u}, v)$ for the system:

$$-\Delta \widetilde{u} = \varepsilon^{-1}(v - f(e^{\widetilde{u}})) - 4\pi \sum_{j=1}^{n} \delta_{p_j} \qquad \text{on } \Sigma \tag{1.6}$$

$$-\Delta v = \varepsilon^{-1}\left[f'(e^{\widetilde{u}})e^{\widetilde{u}}(s-v) - \varepsilon^{-1}(v - f(e^{\widetilde{u}}))\right] \qquad \text{on } \Sigma. \tag{1.7}$$

Here Σ and $p_1, \ldots, p_n$ are as in (1.1)–(1.2), $f = f(t)$, $t \geq 0$ is smooth and *strictly increasing*, $s \in \mathbb{R}$ satisfies $f(0) < s < \sup_{t>0} f(t)$. Without loss of generality, we assume $\mathrm{vol}\Sigma = 1$.

Clearly, when $f(t) = t$ and $s = 1$, system (1.6)–(1.7) reduces to (1.1)–(1.2). On the other hand, setting $v := V - S$, $s := -S$, $\varepsilon^{-1} := 2Q$, system (1.6)–(1.7) reduces to system (1.4)–(1.5) with f defined by $f(t) = (t-1)/(t+1)$.

By a formal analysis of (1.6)–(1.7) we expect that, up to subsequences, $(\widetilde{u}, v)$ should converge to $(\widetilde{u}_\infty, f(e^{\widetilde{u}_\infty}))$, where $\widetilde{u}_\infty$ is a solutions for the equation for the equation:

$$-\Delta\widetilde{u}_\infty = f'(e^{\widetilde{u}_\infty})e^{\widetilde{u}_\infty}(s - f(e^{\widetilde{u}_\infty})) - 4\pi\sum_{j=1}^{n}\delta_{p_j} \qquad \text{on } \Sigma. \tag{1.8}$$

Our main result states that this is indeed the case, with respect to any relevant norm:

THEOREM 1.3 ([**12**]). *Let* $(\widetilde{u}, v)$ *be (distributional) solutions to* (1.6)–(1.7), *with* $\varepsilon \to 0$. *There exists a (distributional) solution* $\widetilde{u}_\infty$ *to* (1.8) *such that a subsequence, still denoted* $(\widetilde{u}, v)$, *satisfies:*

$$(e^{\widetilde{u}}, v) \to \left(e^{\widetilde{u}_\infty}, f(e^{\widetilde{u}_\infty})\right) \qquad \textit{in } C^h(\Sigma)\times C^h(\Sigma),\ \forall h \geq 0.$$

In the rest of this note we shall outline the proof of Theorem 1.3. The detailed proof is contianed in [**12**], although some arguments are proved here in a simpler form. Henceforth we denote by $C > 0$ a general constant independent of ε, which may vary from line to line. Unless otherwise specified, all equations are defined on Σ and all integrals are taken over Σ with respect to the Lebesgue measure.

2. Proof of Theorem 1.3

In order to work in suitable Sobolev spaces, it is standard (see [**14**]) to define a "Green's function" u_0, solution for the problem

$$-\Delta u_0 = 4\pi\left(n - \sum_{j=1}^{n}\delta_{p_j}\right) \qquad \text{on } \Sigma$$

$$\int_\Sigma u_0 = 0$$

(see [**1**] for the unique existence of u_0). Setting $\widetilde{u} = u_0 + u$, we obtain the equivalent system for $(u, v) \in H^1(\Sigma) \times H^1(\Sigma)$:

$$-\Delta u = \varepsilon^{-1}\left(v - f(e^{u_0+u})\right) - 4\pi n \qquad \text{on } \Sigma \tag{2.1}$$

$$-\Delta v = \varepsilon^{-1}\left[f'(e^{u_0+u})e^{u_0+u}(s-v) - \varepsilon^{-1}\left(v - f(e^{u_0+u})\right)\right] \qquad \text{on } \Sigma, \tag{2.2}$$

where e^{u_0} is *smooth.*

The proof is obtained by a priori estimates and an inductive argument. It will be convenient to introduce the spaces $X^k := H^k \cap L^\infty$. By the following inequality, which follows from the well-known Sobolev-Gagliardo-Nirenberg inequality (see e.g. [**10**]):

$$\|D^j u\|_{L^{2k/j}} \le C\|D^k u\|_{L^2}^{j/k}\|u\|_{L^\infty}^{1-j/k} \qquad \forall u \in C^\infty(\Sigma), \tag{2.3}$$

X^k is a *Banach algebra* for every $k \ge 0$, i.e.,

$$\|u_1 u_2\|_{X^k} \le C\|u_1\|_{X^k}\|u_2\|_{X^k}.$$

It will also be convenient to set

$$w := \varepsilon^{-1}(v - f(e^{u_0+u}))$$

and to consider w as a third unknown function. Then the triple (u, v, w) satisfies a system of the following simple form:

$$-\Delta u = w - 4\pi n \tag{2.4}$$

$$-\varepsilon^2\Delta v + [1 + \varepsilon c(x,u)]v = F_\varepsilon(x,u) \tag{2.5}$$

$$-\varepsilon^2\Delta w + [1 + \varepsilon c(x,u)]w = G_\varepsilon(x,u,v,\nabla u) \tag{2.6}$$

where

$$\begin{aligned} c(x,u) =& f'(e^{u_0+u})e^{u_0+u} \\ F_\varepsilon(x,u) =& f(e^{u_0+u}) + s\varepsilon f'(e^{u_0+u})e^{u_0+u} \\ G_\varepsilon(x,u,v,\nabla u) =& f'(e^{u_0+u})e^{u_0+u}(s - v) \\ &+ \varepsilon\left(f''(e^{u_0+u})e^{u_0+u} + f'(e^{u_0+u})\right)e^{u_0+u}|\nabla(u_0+u)|^2. \end{aligned}$$

Furthermore, using (2.3), we obtain:

LEMMA 2.1. *Let $F \in C^\infty(\Sigma \times \mathbb{R})$, $G \in C^\infty(\Sigma \times \mathbb{R} \times \mathbb{R} \times \mathbb{R}^2)$. Then for all $k \ge 0$ there exists constants $C_k = C_k(\|u\|_{L^\infty})$, $C'_k = C'_k(\|u\|_{L^\infty}, \|v\|_{L^\infty}, \|\nabla u\|_{L^\infty})$, such that:*

$$\begin{aligned} &\|F(x,u)\|_{X^k} \le C_k(1 + \|u\|_{X^k}^k) \\ &\|G(x,u,v,\nabla u)\|_{X^{k-1}} \le C'_k(1 + \|u\|_{X^k}^{k-1} + \|v\|_{X^{k-1}}). \end{aligned}$$

The basis for the induction is incuded in the following

PROPOSITION 2.1. *There exists a constant $C > 0$ independent of $\varepsilon \to 0$, such that:*

$$\|u\|_{X^1} + \|v\|_{X^1} + \|w\|_{X^0} \le C \tag{i}$$

$$\|u\|_{L^\infty} \le C \tag{ii}$$

$$\|\nabla u\|_{L^\infty} \le C \tag{iii}$$

In order to prove Proposition 2.1 we need some preliminary estimates.

LEMMA 2.2. *The following estimates hold, pointwise on Σ:*

$$f(0) \leq f(e^{\widetilde{u}}) \leq s \tag{i}$$

$$f(0) \leq v \leq s. \tag{ii}$$

PROOF. By maximum principle. The increasing monotonicity of f is essential here. ∎

As a consequence of Lemma 2.2, the nonlinearity f may be *truncated*. Therefore in what follows, without loss of generality, we assume that:

$$\sup_{t>0}\{|f(t)| + |f'(t)| + |f''(t)|\} \leq C. \tag{2.7}$$

The next identity is the main step in deriving the H^1-estimate for v and the L^2-estimate for w:

LEMMA 2.3. *The following identity holds:*

$$\int |\nabla v|^2 + \int w^2 = \int (s-v)\left(f''(e^{\widetilde{u}})e^{\widetilde{u}} + f'(e^{\widetilde{u}})\right) e^{\widetilde{u}}|\nabla \widetilde{u}|^2. \tag{2.8}$$

PROOF. We compute:

$$\Delta f(e^{\widetilde{u}}) = \left(f''(e^{\widetilde{u}})e^{\widetilde{u}} + f'(e^{\widetilde{u}})\right) e^{\widetilde{u}}|\nabla \widetilde{u}|^2 + f'(e^{\widetilde{u}})e^{\widetilde{u}}\Delta\widetilde{u}.$$

Therefore $f(e^{\widetilde{u}})$ satisfies the equation:

$$-\Delta f(e^{u}) + \varepsilon^{-1} f'(e^{\widetilde{u}})e^{\widetilde{u}} f(e^{\widetilde{u}}) = \varepsilon^{-1} f'(e^{\widetilde{u}})e^{\widetilde{u}}v - \left(f''(e^{\widetilde{u}})e^{\widetilde{u}} + f'(e^{\widetilde{u}})\right) e^{\widetilde{u}}|\nabla \widetilde{u}|^2. \tag{2.9}$$

Integrating (2.9), we obtain

$$\varepsilon^{-1}\int f'(e^{\widetilde{u}})e^{\widetilde{u}}(v - f(e^{\widetilde{u}})) = \int \left(f''(e^{\widetilde{u}})e^{\widetilde{u}} + f'(e^{\widetilde{u}})\right) e^{\widetilde{u}}|\nabla \widetilde{u}|^2 \tag{2.10}$$

Now we multiply (1.7) by $v - f(e^{\widetilde{u}})$ and integrate to obtain:

$$\int -\Delta v(v - f(e^{\widetilde{u}})) = \varepsilon^{-1}\int f'(e^{\widetilde{u}})e^{\widetilde{u}}(s-v)(v - f(e^{\widetilde{u}})) - \varepsilon^{-2}\int (v - f(e^{\widetilde{u}}))^2.$$

Integrating by parts and using (2.9) we find:

$$\int -\Delta v(v - f(e^{\widetilde{u}})) = \int |\nabla v|^2 + \int v\Delta f(e^{\widetilde{u}})$$

$$= \int |\nabla v|^2 - \varepsilon^{-1}\int v f'(e^{\widetilde{u}})e^{\widetilde{u}}(v - f(e^{\widetilde{u}})) + \int v\left(f''(e^{\widetilde{u}})e^{\widetilde{u}} + f'(e^{\widetilde{u}})\right) e^{\widetilde{u}}|\nabla \widetilde{u}|^2.$$

Equating left hand sides in the last two identities, we obtain

$$\int |\nabla v|^2 + \varepsilon^{-2}\int (v - f(e^{\widetilde{u}}))^2 + \int v\left(f''(e^{\widetilde{u}})e^{\widetilde{u}} + f'(e^{\widetilde{u}})\right) e^{\widetilde{u}}|\nabla \widetilde{u}|^2$$

$$= s\varepsilon^{-1}\int f'(e^{\widetilde{u}})e^{\widetilde{u}}(v - f(e^{\widetilde{u}})),$$

and thus identity (2.8) is established. ∎

We shall need some a priori estimates for solutions to

$$-\varepsilon^2\Delta u + (1+\varepsilon c)u = f. \tag{2.11}$$

Indeed, both (2.5) and (2.6) are of the form (2.11).

LEMMA 2.4. *Let $c, f \in X^k$ and suppose that u satisfies: (2.11). For every $k \geq 0$ there exist $\varepsilon_k > 0$, $C_k > 0$ such that*

$$||u||_{X^k} \leq C_k ||f||_{X^k},$$

for all $\varepsilon \leq \varepsilon_k$.

PROOF. The proof is an easy consequence of the following fact Let G_ε be the Green's function for

$$-\varepsilon\Delta_x G_\varepsilon + G_\varepsilon = \delta_y \qquad \text{on } \Sigma.$$

Then $G_\varepsilon(x,y) \to \delta_y$ weakly in the sense of measures. Note that since the operator $-\varepsilon\Delta + 1$ is coercive, the Green's function G_ε is uniquely defined on Σ. By the maximum principle, $G_\varepsilon > 0$ on Σ. Integrating over Σ, we find $\int G_\varepsilon = \int |G_\varepsilon| = 1$. Therefore, there exists a Radon measure μ such that $G_\varepsilon \rightharpoonup \mu$ weakly in the sense of measures. For any $\varphi \in C^\infty(\Sigma)$ we have:

$$\varepsilon \int -\Delta G_\varepsilon \varphi + \int G_\varepsilon \varphi = \varphi(y).$$

Taking limits, we find $\int \varphi \, \mathrm{d}\mu = \varphi(y)$ and the statement of the lemma is established. ∎

Now we can provide the

PROOF OF PROPOSITION 2.1. We begin by establishing

$$\int e^{\widetilde{u}} |\nabla \widetilde{u}|^2 \leq C. \tag{2.12}$$

Proof of the Claim. Multiplying equation (1.6) by $e^{\widetilde{u}}$ and integrating by parts, we obtain

$$\varepsilon^{-1} \int e^{\widetilde{u}}(v - f(e^{\widetilde{u}})) = \int e^{\widetilde{u}} |\nabla \widetilde{u}|^2 \geq 0.$$

By the pointwise estimates in Lemma 2.2, it follows that:

$$\varepsilon \int e^{\widetilde{u}} |\nabla \widetilde{u}|^2 \leq C. \tag{2.13}$$

Multiplying (1.7) by $e^{\widetilde{u}}$ and integrating, we find

$$\varepsilon^{-1} \int e^{\widetilde{u}}(v - f(e^{\widetilde{u}})) = \int e^{2\widetilde{u}} f'(e^{\widetilde{u}})(s - v) + \varepsilon \int e^{\widetilde{u}} \Delta v. \tag{2.14}$$

Integration by parts yields:

$$\varepsilon \int e^{\widetilde{u}} \Delta v = -\int v e^{\widetilde{u}}(v - f(e^{\widetilde{u}})) + \varepsilon \int v e^{\widetilde{u}} |\nabla \widetilde{u}|^2.$$

Hence, by the pointwise estimates as in Lemma 2.2, and taking into account (2.13), we conclude that

$$\varepsilon\left|\int e^{\widetilde{u}}\Delta v\right| \leq C.$$

Inserting into (2.14), recalling Lemma 2.2, we derive that

$$\varepsilon^{-1}\int e^{\widetilde{u}}(v-f(e^{\widetilde{u}}))\leq C,$$

and thus it follows that

$$\int e^{\widetilde{u}}|\nabla\widetilde{u}|^2=\varepsilon^{-1}\int e^{\widetilde{u}}(v-f(e^{\widetilde{u}}))\leq C.$$

(2.12) is established.

Lemma 2.2 readily implies $\|e^{\widetilde{u}}\|_{L^\infty}\leq C$ and $\|v\|_{L^\infty}\leq C$. In order to obtain the H^1-estimate for $e^{\widetilde{u}}$, it suffices to observe that by Lemma 2.2–(i) and by (2.12) we have:

$$\int|\nabla e^{\widetilde{u}}|^2=\int e^{2\widetilde{u}}|\nabla\widetilde{u}|^2\leq C\int e^{\widetilde{u}}|\nabla\widetilde{u}|^2\leq C.$$

Now we estimate $\|\nabla v\|_{L^2}$ and $\|\varepsilon^{-1}(v-f(e^{\widetilde{u}}))\|_{L^2}$. Using identity (2.8), we have:

$$\begin{aligned}\int|\nabla v|^2+\int w^2 \leq&\|s-v\|_\infty\|f''(e^{\widetilde{u}})e^{\widetilde{u}}+f^(\widetilde{u})\|_\infty\int e^{\widetilde{u}}|\nabla\widetilde{u}|^2\\ \leq& C\int e^{\widetilde{u}}|\nabla\widetilde{u}|^2\leq C,\end{aligned}$$

where we again used Lemma 2.2 and (2.12) in order to derive the last step. Proof of (i). Multiplying (2.1) by $u-\int u$ and integrating, we have:

$$\begin{aligned}\int|\nabla u|^2=&q\int(v-f(e^{\widetilde{u}}))(u-\int u)\\ \leq&\|q(v-f(e^{\widetilde{u}}))\|_2\|u-\int u\|_2\leq C\|\nabla u\|_2,\end{aligned}$$

where the last inequality follows by Lemma 2.2 and by the Poincaré inequality. Hence $\|\nabla u\|_2\leq C$. By Lemma 2.2–(ii), we have that $e^{\widetilde{u}}\leq C$, and thus we only have to show that $\int u\geq -C$. To this end, we first observe that integrating (2.1) and (2.2) we obtain:

$$\int f'(e^{u_0+u})e^{u_0+u}(s-v)=q\int(v-f(e^{u_0+u}))=4\pi n.$$

On the other hand, we have in a straightforward manner:

$$\int f'(e^{u_0+u})e^{u_0+u}(s-v)\leq C\int e^{u_0+u}\leq Ce^{\int u}\|e^{u_0}\|_\infty\int e^{u-\int u}\leq C\int e^{u-\int u}.$$

Hence, recalling the Moser-Trudinger inequality (see [**1**]) and the estimate for $\|\nabla u\|_2$, we conclude that

$$4\pi n\leq Ce^{\int u}\int e^{u-\int u}\leq Ce^{\int u}e^{\gamma\int|\nabla u|^2}\leq Ce^{\int u},$$

which establishes (i). Proof of (ii). Since $\|w\|_{L^2} \le C$, by (i) and elliptic regularity we obtain $\|u\|_{H^2} \le C$. Then Sobolev embeddings yield $\|\nabla u\|_{L^p} \le C$, for any $1 \le p < +\infty$ and $\|u\|_{L^\infty} \le C$, which establishes (ii). Proof of (iii). By (2.6), $\|\nabla u\|_{L^p} \le C$ and Lemma 2.4 imply that $\|w\|_{L^p} \le C$, for any $1 \le p < +\infty$. Then (2.4) and Sobolev embeddings yield $\|u\|_{W^{2,p}} \le C$, for any $1 \le p < +\infty$. For $p > 2$, the Sobolev embeddings yield (iii). ∎

PROPOSITION 2.2. *For all $k \ge 0$ there exists a constant $C > 0$ (possibly depending on k) such that:*

$$\|\widetilde{u} - u_0\|_{H^k} + \|v\|_{H^k} \le C.$$

PROOF OF PROPOSITION 2.2. We argue by induction on $k \in \mathbb{N}_0$.
CLAIM: Suppose:

$$\|u\|_{X^k} + \|v\|_{X^k} + \|w\|_{X^{k-1}} \le C_k.$$

Then:

$$\|u\|_{X^{k+1}} + \|v\|_{X^{k+1}} + \|w\|_{X^k} \le C_{k+1}.$$

Indeed,

$$\begin{aligned} \|w\|_{X^{k-1}} \le C \Rightarrow \|u\|_{X^{k+1}} \le C \quad & \text{by (2.4) and standard elliptic regularity} \\ \Rightarrow \|v\|_{X^{k+1}} \le C \quad & \text{by (2.5), Lemma 2.1 and Lemma 2.4} \\ \Rightarrow \|w\|_{X^k} \le C \quad & \text{by (2.6), Lemma 2.1 and Lemma 2.4.} \end{aligned}$$

Now Proposition 2.1, the Claim and a standard induction argument conclude the proof. ∎

Finally, we can prove our main result:

PROOF OF THEOREM 1.3. Let (u, v) be solutions to system (2.1)–(2.2), with $\varepsilon \to 0$. By the a priori estimates as stated in Proposition 2.2 and by standard compactness arguments, there exist u_∞, v_∞ such that up to subsequences $u \to u_\infty$ and $v \to v_\infty$ in C^h, for all $h \ge 0$. We write (2.1) in the form:

$$v = f(e^{u_0+u}) + \varepsilon(-\Delta u + 4\pi n).$$

Taking limits, we find $v_\infty = f(e^{u_0+u_\infty})$. Furthermore, taking limits in (2.2), we obtain

$$\varepsilon^{-1}(v - f(e^{u_0+u})) \to f'(e^{u_0+u_\infty})e^{u_0+u_\infty}(s - f(e^{u_0+u_\infty})),$$

where the convergence holds in C^h, for any $h \ge 0$. Consequently, taking limits in (2.1), we find that u_∞ satisfies:

$$-\Delta u_\infty = f'(e^{u_0+u_\infty})e^{u_0+u_\infty}(s - f(e^{u_0+u_\infty})) - 4\pi \sum_{j=1}^{n} \delta_{p_j}. \tag{2.15}$$

Setting $\widetilde{u}_\infty = u_0 + u_\infty$, we conclude the proof of Theorem 1.3. ∎

Acknowledgement

I should like to thank the Organizers of the ICM2002 Satellite Conference on Nonlinear Functional Analysis, Taiyuan, August 14–18, as well as Professor Yanyan Li, for the invitation.

References

[1] Th. Aubin, *Nonlinear Analysis on Manifolds: Monge-Ampère Equations*, Springer-Verlag, New York, Berlin, 1982

[2] D. Chae and N. Kim, Vortex Condensates in the Relativistic Self-Dual Maxwell-Chern-Simons-Higgs System, preprint

[3] D. Chae and H.S. Nam, On the Condensate Multivortex Solutions of the Self-Dual Maxwell-Chern-Simons $CP(1)$ Model, Ann. H. Poincaré **2** (2001), 887–906

[4] K. Kimm, K. Lee and T. Lee, Anyonic Bogomol'nyi Solitons in a Gauged $O(3)$ Sigma Model, Phys. Rev. D **53** (1996), 4436–4440

[5] G. Dunne, *Self-Dual Chern-Simons Theories*, Lecture Notes in Physics, vol. M36, Springer-Verlag, Berlin, New York, 1995

[6] J. Hong, Y. Kim and P.Y. Pac, Multivortex solutions of the Abelian Chern-Simons theory, Phys. Rev. Lett. **64** (1990), 2230–2233

[7] R. Jackiw and E. Weinberg, Self-dual Chern–Simons vortices, Phys. Rev. Lett. **64** (1990), 2234–2237

[8] A. Jaffe, C. Taubes, *Vortices and Monopoles*, Birkhäuser, Boston, 1980

[9] C. Lee, K. Lee and H. Min, Self-Dual Maxwell Chern-Simons Solitons, Phys. Lett. B, **252** (1990), 79–83

[10] L. Nirenberg, On elliptic partial differential equations, Ann. Sc. Norm. Sup. Pisa, **13** (1959), 116–162

[11] T. Ricciardi, Asymptotics for Maxwell-Chern-Simons multivortices, Nonlin. Anal. T.M.A. **50** No. 8 (2002), 1093–1106.

[12] T. Ricciardi, On a nonlinear elliptic system from Maxwell-Chern-Simons vortex theory, preprint.

[13] T. Ricciardi and G. Tarantello, Vortices in the Maxwell-Chern-Simons Theory, Comm. Pure Appl. Math. **53** no. 7 (2000), 811–851

[14] Y. Yang, *Solitons in field theory and nonlinear analysis,* Springer Monographs in Mathematics, Springer-Verlag, New York, 2001

Multi-parameter bifurcation and applications

Junping Shi

Department of Mathematics, College of William and Mary
Williamsburg, VA 23185, USA

and

School of Mathematics, Harbin Normal University
Harbin, Heilongjiang, P.R.China

1. Introduction

In this article we discuss some new abstract bifurcation theorems and their applications to the global bifurcation problem of a semilinear boundary value problem:

$$\Delta u + \lambda f(\varepsilon, u) = 0 \text{ in } \Omega, \quad u = 0 \text{ on } \partial\Omega, \tag{1.1}$$

where $\lambda, \varepsilon > 0$, f is a nonlinear function, and Ω is a smooth bounded domain in $\mathbf{R}^n$. The global bifurcation diagrams and exact multiplicity of equation (1.1) have been studied in many recent work. In particular, a systematic approach for the positive radially symmetric solutions of (1.1) when Ω is a ball was presented in Ouyang and Shi [**OS1**], [**OS2**] based on previous results by Korman, Li and Ouyang [**KLO1**], [**KLO2**]. The techniques we developed in this article are extensions of the approach in these work, but emphasizing on different aspects of the bifurcation problems.

First in many bifurcation problems in application, "imperfect bifurcations" occur due to the small perturbations or noises in the physical system. For example, in a pitch-fork bifurcation, an imperfect bifurcation diagram as in Fig. 1b appears often in the literature of bifurcation theory:

Such diagrams can be rigorously derived by using Lyapunov-Schmidt reduction method, see for example [**GS**]. We will use bifurcation theorems by Crandall and Rabinowitz [**CR1**], [**CR2**] to give a more direct approach. On the other hand, there is a considerable interest in the evolution of global bifurcation diagrams in (λ, u) space when another parameter ε changes. For example, it has been conjectured that for $f(\varepsilon, u) = e^{u/(1+\varepsilon u)}$ (this equation arises from combustion theory, and it is

e-mail: `shij@math.wm.edu`.

FIGURE 1. (a) Pitchfork bifurcation; (b) Imperfect pitchfork bifurcation

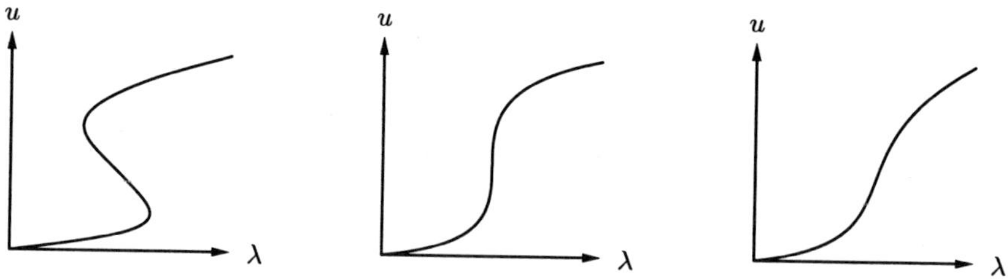

(a) $0 < \varepsilon < \varepsilon_0$ (b) $\varepsilon = \varepsilon_0$ (c) $\varepsilon > \varepsilon_0$

FIGURE 2. Evolution of bifurcation diagrams

usually called perturbed Gelfand's equation), the bifurcation diagrams of (1.1) are as in Fig. 2. We will also show that our abstract results can be applied to these evolution problems of bifurcation diagrams.

In Section 2, we introduce some abstract multi-parameter bifurcation results. We apply the abstract imperfect bifurcation results to an equation from natural resource management in Section 3, and in Section 4, we give some applications of fold and cusp bifurcations. Some open problems and conjectures are mentioned at the end of Section 4. For a linear operator L, we use $N(L)$ to denote the null space of L, and $R(L)$ to denote the range of L.

2. Abstract Theory

In this section, we recall some abstract bifurcation results from [**S**]. We consider an equation

$$F(\varepsilon, \lambda, u) = 0, \tag{2.1}$$

where $F : M \equiv \mathbf{R} \times \mathbf{R} \times X \to Y$ is a nonlinear differentiable map and X, Y are Banach spaces. We call $(\varepsilon_0, \lambda_0, u_0)$ a *degenerate* solution of (2.1) if $F_u(\varepsilon_0, \lambda_0, u_0)$ is not invertible.

First we consider the case when $F(\varepsilon_0, \lambda, u_0) \equiv 0$ for all $\lambda \in \mathbf{R}$, a bifurcation from the trivial solution $u = u_0$ occurs if $F_u(\varepsilon_0, \lambda_0, u_0)$ is not invertible. In [**CR1**], Crandall and Rabinowitz proved:

THEOREM 2.1. ([**CR1**] Theorem 1.7) *Let $F : M \to Y$ be continuously differentiable. Suppose that $F(\varepsilon_0, \lambda, u_0) = 0$ for $\lambda \in \mathbf{R}$, the partial derivative $F_{\lambda u}$ exists and is continuous. At $(\varepsilon_0, \lambda_0, u_0) \in M$, F satisfies*

(F1) *$\dim N(F_u(\varepsilon_0, \lambda_0, u_0))=\operatorname{codim} R(F_u(\varepsilon_0, \lambda_0, u_0)) = 1$, and $N(F_u(\varepsilon_0, \lambda_0, u_0)) = \operatorname{span}\{w_0\}$,*

and

(F2) *$F_{\lambda u}(\varepsilon_0, \lambda_0, u_0)[w_0] \notin R(F_u(\varepsilon_0, \lambda_0, u_0))$.*

Then for fixed $\varepsilon = \varepsilon_0$, the solutions of (2.1) near (λ_0, u_0) consists precisely of the curves $u = u_0$ and $(\lambda(s), u(s))$, $s \in I = (-\delta, \delta)$, where $(\lambda(s), u(s))$ are C^1 functions such that $\lambda(0) = \lambda_0$, $u(0) = u_0$, $u'(0) = w_0$. Moreover, if F is C^2 in u, then

$$\lambda'(0) = -\frac{\langle l, F_{uu}(\varepsilon_0, \lambda_0, u_0)[w_0, w_0]\rangle}{2\langle l, F_{\lambda u}(\varepsilon_0, \lambda_0, u_0)[w_0]\rangle}, \tag{2.2}$$

where $l \in Y^*$ satisfying $N(l) = R(F_u(\varepsilon_0, \lambda_0, u_0))$, and Y^* is the dual space of Y.

In Theorem 2.1, if

(F3) $F_{uu}(\varepsilon_0, \lambda_0, u_0)[w_0, w_0] \notin R(F_u(\varepsilon_0, \lambda_0, u_0))$

is satisfied, then $\lambda'(0) \neq 0$, and a *transcritical* bifurcation occurs at (λ_0, u_0); and if $\lambda'(0) = 0$ and $\lambda''(0) \neq 0$, then a *pitchfork* bifurcation occurs at (λ_0, u_0).

FIGURE 3. (a) Transcritical bifurcation; (b) Pitchfork bifurcation

By studying the degenerate solutions of (2.1), we proved the following result of imperfect bifurcation:

THEOREM 2.2. ([**S**] Theorem 2.5) *Assume the conditions in Theorem 2.1 are satisfied. In addition we assume that (F3) and*

(F4) *$F_\varepsilon(\varepsilon_0, \lambda_0, u_0) \notin R(F_u(\varepsilon_0, \lambda_0, u_0))$*

are satisfied. Then there exists $\rho_1, \delta_1, \delta_2 > 0$, such that for $N = \{(\lambda, u) \in \mathbf{R} \times X : |\lambda - \lambda_0| \leq \delta_1, ||u|| \leq \delta_2\}$,

(A) *for $\varepsilon \in (\varepsilon_0 - \rho_1, \varepsilon_0)$, (or $\varepsilon \in (\varepsilon_0, \varepsilon_0 + \rho_1)$),*

$$F^{-1}(0) \bigcap N = \Sigma_\varepsilon^1 \bigcup \Sigma_\varepsilon^2,\ \Sigma_\varepsilon^i = \{(\lambda, \overline{u_i}(\lambda)) : \lambda \in [\lambda_0 - \delta_1, \lambda_0 + \delta_1]\},\ i = 1, 2, \tag{2.3}$$

and $\overline{u_i}'(\lambda) > 0$ for $\lambda \in [\lambda_0 - \delta_1, \lambda_0 + \delta_1]$, $i = 1, 2$;

(B) *for $\varepsilon = \varepsilon_0$,*

$$F^{-1}(0) \bigcap N = \{(\lambda, 0) : \lambda \in [\lambda_0 - \delta_1, \lambda_0 + \delta_1]\} \bigcup \Sigma_0,\ \Sigma_0 = \{(\overline{\lambda}(t), \overline{u}(t)) : t \in [-\eta, \eta]\}, \tag{2.4}$$

and $\overline{u}'(t) > 0$ *for* $t \in [-\eta, \eta]$;
(C) *for* $\varepsilon \in (\varepsilon_0, \varepsilon_0 + \rho_1)$, *(or* $\varepsilon \in (\varepsilon_0 - \rho_1, \varepsilon_0)$*),*

$$F^{-1}(0) \bigcap N = \Sigma_\varepsilon^+ \bigcup \Sigma_\varepsilon^-, \ \Sigma_\varepsilon^\pm = \{(\overline{\lambda}_\pm(t), \overline{u}_\pm(t)) : t \in [-\eta, \eta]\}, \tag{2.5}$$

$\overline{\lambda}_+(\pm\eta) = \lambda_0 + \delta_1$, $\overline{\lambda}_-(\pm\eta) = \lambda_0 - \delta_1$, $\overline{\lambda}'_\pm(0) = 0$, $\overline{\lambda}''_+(0) < 0$, $\overline{\lambda}''_-(0) > 0$, *and there are exactly one turning point on each component* $\Sigma_\varepsilon^\pm$. *(See Fig. 4.)*

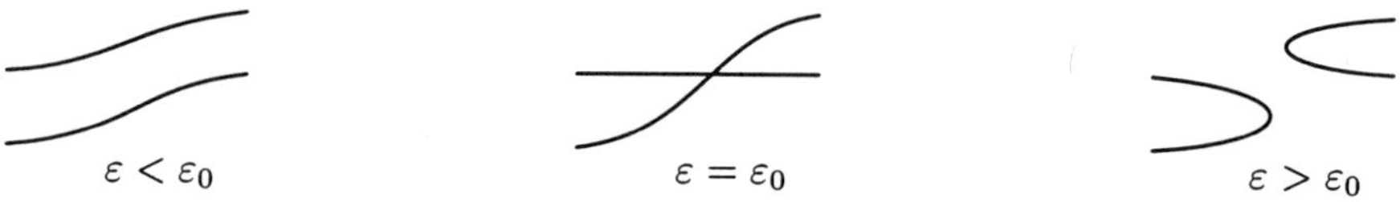

FIGURE 4. Imperfect bifurcation in a transcritical bifurcation

Secondly, we consider the case when the degenerate solution is a turning point on the bifurcation diagram. The following result was proved by Crandall and Rabinowitz [**CR2**]:

THEOREM 2.3. ([**CR2**] Theorem 3.2) *Let* $F : M \to Y$ *be continuously differentiable. At* $(\varepsilon_0, \lambda_0, u_0) \in M$, $F(\varepsilon_0, \lambda_0, u_0) = 0$, F *satisfies (F1) and*
(F5) $F_\lambda(\varepsilon_0, \lambda_0, u_0) \notin R(F_u(\varepsilon_0, \lambda_0, u_0))$.
Then for fixed $\varepsilon = \varepsilon_0$, *the solutions of* (2.1) *near* (λ_0, u_0) *form a* C^1 *curve* $(\lambda(s), u(s))$, $\lambda(0) = \lambda_0$, $u(0) = u_0$, $\lambda'(0) = 0$ *and* $u'(0) = w_0$. *Moreover, if* F *is* C^2 *in* u, *then*

$$\lambda''(0) = -\frac{\langle l, F_{uu}(\varepsilon_0, \lambda_0, u_0)[w_0, w_0]\rangle}{\langle l, F_\lambda(\varepsilon_0, \lambda_0, u_0)\rangle}, \tag{2.6}$$

where $l \in Y^*$ *satisfying* $N(l) = R(F_u(\varepsilon_0, \lambda_0, u_0))$, *and* Y^* *is the dual space of* Y.

In Theorem 2.3, if (F3) is also satisfied, then $\lambda''(0) \neq 0$, and a *saddle-node* bifurcation occurs at (λ_0, u_0). In singularity theory, a degenerate solution with $\lambda''(0) \neq 0$ is called a *fold*, and one with $\lambda''(0) = 0$ but $\lambda'''(0) \neq 0$ is a *cusp*. (For applications of singularity theory in partial differential equations, see [**CT**].) In [**S**], we proved

THEOREM 2.4. ([**S**] Theorem 2.3)

1. *If* F *satisfies (F1), (F3) and (F5) at* $(\lambda_0, \varepsilon_0, u_0)$, *then the saddle-node degenerate solution persists for* ε *near* ε_0.
2. *If* F *satisfies (F1), (F5),*
   (F3') $F_{uu}(\varepsilon_0, \lambda_0, u_0)[w_0, w_0] \in R(F_u(\varepsilon_0, \lambda_0, u_0))$,
   and
   (F6) $D_\varepsilon \begin{pmatrix} F(\cdot) \\ F_u(\cdot)[w_0] \end{pmatrix} \notin R\left(D_{(\lambda,u,w)} \begin{pmatrix} F(\cdot) \\ F_u(\cdot)[w_0] \end{pmatrix}\right)$,
   where $\cdot = (\varepsilon_0, \lambda_0, u_0)$. *In addition we assume that* $F \in C^3(M)$ *and*
   (F7) $F_{uuu}(\varepsilon_0, \lambda_0, u_0)[w_0, w_0, w_0] + 3F_{uu}(\varepsilon_0, \lambda_0, u_0)[\theta, w_0] \notin R(F_u(\varepsilon_0, \lambda_0, u_0))$, *where* θ *is the solution of* $F_u(\cdot)[\theta] + F_{uu}(\cdot)[w_0, w_0] = 0$.

Then there exists $\rho_1 > 0$ *such that for* $\varepsilon \in (\varepsilon_0 - \rho_1, \varepsilon_0 + \rho_1)$, *all the solutions of* (2.1) *near* (λ_0, u_0) *are on a curve* $\Sigma_\varepsilon = (\overline{\lambda}(t), \overline{u}(t))$, *where* $t \in I = (-\eta, \eta)$ *for* $\eta = \eta(\varepsilon)$. *Moreover,*

(A) *for* $\varepsilon \in (\varepsilon_0 - \rho_1, \varepsilon_0)$, *(or* $\varepsilon \in (\varepsilon_0, \varepsilon_0 + \rho_1)$*),* $\overline{\lambda}'(t) > 0$ *for* $t \in I$*;*

(B) *for* $\varepsilon = \varepsilon_0$, $\overline{\lambda}(0) = \lambda_0$, $\overline{\lambda}'(0) = \overline{\lambda}''(0) = 0$, $\overline{\lambda}'''(0) > (<)0$ *and* $\overline{\lambda}'(t) > (<)0$ *for* $t \in I\backslash\{0\}$*;*

(C) *for* $\varepsilon \in (\varepsilon_0, \varepsilon_0 + \rho_1)$, *(or* $\varepsilon \in (\varepsilon_0 - \rho_1, \varepsilon_0)$*), there exists* $t_1, t_2 \in I$ *such that* $\overline{\lambda}'(t_1) = \overline{\lambda}'(t_2) = 0$, $\overline{\lambda}''(t_1) < (>)0$, $\overline{\lambda}''(t_2) > (<)0$, $\overline{\lambda}'(t) > (<)0$ *in* $(-\eta, t_1) \bigcup (t_2, \eta)$ *and* $\overline{\lambda}'(t) < (>)0$ *in* (t_1, t_2). *(See Figure 5.)*

FIGURE 5. Cusp Bifurcation

The proof of Theorems 2.2 and 2.4 are based on implicit function theorem and bifurcation theorems in [**CR1, CR2**] (which are also based on implicit function theorems). For details of proof, we refer to [**S**].

3. Application to a Harvesting Problem

As applications of the abstract results established in Section 2, we consider a diffusive logistic equation with harvesting, which arises from population biology:

$$\frac{\partial u}{\partial t} = D\Delta u + au\left(1 - \frac{u}{N}\right) - c \cdot h(x), \qquad (t, x) \in (0, T) \times \Omega, \tag{3.1}$$

with the initial and boundary conditions:

$$\begin{aligned} u(t, x) &= 0 \qquad (t, x) \in (0, T) \times \partial\Omega, \\ u(0, x) &= u_0(x) \geq 0, \qquad x \in \Omega, \end{aligned} \tag{3.2}$$

where $D > 0$ is the diffusion coefficient, $a > 0$ is the linear reproduction rate and $N > 0$ is the carrying capacity of the environment. (3.1) and (3.2) arise from the studies of population biology of one species. Here $u(x, t)$ is the concentration of a substance or the density of a population. We assume that (a) the individuals of the species disperse randomly in a bounded habitat Ω; (b) the reproduction of the species follows the logistic growth; (c) the boundary of the environment $\partial\Omega$ is hostile to the species; (d) the habitat Ω is homogeneous, (*i.e.* the diffusion and the growth do not depend on x); and (e) the population is harvested at a constant rate $c \cdot h(x)$, where $c > 0$ is a parameter which represents the level of harvesting, and $h(x)$ represents the natural heterogeneous fishing rate with $h(x) > 0$ for $x \in \Omega$, and $||h||_\infty = 1$. Such harvesting pattern arises naturally from the fishery management problems, where $c \cdot h(x)$ is related to the fishing quota imposed by fishing regulating authorities. The equation (3.1) with (3.2) are generalization of well-known ordinary differential equation logistic model with constant yield harvesting $u' = au(1 - u/N) - c$ (see [**Cl**], [**BC**].)

We study the steady state solutions of (3.1) and (3.2). In fact, we consider a more general nonlinearity. We consider a semilinear elliptic equation:

$$\Delta u + \lambda f(u) - ch(x) = 0 \quad \text{in } \Omega, \quad u = 0 \quad \text{on } \partial\Omega, \tag{3.3}$$

where $\lambda, c > 0$, $h(x)$ is defined as in last paragraph, and $f(u)$ satisfies

(f1) $f \in C^2(\mathbf{R})$, $f(0) = 0$, $f'(0) > 0$, $f(u) > 0$ for $u \in (0, M)$, where either $M = \infty$ or $M < \infty$, $f(M) = 0$ and $f'(M) < 0$;

(f2) $f''(u) < 0$ for $u \in \mathbf{R}$;

(f3) If $M = \infty$, then $\lim_{u\to\infty} \frac{f(u)}{u} = 0$.

An example for the case of $M < \infty$ in (f1) is the logistic growth $f(u) = au - bu^2$ for some $a, b > 0$, and examples for the case of $M = \infty$ are $g(u) = au/(1 + bu)$ for $a, b > 0$ and $g(u) = 1 - e^{-au}$ for $a > 0$, which are sublinear functions modeling the saturating effect. We denote by λ_k the k-th eigenvalue of

$$\Delta\phi + \lambda\phi = 0 \quad \text{in } \Omega, \quad \phi = 0 \quad \text{on } \partial\Omega. \tag{3.4}$$

It is well-known that λ_1 is simple, and its eigenfunction does not change sign. We define $\lambda_k^0 = \lambda_k / f'(0)$. It is easy to show that when $\lambda \le \lambda_1^0$, then (3.3) has no non-negative solutions if $c \ge 0$. General existence results via comparison and bifurcation methods were obtained in [**OSS**], and we also proved an exact multiplicity result which can be obtained by using the abstract bifurcation in Section 2:

THEOREM 3.1. ([**OSS**] Theorem 4.1) *Suppose that f satisfies (f1)-(f3). There exists $c_2, \delta_2 > 0$ such that when $\lambda \in (\lambda_*, \lambda_* + \delta_2)$,*

1. *(3.3) has exactly two positive solutions $u_1(\cdot, c)$ and $u_2(\cdot, c)$ for $c \in [0, c_2)$, exactly one positive solution $u_1(\cdot, c)$ for $c = c_2$, and no positive solution for $c > c_2$;*
2. *$u_1(\cdot, c)$ is stable and $u_2(\cdot, c)$ is unstable with Morse index is 1 for $c \in [0, c_2)$, $u_1(\cdot, c_2)$ is degenerate with and neutrally stable;*
3. *All solutions lie on a smooth curve Σ. On (c, u) space, Σ starts from $(0, 0)$, continues to the right, reaches the unique turning point at $c = c_2$ where it turns back, then continues to the left without any turnings until it reaches $(0, v_a)$, where v_a is the unique positive solution of (3.3) with $c = 0$. (see Fig. 6).*

In Theorem 3.1, we use the harvesting rate c as the bifurcation parameter. Note that in this case, the bifurcation diagram is exactly same as that of ODE: $u' = au(1 - u/N) - c$, but the dynamics in function space is much more complicated. Next we use the recipcle of the diffusion coefficient as the bifurcation parameter, and we obtain a local exact multiplicity result. Here we consider

$$\Delta u + \lambda[f(u) - \varepsilon] = 0 \quad \text{in } \Omega, \quad u = 0 \quad \text{on } \partial\Omega, \tag{3.5}$$

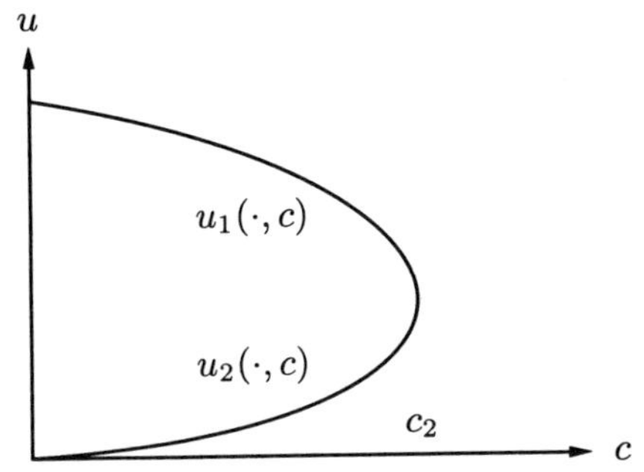

FIGURE 6. Precise Bifurcation Diagram for Diffusive Logistic Equation with Harvesting

where $\varepsilon, \lambda > 0$, and $f(u)$ satisfies (f1)-(f3).

THEOREM 3.2. ([**SS**] Theorem 1.2) *Suppose that $f(u)$ satisfies (f1)-(f3). We assume that λ_2 is a simple eigenvalue of* (3.4) *and*

$$\int_\Omega \phi_2(x)dx \cdot \int_\Omega \phi_2^3(x)dx > 0, \tag{3.6}$$

where ϕ_2 is an eigenfunction corresponding to λ_2. Let $\Sigma = \{(\lambda, u) \in \mathbf{R} \times X : (\lambda, u) \text{ solves } (3.5)\}$, and $T(a,b,c) = \{(\lambda, u) : a < \lambda < b, ||u||_X < c\}$. Then for any small $\delta_1, \delta_2 > 0$, there exists $\varepsilon_1 = \varepsilon_1(\delta_1, \delta_2, g) > 0$ such that for any $\varepsilon \in (0, \varepsilon_1)$,

$$\Sigma_0 \equiv \Sigma \bigcap T(\lambda_1^0 - \delta_1, \lambda_2^0 + \delta_1, \delta_2) = \bigcup_{i=1}^{3} \Sigma_i,$$

where Σ_i is a connected component of Σ_0, $(i = 1, 2, 3)$. Moreover,

(A) *Each Σ_i $(i = 1, 2, 3)$ is a smooth curve in $\mathbf{R} \times X$;*
(B) *Σ_1 is exactly $\supset$-shaped, there is a unique degenerate solution on Σ_1, and each solution on Σ_1 is negative;*
(C) *Σ_3 is exactly $\subset$-shaped, there is a unique degenerate solution on Σ_3, and each solution on Σ_3 is sign-changing;*
(D) *Σ_2 is exactly S-shaped, and there are exactly two degenerate solutions on Σ_2; Σ_2 can be parameterized as $(\lambda(s), u(s))$, $s \in (s_1, s_4)$, such that for $s \in (s_1, s_2)$, $u(s)$ is positive, and for $s \in (s_3, s_4)$, $u(s)$ is sign-changing, where $s_1 < s_2 < s_3 < s_4$; The portion of Σ_2 with $s \in (s_1, s_2)$ contains the degenerate solution on the left, and the portion of Σ_2 with $s \in (s_3, s_4)$ contains the degenerate solution on the right (see Fig. 7);*
(E) *There exist $\delta_i = \delta_i(\varepsilon) > 0$, $(i = 3, 4, 5, 6)$ such that*

$$proj\Sigma_1 = [\lambda_1^0 - \delta_1, \lambda_1^0 - \delta_3],\ proj\Sigma_2 = [\lambda_1^0 + \delta_4, \lambda_2^0 - \delta_5],\ proj\Sigma_3 = [\lambda_2^0 + \delta_6, \lambda_2^0 + \delta_1],$$

where $proj\Sigma_i$ is the projection of Σ_i into $\mathbf{R} = (\lambda)$.

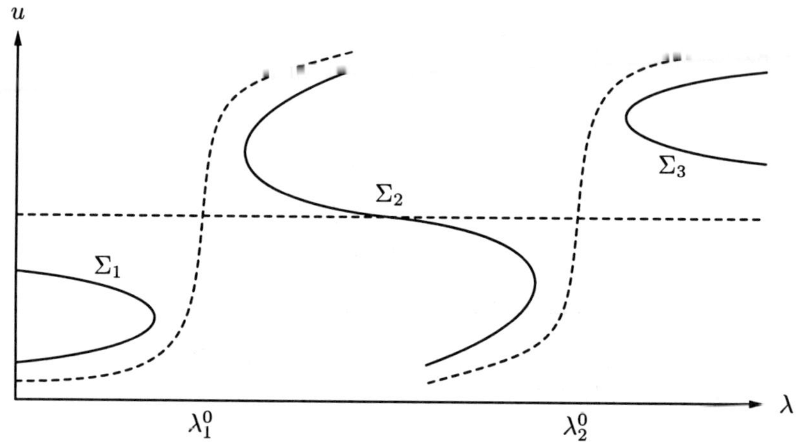

FIGURE 7. Bifurcation diagram in Theorem 3.2; solid curve: small $\varepsilon > 0$, dashed curve: $\varepsilon = 0$

The bifurcation diagrams in Theorems 3.1 and 3.2 are both of exact shape (there is no wiggling of the diagrams), and hold for very general domains. (In Theorem 3.1, there is no restriction on Ω; and in Theorem 3.2, (3.6) is needed. When the $<$ sign in (3.6) is replaced by $>$, a similar diagram can be shown, see details in [**SS**]; and when $<$ sign is $=$, some more interesting bifurcation occurs, also see details in [**SS**].)

4. Fold and Cusp

In this section, we apply Theorem 2.4 to a few application problems. Our first example is

$$\Delta u + \lambda(1 + u + u^2 - \varepsilon u^3) = 0 \text{ in } \Omega, \quad u = 0 \text{ on } \partial\Omega, \tag{4.1}$$

which was studied by Crandall and Rabinowitz [**CR2**], and Brown, Ibrahim and Shivaji [**BIS**].

THEOREM 4.1. *The bifurcation diagram of* (4.1) *is exactly* S*-shaped when* $\varepsilon > 0$ *is small and* Ω *is a ball* B^n *in* $\mathbf{R}^n$ *with* $1 \le n \le 6$.

We sketch a proof of Theorem 4.1. The proof is helped by two auxiliary equations: (A) $\Delta u + \lambda(1 + u + u^2) = 0$ in Ω, $u = 0$ on $\partial\Omega$; and (B) $\Delta v + \mu(v^2 - v^3) = 0$ in Ω, $u = 0$ on $\partial\Omega$. The second equation can be obtained by letting $v = \varepsilon u$, $\mu = \varepsilon^{-1}\lambda$, and then setting $\varepsilon = 0$. The basic idea is to use solutions of (A) to approximate solutions of (4.1) when $u(0)$ is small, and to use (B) when $u(0)$ is in the scale of $O(\varepsilon^{-1})$. The exact multiplicity of solutions of (A) and (B) are known from previous results: the bifurcation diagram of (A) is exactly $\supset$-shaped (see for example [**OS2**], Fig. 8a); and that of (B) is exactly $\subset$-shaped (see [**OS1**], Fig. 8b.) To get $\supset$-shaped curve in (A), we need the condition $n \le 6$ to make u^2 to be subcritical. We can show that when $0 < u(0) < (3\varepsilon)^{-1}$, the portion of the bifurcation diagram is exactly $\supset$-shaped since for $f(u) = 1 + u + u^2 - \varepsilon u^3$, $f''(u) > 0$ for $u \in (0, (3\varepsilon)^{-1})$, then a standard bifurcation argument in **[OS2]** can be used. When $((3+\delta)\varepsilon)^{-1} < u(0) < ((1+\delta)\varepsilon)^{-1}$, for a small $\delta > 0$, the portion of the bifurcation diagram is exactly $\subset$-shaped since in the (μ, v) coordinate, it is a perturbation of a compact portion of bifurcation diagram of (B) which is of that shape, and there is exactly one degenerate solution on that portion (see **[Da]** for perturbation of a compact portion of a curve of solutions). Finally we can prove that when $u(0) > ((1+\delta)\varepsilon)^{-1}$, the diagram is monotone with respect to λ. Thus the bifurcation diagram is exactly S-shaped as in Fig. 8c.

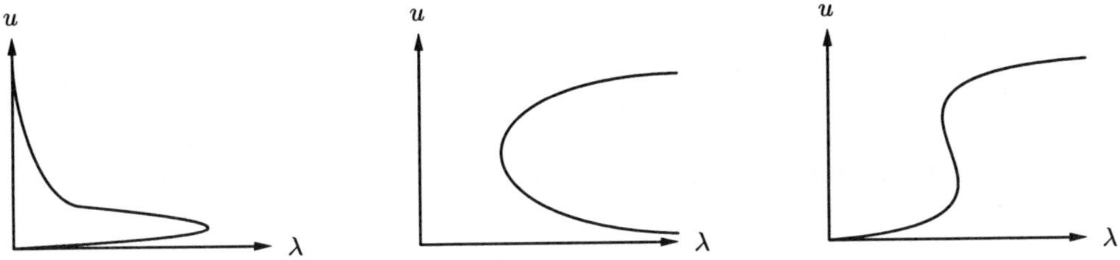

FIGURE 8. (a) Equation (A); (b) Equation (B); (c) Equation (4.1)

Our second example is the perturbed Gelfand equation:

$$\Delta u + \lambda \exp[-1/(u+\varepsilon)] = 0 \text{ in } B^n, \; u = 0 \text{ on } \partial B^n, \tag{4.2}$$

where B^n is a ball in $\mathbf{R}^n$. The original perturbed Gelfand equation is $\Delta v + \mu \exp[v/(\varepsilon v + 1)] = 0$ in B^n, $v = 0$ on ∂B^n. (4.2) is obtained by a change of variables: $u = \varepsilon^2 v$ and $\lambda = \varepsilon^2 \exp(1/\varepsilon)\mu$, which was introduced by Du and Lou **[DL]**. Let Σ_ε be the solution set of (4.2). It has been a long-time conjecture that when $n = 1$ or 2, there exists $\varepsilon_0 > 0$ such that when $\varepsilon \in (0, \varepsilon_0)$, Σ_ε is exactly S-shaped; when $\varepsilon \in (\varepsilon_0, \infty)$, Σ_ε is monotone increasing; and when $\varepsilon = \varepsilon_0$, there is a cusp type degenerate solution. (See Fig. 2.)

It is easy to show that when $\varepsilon \geq 1/4$, Σ_ε is monotone increasing. The proof of exact S-shaped curve for small $\varepsilon > 0$ has been studied by several authors. Dancer **[Da]** first proved (4.2) has exactly three solutions for $\lambda \in (\lambda_0, \lambda_1(\varepsilon))$, $\varepsilon \in (0, \varepsilon_1)$ and $n = 1, 2$. Hastings and McLeod **[HM]** proved Σ_ε is exactly S-shaped for $n = 1$, $\varepsilon \in (0, \varepsilon_1)$ using quadratures. Recently, Wang **[W2]** showed that for $n = 1$ and $\varepsilon \in (0, 1/4.4967]$, Σ_ε is exactly S-shaped using a different quadrature method, and the upper bound was improved to $1/4.35$ by Korman and Li **[KL]** using the bifurcation method. Du and Lou **[DL]** proved Σ_ε is exactly S-shaped for $n = 2$ and $\varepsilon \in (0, \varepsilon_1)$ for some small $\varepsilon_1 > 0$ using a combination of bifurcation method and a perturbation argument.

Using the proof of Theorem 4.1 above, we can prove the result of **[DL]** in a more direct way. We point out that, to completely resolve the conjecture for all

$\varepsilon > 0$, we need to prove the following estimates at a cusp degenerate solution of (4.2)

$$3\int_{B^n} f_{uu}(\varepsilon, u(x))w^2(x)\theta(x)dx + \int_{B^n} f_{uuu}(\varepsilon, u(x))w^4(x)dx < 0. \qquad (4.3)$$

where θ is the solution of

$$\Delta\theta + \lambda_0 f'(\varepsilon_0, u_0)\theta + \lambda_0 f''(\varepsilon_0, u_0)w_0^2 = 0 \text{ in } B^n, \;\; u = 0 \text{ on } \partial B^n. \qquad (4.4)$$

Finally we mention that another long-standing open conjecture is the bifurcation diagram of

$$u'' + \lambda(u-\varepsilon)(u-b)(c-u) = 0, \text{ in } (0,1), \;\; u(0) = u(1) = 0, \qquad (4.5)$$

where $c > b > \varepsilon > 0$, and $c - b > b - \varepsilon$. The equation arises from the studies of dynamics of FitzHugh-Nagumo equation and population biology. When $\varepsilon = 0$, Smoller and Wasserman [**SW**] proved that the bifurcation diagram is exactly $\subset$-shaped (see Fig. 9a) by quadrature methods, and later it was also proved by Korman, Li and Ouyang [**KLO1**] using bifurcation method. (The higher dimensional analog for radially symmetric solutions was proved by Korman, Li and Ouyang [**KLO2**] for two-dimensional ball, and all dimensions by Ouyang and Shi [**OS1**].

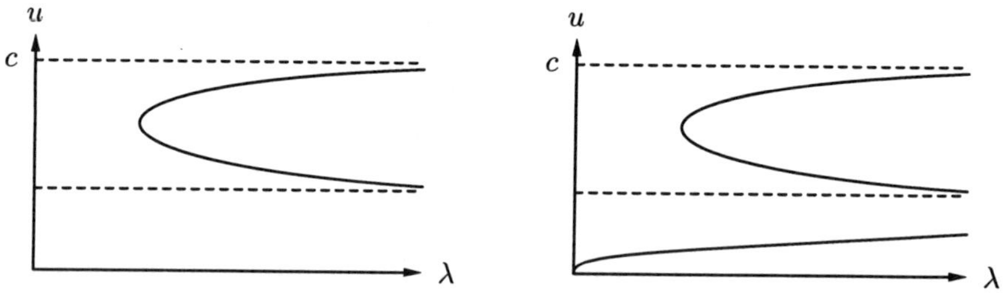

FIGURE 9. (a) Bifurcation diagram when $\varepsilon = 0$; (b) Bifurcation diagram when $\varepsilon > 0$.

For $\varepsilon > 0$, Smoller and Wasserman [**SW**] claimed that the bifurcation diagram is as Fig. 9b, but Wang [**W1**] found an error in their proof. The exact multiplicity result as in Fig. 9b was proved by Wang [**W1**], Ye and Li [**YL**] and Korman, Li and Ouyang [**KLO1**] but all of them need some extra conditions on $f(u)$. An approach using the ideas in this paper is possible for a complete solution for the problem: here one need to show a cusp bifurcation will not occur when ε increases from 0 to ∞. This was pursued by Korman, Li and Ouyang [**KLO3**] recently, but still some extra conditions are needed.

References

[BC] Brauer, Fred; Castillo-Chávez, Carlos, *Mathematical models in population biology and epidemiology. Texts in Applied Mathematics*, **40**. Springer-Verlag, New York, (2001).

[BIS] Brown, K. J.; Ibrahim, M. M. A.; Shivaji, R., *S*-shaped bifurcation curves. *Nonlinear Anal.* **5** (1981), no. 5, 475–486.

[CT] Church, P. T.; Timourian, J. G., Global structure for nonlinear operators in differential and integral equations. I. Folds. II. Cusps. Topological nonlinear analysis, II (Frascati, 1995), 109–160, 161–245, Progr. Nonlinear Differential Equations Appl., 27, Birkhaüser Boston, Boston, MA, 1997.

[Cl] Clark, Colin W., Mathematical Bioeconomics, *The Optimal Management of Renewable Resources*, John Wiley & Sons, Inc. New York (1991).

[CR1] Crandall, Michael G.; Rabinowitz, Paul H, Bifurcation from simple eigenvalues. *J. Functional Analysis*, **8** (1971), 321–340.

[CR2] Crandall, Michael G.; Rabinowitz, Paul H., Bifurcation, perturbation of simple eigenvalues and linearized stability. *Arch. Rational Mech. Anal.* **52** (1973), 161–180.

[Da] Dancer, E. N., On the structure of solutions of an equation in catalysis theory when a parameter is large. *J. Differential Equations* **37** (1980), no. 3, 404–437.

[DL] Du, Yihong; Lou, Yuan, Proof of a conjecture for the perturbed Gelfand equation from combustion theory. *J. Differential Equations* **173** (2001), no. 2, 213–230.

[GS] Golubitsky, Martin; Schaeffer, David G., Singularities and groups in bifurcation theory. Vol. I. *Applied Mathematical Sciences*, **51**. Springer-Verlag, New York, 1985.

[HM] Hastings, S. P.; McLeod, J. B., The number of solutions to an equation from catalysis. *Proc. Roy. Soc. Edinburgh Sect. A* **101** (1985), no. 1-2, 15–30.

[KL] Korman, Philip; Li, Yi, On the exactness of an S-shaped bifurcation curve. *Proc. Amer. Math. Soc.*, **127** (1999), no. 4, 1011–1020.

[KLO1] Korman, Philip; Li, Yi; Ouyang, Tiancheng, Exact multiplicity results for boundary value problems with nonlinearities generalising cubic. *Proc. Roy. Soc. Edinburgh Sect. A* **126** (1996), no. 3, 599–616.

[KLO2] Korman, Philip; Li, Yi; Ouyang, Tiancheng, An exact multiplicity result for a class of semilinear equations. *Comm. Partial Differential Equations* **22** (1997), no. 3-4, 661–684.

[KLO3] Korman, Philip; Li, Yi; Ouyang, Tiancheng, preprint (2002).

[OSS] Oruganti, Shobha, Shi, Junping and Shivaji, Ratnasingham, Diffusive Logistic equation with constant effort harvesting, I: Steady States. *Trans. Amer. Math. Soc.*, **354** (2002), no. 9, 3601–3619.

[OS1] Ouyang, Tiancheng and Shi, Junping, Exact multiplicity of positive solutions for a class of semilinear problem. *Jour. Diff. Equa.* **146** (1998), no. 1, 121–156.

[OS2] Ouyang, Tiancheng and Shi, Junping, Exact multiplicity of positive solutions for a class of semilinear problem:II. *Jour. Diff. Equa.* **158**, (1999), no. 1, 94–151.

[S] Shi, Junping, Persistence and bifurcation of degenerate solutions, *Jour. Func. Anal.* **169**, (1999), No. 2, 494–531.

[SS] Shi, Junping; Shivaji, Ratnasingham, *Global bifurcation for concave semipositon problems*. To appear in *Recent Advances in Evolution Equations*, (2002).

[SW] Smoller, J.; Wasserman, A., Global bifurcation of steady-state solutions. *J. Differential Equations* **39** (1981), no. 2, 269–290.

[W1] Wang, Shin-Hwa, A correction for a paper by J. Smoller and A. G. Wasserman: "Global bifurcation of steady-state solutions". *J. Differential Equations* **77** (1989), no. 1, 199–202.

[W2] Wang, Shin Hwa, On **S**-shaped bifurcation curves. *Nonlinear Anal.* **22** (1994), no. 12, 1475–1485.

[YL] Ye, Qi Xiao; Li, Zheng Yuan, Introduction to reaction-diffusion equations (in Chinese). Science Press, Beijing, (1990).

A semilinear Schrödinger equation with magnetic field

Andrzej Szulkin

Department of Mathematics, Stockholm University
106 91 Stockholm, Sweden

1. Introduction

In this note we describe some recent results by Gianni Arioli and the author [2] on the existence of nontrivial solutions of the semilinear stationary Schrödinger equation in a magnetic field

$$(-i\nabla + A)^2 u + V(x)u = g(x, |u|)u, \quad x \in \mathbb{R}^N. \tag{1.1}$$

Here $u : \mathbb{R}^N \to \mathbb{C}$ ($N \geq 2$), $V : \mathbb{R}^N \to \mathbb{R}$ is the scalar (or electric) potential and $A = (A_1, \ldots, A_N) : \mathbb{R}^N \to \mathbb{R}^N$ is the vector (or magnetic) potential. Let $B := \operatorname{curl} A$. For $N = 3$ this is the usual curl operator and for general N, $B = (B_{jk})$, $1 \leq j, k \leq N$, where $B_{jk} := \partial_j A_k - \partial_k A_j$. B represents the external magnetic field whose source is A. In what follows we always assume $A \in L^2_{loc}(\mathbb{R}^N, \mathbb{R}^N)$, $V \in L^1_{loc}(\mathbb{R}^N)$ and V is bounded below. Denote

$$\nabla_A u = (\nabla + iA)\, u$$

and let $2^* := 2N/(N-2)$. Appropriate spaces in which we shall consider the problem (1.1) are

$$H^1_A(\mathbb{R}^N) := \{u \in L^2(\mathbb{R}^N) : \nabla_A u \in L^2(\mathbb{R}^N)\}$$

and, for $N \geq 3$,

$$\mathcal{D}^{1,2}_A(\mathbb{R}^N) := \{u \in L^{2^*}(\mathbb{R}^N) : \nabla_A u \in L^2(\mathbb{R}^N)\}.$$

Both $H^1_A(\mathbb{R}^N)$ and $\mathcal{D}^{1,2}_A(\mathbb{R}^N)$ are Hilbert spaces with inner product respectively

$$\int_{\mathbb{R}^N} \nabla_A u \cdot \overline{\nabla_A v} + u\overline{v}$$

Supported in part by the Swedish Research Council.

and

$$\int_{\mathbb{R}^N} \nabla_A u \cdot \overline{\nabla_A v}$$

(the bar denotes complex conjugation). By Section 2 of [**6**] and Theorem 7.22 of [**10**], $C_0^\infty(\mathbb{R}^N)$ is dense in $H_A^1(\mathbb{R}^N)$ and $\mathcal{D}_A^{1,2}(\mathbb{R}^N)$ (in [**6**] $\mathcal{D}_A^{1,2}(\mathbb{R}^N)$ has in fact been defined as the closure of $C_0^\infty(\mathbb{R}^N)$ with respect to the norm corresponding to the inner product above).

Let $|g(x,|u|)| \le c(1+|u|^{2^*-2})$, $F(x,|u|) := \int_0^{|u|} g(x,s)s\,ds$ and consider the functional

$$J(u) := \frac{1}{2}\int_{\mathbb{R}^N} |\nabla_A u|^2 + V|u|^2 - \int_{\mathbb{R}^N} F(x,|u|). \tag{1.2}$$

Suppose $u \in H_A^1(\mathbb{R}^N)$. By the diamagnetic inequality $\big|\nabla|u|(x)\big| \le |\nabla_A u(x)|$ a.e. in $\mathbb{R}^N$ (see e.g. Theorem 7.21 in [**10**]), $|u| \in H^1(\mathbb{R}^N)$ and therefore $u \in L^p(\mathbb{R}^N)$ for any $p \in [2, 2^*]$ (for any $p \in [2,+\infty)$ if $N = 2$). It follows that whenever $V \in L^\beta(\mathbb{R}^N)$, where $\beta \ge N/2$ ($\beta > 1$ if $N = 2$), then $J \in C^1(H_A^1(\mathbb{R}^N), \mathbb{R})$ and critical points of J are weak solutions of (1.1). We note also that $J(e^{i\theta}u) = J(u)$ for any $\theta \in \mathbb{R}$, hence J is S^1-invariant.

Suppose A, $\tilde{A} \in L^\alpha_{loc}(\mathbb{R}^N, \mathbb{R}^N)$ for some $\alpha \in [1,+\infty)$ and $\operatorname{curl} A = B = \operatorname{curl} \tilde{A}$ (in the sense of distributions). Then $\tilde{A} - A = \nabla\varphi$ for some $\varphi \in W^{1,\alpha}_{loc}(\mathbb{R}^N)$, see Lemma 1.1 of [**9**]. It is easy to see that if $\tilde{u} = e^{-i\varphi}u$, then $\nabla_{\tilde{A}}\tilde{u} = e^{-i\varphi}\nabla_A u$ and hence

$$\int_{\mathbb{R}^N} |\nabla_{\tilde{A}}\tilde{u}|^2 = \int_{\mathbb{R}^N} |\nabla_A u|^2 .$$

It follows that if $u \in H_A^1(\mathbb{R}^N)$, then $\tilde{u} \in H_{\tilde{A}}^1(\mathbb{R}^N)$ and if u satisfies (1.1), then so does $\tilde{u}$ with A replaced by $\tilde{A}$. The above properties are called the gauge invariance and they reflect the fact that the magnetic field B and not the particular choice of the vector potential A should be essential. The transformation $u \mapsto \tilde{u}$ is called the change of gauge. Note that there is a trivial change of gauge $u \mapsto \tilde{u} = e^{-i\theta}u$, where θ is a constant. Then $\tilde{A} = A$ and it is a consequence of this property that J is S^1-invariant. While there is a vast literature concerning the Schrödinger equation (1.1) with $A = 0$, to the best of our knowledge there are very few papers dealing with the magnetic case [**6, 11, 12**]. Also in [**5, 8**] the magnetic case has been considered, but from a very different point of view (semiclassical limits and related concentration phenomena).

Denote $-\Delta_A := (-i\nabla + A)^2$. Since V is bounded below, so is the spectrum $\sigma(-\Delta_A + V)$ in $L^2(\mathbb{R}^N)$. Suppose $F \ge 0$. If $0 \notin \sigma(-\Delta_A + V)$, then either $\sigma(-\Delta_A + V) \subset (0,+\infty)$ and the functional has a mountain pass geometry, or $\sigma(-\Delta_A + V) \cap (-\infty, 0) \ne \emptyset$ and then J has a geometry of linking type.

In what follows $\|\cdot\|_p$ will denote the usual L^p-norm in $\mathbb{R}^N$, $\sigma(-\Delta_A + V)$ the spectrum of $-\Delta_A + V$ in $L^2(\mathbb{R}^N)$ and $B(a,r)$ the open ball centered at a and having radius r.

2. The results

First we consider a minimization problem in $\mathbb{R}^N$, $N \geq 3$. Let

$$\bar{S} := \inf_{u \in \mathcal{D}_A^{1,2}(\mathbb{R}^N)\setminus\{0\}} \frac{\int_{\mathbb{R}^N} |\nabla_A u|^2 + V|u|^2}{||u||_{2^*}^2}. \tag{2.1}$$

Our first result is a slight generalization of Theorem 3.7 in [**6**].

THEOREM 2.1. *If $V \geq 0$, $V \in L_{loc}^{N/2}(\mathbb{R}^N)$ and $A \in L_{loc}^N(\mathbb{R}^N, \mathbb{R}^N)$, then the infimum in (2.1) is attained if and only if $V \equiv 0$ and $B = \operatorname{curl} A \equiv 0$.*

Note that since V is only in $L_{loc}^{N/2}(\mathbb{R}^N)$, $\int_{\mathbb{R}^N} V|u|^2$ need not be finite for all $u \in \mathcal{D}_A^{1,2}(\mathbb{R}^N)$. However, it is finite for $u \in C_0^\infty(\mathbb{R}^N)$ and therefore the minimization problem (2.1) makes sense. Note also that if $\bar{S}$ is attained at some $u \in \mathcal{D}_A^{1,2}(\mathbb{R}^N)$, then u is a solution of (1.1) with $g(x,|u|) = \bar{S}|u|^{2^*-2}$; hence $v = \bar{S}^{1/(2^*-2)}u$ solves (1.1) with $g(x,|u|) = |u|^{2^*-2}$.

THEOREM 2.2. *Suppose $N \geq 4$, $V \in L_{loc}^1(\mathbb{R}^N)$, $V^- := \max\{-V, 0\} \in L^{N/2}(\mathbb{R}^N)$, $A \in L_{loc}^2(\mathbb{R}^N, \mathbb{R}^N)$ and $\sigma(-\Delta_A + V) \subset (0, +\infty)$. If there exists $\bar{x} \in \mathbb{R}^N$ such that $V(x) \leq -c < 0$ in a neighborhood of $\bar{x}$ and A is continuous at $\bar{x}$, then the infimum of (2.1) is attained for some $u \in H_A^1(\mathbb{R}^N) \setminus \{0\}$.*

Since $\sigma(-\Delta_A + V) \subset (0, +\infty)$, it follows that if $\bar{S}$ is attained, then it is positive. Indeed, $\bar{S} \geq 0$ and if $\bar{S} = 0$ is attained at some $u \neq 0$, then u is an eigenvalue of $-\Delta_A + V$ which is impossible.

In the next theorems we shall need the following assumptions:

A1: $V \in L^\infty(\mathbb{R}^N)$, $g \in C(\mathbb{R}^N \times \mathbb{R}^+, \mathbb{R})$ and $A \in L_{loc}^2(\mathbb{R}^N, \mathbb{R}^N)$.

A2: V, g and $B = \operatorname{curl} A$ are 1-periodic in x_j, $1 \leq j \leq N$.

A3: $g(x, 0) = 0$.

A4: There are constants $C > 0$ and $p \in (2, 2^*)$ ($p > 2$ if $N = 2$) such that $|g(x,|u|)| \leq C(1 + |u|^{p-2})$ for all x, u.

A5: There is a constant $\mu > 2$ such that $0 < \mu F(x,|u|) \leq g(x,|u|)|u|^2$ whenever $u \neq 0$.

A6: There are constants $\bar{C}, \varepsilon_0 > 0$ such that

$$|g(x, |u+v|)(u+v) - g(x,|u|)u| \leq \bar{C}|v|(1 + |u|^{p-1})$$

whenever $|v| \leq \varepsilon_0$, where p is as in (A4).

Note that in view of the definition of F, (A5) is the usual superlinearity condition. Since $B_{jk} = \partial_j A_k - \partial_k A_j$ in the sense of distributions, the periodicity of B should be interpreted as $B(\cdot) - B(\cdot + e_j)$ being the zero distribution for any element e_j of the standard basis in $\mathbb{R}^N$. It is also clear that according to (A3), (1.1) has the trivial solution $u = 0$.

THEOREM 2.3. *If $0 \notin \sigma(-\Delta_A + V)$ and conditions (A1)-(A5) are satisfied, then equation (1.1) has a nontrivial solution $u \in H_A^1(\mathbb{R}^N)$.*

A corresponding result is well-known for the Schrödinger equation with $A = 0$ (see e.g. [**7, 14**] and the references there).

Finally we shall exploit the S^1-invariance of J in order to show the existence of infinitely many solutions of (1.1).

THEOREM 2.4. *If $0 \notin \sigma(-\Delta_A + V)$ and conditions (A1)-(A6) are satisfied, then equation (1.1) has infinitely many geometrically distinct solutions.*

By geometrically distinct we mean such u, v that $v \neq e^{i\theta}u$ for any $\theta \in \mathbb{R}$ and $v \neq T_z u$ for any $z \in \mathbb{Z}^N$, where T_z is a certain operator corresponding to the translation by elements of $\mathbb{Z}^N$ in the nonmagnetic case. A more precise definition will be given in the next section.

The above result should be compared to the one contained in [**3, 7**], where A was equal to 0.

We would also like to mention that each of the following conditions is sufficient for $\sigma(-\Delta_A + V)$ to be contained in $(0, +\infty)$ (see [**2**]):

P1: $A \in L^N_{loc}(\mathbb{R}^N)$, $V \in L^1_{loc}(\mathbb{R}^N)$, where $N \geq 3$, and there exists a bounded set $\Omega \subset \mathbb{R}^N$ and a constant $c > 0$ such that $V(x) \geq c$ for all $x \notin \Omega$ and $\inf_{x\in\Omega} V(x) > -S\mu(\Omega)^{-2/N}$.

P2: $A \in W^{1,\infty}_{loc}(\mathbb{R}^N, \mathbb{R}^N)$, $V \in L^1_{loc}(\mathbb{R}^N)$ and there exists a constant $c > 0$ such that $\inf_{x\in\mathbb{R}^N} V(x) > -c$ and either $\partial_j A_k - \partial_k A_j \geq c$ a.e. in $\mathbb{R}^N$ or $\partial_j A_k - \partial_k A_j \leq -c$ a.e. in $\mathbb{R}^N$ for some $j, k \in \{1, \ldots, N\}$.

Here

$$S := \inf_{u\in\mathcal{D}^{1,2}(\mathbb{R}^N)\setminus\{0\}} \frac{\int_{\mathbb{R}^N} |\nabla u|^2 + V|u|^2}{||u||^2_{2^*}} \tag{2.2}$$

is the Sobolev constant for the embedding $\mathcal{D}^{1,2}(\mathbb{R}^N) \hookrightarrow L^{2^*}(\mathbb{R}^N)$ and $\mu(\Omega)$ is the measure of Ω.

3. Outline of proofs

An important role in the proofs is played by the following two results:

PROPOSITION 3.1. (Diamagnetic inequality [**10**, Theorem 7.21]) *If $u \in H^1_A(\mathbb{R}^N)$ (resp. $u \in \mathcal{D}^{1,2}_A(\mathbb{R}^N)$), then $|u| \in H^1(\mathbb{R}^N)$ (resp. $|u| \in \mathcal{D}^{1,2}(\mathbb{R}^N)$) and*

$$\left|\nabla\, |u|\, (x)\right| \leq |\nabla u(x) + iA(x)u(x)| \quad \textit{for a.e. } x \in \mathbb{R}^N.$$

Proof Since A is real-valued,

$$\left|\nabla\, |u|\, (x)\right| = \left|\mathrm{Re}\left(\nabla u \frac{\bar{u}}{|u|}\right)\right| = \left|\mathrm{Re}\left((\nabla u + iAu)\frac{\bar{u}}{|u|}\right)\right| \leq |\nabla u + iAu|. \tag{3.1}$$

See [**10**] for more details. □

PROPOSITION 3.2. *Let $A \in L^2_{loc}(\mathbb{R}^N, \mathbb{R}^N)$ and suppose $u_n \rightharpoonup u$ in $\mathcal{D}^{1,2}_A(\mathbb{R}^N)$. Then, up to a subsequence, $u_n \to u$ a.e. in $\mathbb{R}^N$ and $u_n \to u$ in $L^q_{loc}(\mathbb{R}^N)$ for any $q \in [2, 2^*)$. The same conclusion holds if $u_n \rightharpoonup u$ in $H^1_A(\mathbb{R}^N)$.*

Proof By the diamagnetic inequality the injection $\mathcal{D}_A^{1,2}(\mathbb{R}^N) \hookrightarrow L^{2^*}(\mathbb{R}^N)$ is continuous, hence $u_n \rightharpoonup u$ in $L^{2^*}(\mathbb{R}^N)$. Moreover, $|u_n - u|$ is bounded in $\mathcal{D}^{1,2}(\mathbb{R}^N)$. So passing to a subsequence, $u_n \to u$ a.e. and $|u_n - u| \rightharpoonup 0$ in $\mathcal{D}^{1,2}(\mathbb{R}^N)$. It follows from the Rellich-Kondrachov theorem that $u_n \to u$ in $L^q_{loc}(\mathbb{R}^N)$. The second part of the lemma is proved similarly. □

Below we shall prove Theorem 2.1 and briefly sketch the proofs of the other theorems.

Proof of Theorem 2.1 *Necessary condition.* We first show that $\bar{S} = S$. By the Sobolev and the diamagnetic inequalities,

$$S \le \frac{\int_{\mathbb{R}^N} |\nabla|u||^2 + V|u|^2}{\|u\|_{2^*}^2} \le \frac{\int_{\mathbb{R}^N} |\nabla_A u|^2 + V|u|^2}{\|u\|_{2^*}^2},$$

hence $S \le \bar{S}$. Let

$$U_\varepsilon(x) = (N(N-2))^{(N-2)/4} \left(\frac{\varepsilon}{\varepsilon^2 + |x|^2} \right)^{(N-2)/2} \tag{3.2}$$

and $u_\varepsilon(x) = \psi(x) U_\varepsilon(x)$, where $\psi \in C_0^\infty(\mathbb{R}^N, [0,1])$, $\psi = 1$ on $B(0,1/2)$ and $\psi = 0$ on $\mathbb{R}^N \setminus B(0,1)$. Then

$$\|\nabla(\psi U_\varepsilon)\|_2^2 \equiv \|\nabla u_\varepsilon\|_2^2 = S^{N/2} + O(\varepsilon^{N-2}) \quad \text{and} \quad \|u_\varepsilon\|_{2^*}^{2^*} = S^{N/2} + O(\varepsilon^N) \tag{3.3}$$

(see e.g. [**14**], p. 35). Since u_ε is bounded in $L^{2^*}(\mathbb{R}^N)$ and $u_\varepsilon \to 0$ a.e., $u_\varepsilon \rightharpoonup 0$ in $L^{2^*}(\mathbb{R}^N)$ as $\varepsilon \to 0$. Therefore $\int_{\mathbb{R}^N} V|u_\varepsilon|^2 \to 0$ and $\int_{\mathbb{R}^N} |Au_\varepsilon|^2 \to 0$ as $\varepsilon \to 0$ (recall $V \in L^{N/2}_{loc}(\mathbb{R}^N)$ and $A \in L^N_{loc}(\mathbb{R}^N, \mathbb{R}^N)$). Since u_ε is real-valued,

$$\frac{\int_{\mathbb{R}^N} |\nabla_A u_\varepsilon|^2 + V|u_\varepsilon|^2}{\|u_\varepsilon\|_{2^*}^2} = \frac{\int_{\mathbb{R}^N} |\nabla u_\varepsilon|^2 + |Au_\varepsilon|^2 + V|u_\varepsilon|^2}{\|u_\varepsilon\|_{2^*}^2} \to S$$

as $\varepsilon \to 0$ and $\bar{S} \le S$.

Now assume that u is a minimizer normalized by $\|u\|_{2^*} = 1$. Then

$$S = \int_{\mathbb{R}^N} |\nabla_A u|^2 + V|u|^2 \ge \int_{\mathbb{R}^N} |\nabla_A u|^2 \ge \int_{\mathbb{R}^N} |\nabla|u||^2 \ge S$$

and it follows that $|u(x)| = U_\varepsilon(x-a)/\|U_\varepsilon\|_{2^*}$ for some $a \in \mathbb{R}^N$ (that the minimizer in (2.2) is unique up to translation and dilation may be seen e.g. from the proof of Theorem 1.42 in [**14**]). In particular, $|u| > 0$ for all x and therefore $V \equiv 0$. Moreover, the inequality of Proposition 3.1 must be an equality a.e. So by (3.1), the imaginary part of $(\nabla u + iAu)\bar{u}$ must be zero which is equivalent to $A = -\mathrm{Im}\,(\nabla u/u)$. An easy computation shows that $\mathrm{curl}\,(\nabla u/u) = 0$.

Sufficient condition. Assume $V \equiv 0$ and $\mathrm{curl}\, A = 0$. Then $A = \nabla\varphi$ for some $\varphi \in W^{1,N}_{loc}(\mathbb{R}^N)$ according to [**9**] and it is easy to verify that $u = U_\varepsilon e^{-i\varphi}$ is a minimizer for (2.1) for any $\varepsilon > 0$. □

Proof of Theorem 2.2 Assume without loss of generality that $\bar{x} = 0$. Let $\theta(x) := -\sum_{j=1}^N A_j(0)x_j$. Then $(A + \nabla\theta)(0) = 0$ and by continuity $|(A + \nabla\theta)(x)|^2 \le c' < c$

for all $|x| < \delta$ provided δ is small enough. Choosing a smaller δ if necessary we may also assume $V(x) \le -c$ whenever $|x| < \delta$. Let U_ε be as in (3.2) and let $v_\varepsilon(x) := \psi(x)U_\varepsilon(x)e^{i\theta(x)}$, where $\psi \in C_0^\infty(\mathbb{R}^N, [0,1])$, $\psi(x) = 1$ in $B(0,\delta/2)$ and $\psi(x) = 0$ in $\mathbb{R}^N \setminus B(0,\delta)$. By (3.3),

$$\int_{\mathbb{R}^N} |\nabla_A v_\varepsilon|^2 + V|v_\varepsilon|^2 \le \int_{\mathbb{R}^N} |\nabla(\psi U_\varepsilon)|^2 + \psi^2 U_\varepsilon^2 |\nabla\theta + A|^2 - c\psi^2 U_\varepsilon^2$$
$$\le S^{\frac{N}{2}} + (c' - c)\int_{B(0,\delta/2)} U_\varepsilon^2 + O(\varepsilon^{N-2})$$

and $\|v_\varepsilon\|_{2^*}^2 = S^{(N-2)/2} + O(\varepsilon^N)$. Since

$$\int_{B(0,\delta/2)} U_\varepsilon^2 \ge \begin{cases} C\varepsilon^2|\log\varepsilon| & \text{if } N = 4 \\ C\varepsilon^2 & \text{if } N \ge 5 \end{cases}$$

for some $C > 0$ and all small $\varepsilon > 0$ (cf. e.g. [**14**], p. 35), an easy computation shows that

$$\bar{S} = \inf_{u \in \mathcal{D}_A^1(\mathbb{R}^N)\setminus\{0\}} \frac{\int_{\mathbb{R}^N} |\nabla_A u|^2 + V|u|^2}{\|u\|_{2^*}^2} < S.$$

Having this, a usual argument based on the concentration-compactness lemma [**14**, Lemma 1.40] shows that if $\{u_n\}$ is a minimizing sequence such that $\|u_n\|_{2^*} = 1$, then $u_n \to u$ in $L^{2^*}(\mathbb{R}^N)$ and in $\mathcal{D}_A^{1,2}(\mathbb{R}^N)$, possibly after passing to a subsequence. Hence u is a minimizer for (2.1). Moreover, since $\sigma(-\Delta_A + V) \subset (0, +\infty)$, $\int_{\mathbb{R}^N} |\nabla_A u|^2 + V|u|^2 \ge \varepsilon\|u\|_2^2$ for some $\varepsilon > 0$; therefore $u \in \mathcal{D}_A^{1,2}(\mathbb{R}^N) \cap L^2(\mathbb{R}^N)$, i.e. $u \in H_A^1(\mathbb{R}^N)$.

Finally we would like to point out that Lemma 1.40 in [**14**] must be adapted to the $\mathcal{D}_A^{1,2}(\mathbb{R}^N)$-setting. However, this is easily done by following the proof in [**14**] and employing Propositions 3.1 and 3.2 above. □

Proof of Theorem 2.3 Let $E := H_A^1(\mathbb{R}^N)$ and let J be as in (1.2). Then $J \in C^1(E, \mathbb{R}^N)$ and $J'(u) = 0$ if and only if u is a solution of (1.1). If $\sigma(-\Delta_A + V) \subset (0, +\infty)$, then the quadratic form $Q(u) := \int_{\mathbb{R}^N} |\nabla_A u|^2 + V|u|^2$ is positive definite on E, otherwise E can be decomposed into the direct sum of two subspaces, E^+ and E^-, invariant with respect to $-\Delta_A + V$ and such that Q is positive definite on E^+ and negative definite on E^- (cf. [**13**], Section 8). In the first case the functional J has the mountain pass geometry, and in the second one it has a geometry of linking type as described e.g. in [**7**]. Hence there exists a Palais-Smale sequence $\{u_n\}$ at some level $c > 0$ (cf. [**7**], Theorem 3.4 and [**14**], Theorems 2.9, 2.10, 6.10). Moreover, $\{u_n\}$ is bounded, so $u_n \rightharpoonup u$ after passing to a subsequence. By Lemma 1.7 in [**7**], either $u_n \to 0$ in E (up to a subsequence) which is impossible because $J(u_n) \to c > 0$, or there exists a sequence $\{z_n\}$ in $\mathbb{Z}^N$ and $r, \eta > 0$ such that $\int_{B(z_n,r)} |u_n(x)|^2 \ge \eta$.

We shall now use (A2) in order to construct a Palais-Smale sequence $\{v_n\}$ such that $v_n \rightharpoonup v \ne 0$. Since $J'(v_n) \rightharpoonup J'(v)$, v is a critical point of J. By (A2),

$$\operatorname{curl} A(x+z) - \operatorname{curl} A(x) = B(x+z) - B(x) = 0$$

for all $z \in \mathbb{Z}^N$. It follows from Lemma 1.1 in [**9**] that

$$A(x+z) - A(x) = \nabla \varphi_z(x) \tag{3.4}$$

for some $\varphi_z \in H^1_{loc}(\mathbb{R}^N)$. Let $(T_z u)(x) := u(x+z)e^{i\varphi_z(x)}$. An explicit computation using (3.4) shows that T_z is an isometry on E, $J(T_z u) = J(u)$ and $J'(T_z u) = T_z J'(u)$ for each $z \in \mathbb{Z}^N$. Denote $v_n := T_{z_n} u_n$. Then $J(v_n) \to c$ and $J'(v_n) \to 0$. Passing to a subsequence, $v_n \rightharpoonup v$ in E and, according to Proposition 3.2, $v_n \to v$ in $L^2_{loc}(\mathbb{R}^N)$. Hence v is a critival point, and $v \neq 0$ because

$$\int_{B(0,r)} |v_n(x)|^2 = \int_{B(z_n,r)} |u_n(x)|^2 \geq \eta.$$

□

Proof of Theorem 2.4 Let J and E, E^+, E^- be as in the preceding proof ($E^- = \{0\}$ if $\sigma(-\Delta_A + V) \subset (0,+\infty)$). Denote $S_\theta u = e^{i\theta}u$, $\theta \in S^1 = \mathbb{R}/2\pi\mathbb{Z}$. Then $J(S_\theta u) = J(u)$ and (as we already have seen) $J(T_z u) = J(u)$ for all $\theta \in S^1$, $z \in \mathbb{Z}^N$, $u \in E$. Note that $S_\theta T_z = T_z S_\theta$. Let

$$\mathcal{O}_{S^1 \times \mathbb{Z}^N}(u) := \{S_\theta T_z u : \theta \in S^1,\ z \in \mathbb{Z}^N\}.$$

Two solutions u, v of (1.1) are called *geometrically distinct* if they belong to different orbits, i.e. if $\mathcal{O}_{S^1\times\mathbb{Z}^N}(u) \neq \mathcal{O}_{S^1\times\mathbb{Z}^N}(v)$.

The proof of Theorem 2.4 is a straightforward adaptation of that in [**1**]. Suppose (1.1) has only finitely many geometrically distinct solutions. Denote the set of critical points of J by K_J, let $\mathcal{C}$ be a set consisting of arbitrarily chosen representatives of the orbits $\mathcal{O}_{S^1\times\mathbb{Z}^N}(u)$, $u \in K_J \setminus \{0\}$ and $\mathcal{K} := \mathcal{O}_{S^1}(\mathcal{C}) = \{S_\theta u : \theta \in S^1,\ u \in \mathcal{C}\}$. Then $\mathcal{K}$ is a compact set,

$$K_J \setminus \{0\} = \mathcal{O}_{\mathbb{Z}^N}(\mathcal{K}) \tag{3.5}$$

and if $\mathcal{F} := P_{E^+}(\mathcal{K})$, where P_{E^+} is the orthogonal projection on E^+, then

$$T_{z_1}\mathcal{F} \cap T_{z_2}\mathcal{F} = \emptyset \text{ whenever } z_1, z_2 \in \mathbb{Z}^N,\ z_1 \neq z_2. \tag{3.6}$$

These conditions correspond to (9) and (10) in [**1**]. Clearly, J is even and $\mathcal{K}$, $\mathcal{F}$ are symmetric (i.e. $\mathcal{K} = -\mathcal{K}$, $\mathcal{F} = -\mathcal{F}$). *From now on we consider J as an even functional and disregard the S^1-invariance.* Let

$$\mathcal{U}_\delta := E^- \oplus \bigcup_{z\in\mathbb{Z}^N} \{u^+ \in E^+ : d(u^+, T_z\mathcal{F}) < \delta\},$$

where $d(u, A)$ denotes the distance from u to the set A and let $\mathcal{H}$ be the class of mappings $f : E \to E$ such that f is a homeomorphism, $f(-u) = -f(u)$ for all u and $f(J^c) \subset J^c$ for all $c \geq -1$ ($J^c := \{u \in E : J(u) \leq c\}$). One can show that if J satisfies (3.5), (3.6) and $c \geq \inf_{K_J\setminus\{0\}} J$, then there exists a mapping $f \in \mathcal{H}$ such that $f(J^{c+\varepsilon} \setminus \mathcal{U}_\delta) \subset J^{c-\varepsilon}$ provided δ and ε are small enough. This is a variant of the deformation lemma which will be needed in the minimax argument below. Fix a small $\rho > 0$. For a closed and symmetric A, define

$$\gamma^*(A) = \min_{f\in\mathcal{H}} \gamma(f(A) \cap \partial B(0,\rho) \cap E^+),$$

where γ denotes the Krasnoselskii genus (γ^* is a variant of Benci's pseudoindex [4]). Let

$$d_k := \inf_{\gamma^*(A)\geq k} \sup_{u\in A} J(u).$$

If ρ is small enough, then $J|_{\partial B(0,\rho)\cap E^+} \geq d > 0$ and it follows that $d_k \geq d$. Moreover, it can be shown that there are sets of arbitrarily large pseudoindex, so d_k is defined for all $k \geq 1$. Since $d_0 := \sup_{\mathcal{U}_\delta} J < \infty$, it follows from the deformation lemma that $d_k \leq d_0$ for all k. Hence $d_k \nearrow \bar{d} \leq d_0$. Since $\gamma(S^1) = 2$, it is easy to see that $\gamma(\mathcal{F}) = 2$ and $\gamma(\bar{\mathcal{U}}_\delta) = 2$ provided δ is small enough. Using the deformation lemma once more we obtain

$$k \leq \gamma^*(J^{d_k+\varepsilon}) \leq \gamma^*(J^{d_k+\varepsilon} \setminus \mathcal{U}_\delta) + \gamma(\bar{\mathcal{U}}_\delta) \leq \gamma^*(J^{d_k-\varepsilon}) + 2.$$

Therefore $\gamma^*(J^{d_k-\varepsilon}) \geq k-2$, so $d_k - \varepsilon \geq d_{k-2}$ and $\bar{d} - \varepsilon \geq \bar{d}$, a contradiction. Hence there is no compact set $\mathcal{K}$ satisfying (3.5) and (3.6) and the number of geometrically distinct solutions of (1.1) must be infinite. □

References

[1] G. Arioli, A. Szulkin, *Homoclinic solutions of Hamiltonian systems with symmetry*, J. Diff. Eq. 158 (1999), 291-313.

[2] G. Arioli, A. Szulkin, *A semilinear Schrödinger equation in the presence of a magnetic field*, Preprint.

[3] T. Bartsch, Y. Ding, *On a nonlinear Schrödinger equation with periodic potential*, Math. Ann. 313 (1999), 15-37.

[4] V. Benci, *On critical point theory for indefinite functionals in the presence of symmetries*, Trans. Amer. Math. Soc. 274 (1982), 533-572.

[5] S. Cingolani, S. Secchi, *Semiclassical limit for nonlinear Schrödinger equation with electromagnetic fields*, Preprint.

[6] M.J. Esteban, P.L. Lions, *Stationary solutions of nonlinear Schrödinger equations with an external magnetic field.* In: Partial Differential Equations and the Calculus of Variations, Vol. 1, F. Colombini, A. Marino, L. Modica and S. Spagnolo eds., Birkhäuser (1989), pp. 401-449.

[7] W. Kryszewski, A. Szulkin, *Generalized linking theorem with an application to semilinear Schrödinger equation*, Adv. Diff. Eq. 3 (1998), 441-472.

[8] K. Kurata, *Existence and semi-classical limit of the least energy solution to a nonlinear Schrödinger equation with electromagnetic fields,* Nonlinear Analysis 41 (2000), 763-778.

[9] H. Leinfelder, *Gauge invariance of Schrödinger operators and related spectral properties*, J. Operator Theory 9 (1983), 163-179.

[10] E.H. Lieb, M. Loss, *Analysis*, Amer. Math. Soc., Providence, R.I., 1997.

[11] A.A. Pankov, *On nontrivial solutions of nonlinear Schrödinger equation with external magnetic field*, Preprint.

[12] I. Schindler, K. Tintarev, *A nonlinear Schrödinger equation with external magnetic field*, Rostock. Math. Kolloq. 56 (2002), 49-54.

[13] C. Stuart, *Bifurcation into spectral gaps*, Bull. Belg. Math. Soc., Supplement (1995).

[14] M. Willem, *Minimax Theorems*, Birkhäuser, Boston, 1996.

An Exact multiplicity theorem of a P-Laplacian Dirichlet problem and its application

Shin-Hwa Wang and Tzung-Shin Yeh

Department of Mathematics, National Tsing Hua University
Hsinchu, Taiwan 300

Abstract. We study the exact multiplicity of positive solutions of the p-Laplacian Dirichlet problem

$$-\left(\varphi_p\left(u'(x)\right)\right)' = \lambda f(u),\ -1 < x < 1,$$
$$u(-1) = u(1) = 0,$$

where $p > 1$, $\varphi_p(y) = |y|^{p-2}y$, $(\varphi_p(u'))'$ is the one-dimensional p-Laplacian, and $\lambda > 0$ is a bifurcation parameter. Assuming that $f \in C[0,\infty) \cap C^2(0,\infty)$ satisfies (F1)-(F4), we show that the bifurcation curve has exactly one critical point, a maximum, on the $(\|u\|_\infty, \lambda)$-plane. Thus we are able to determine the exact multiplicity of positive solutions. We give an interesting application for a nonlinear Dirichlet problem of polynomial nonlinearities with positive coefficients.

1. INTRODUCTION

We study the exact multiplicity of positive solutions of the p-Laplacian Dirichlet problem

$$-\left(\varphi_p\left(u'(x)\right)\right)' = \lambda f(u),\ -1 < x < 1, \qquad u(-1) = u(1) = 0, \tag{1.1}$$

where $f \in C[0,\infty) \cap C^2(0,\infty)$ satisfies $f(u) > 0$ for $u > 0$, $p > 1$, $\varphi_p(y) = |y|^{p-2}y$, $(\varphi_p(u'))'$ is the one-dimensional p-Laplacian, and $\lambda > 0$ is a bifurcation parameter. This problem arises in the study of non-Newtonian fluids and nonlinear diffusion

1991 *Mathematics Subject Classification.* 34B15, 74G35.

Key words and phrases. Exact multiplicity, solution curve, bifurcation, p-Laplacian, time map.

problems (see, e.g. Diaz [4] and its bibliography). The motivating example is the nonlinear Dirichlet problem of polynomial nonlinearities with positive coefficients:

$$\begin{gathered} -\left(\varphi_p\left(u'(x)\right)\right)' = \lambda\left[\sum_{i=1}^{m} a_i u^{q_i} + \sum_{j=1}^{n} b_j u^{r_j}\right], \ -1 < x < 1, \\ u(-1) = u(1) = 0, \end{gathered} \tag{1.2}$$

where $f(u) = \sum_{i=1}^{m} a_i u^{q_i} + \sum_{j=1}^{n} b_j u^{r_j}$ satisfies

$$\begin{gathered} 0 < q_1 < q_2 < \cdots < q_m < p-1 \le r_1 < r_2 < \cdots < r_n,\ m,n \ge 1,\ r_n > p-1, \\ a_i > 0 \text{ for } i = 1,2,...,m \text{ and } b_j > 0 \text{ for } j = 1,2,...,n. \end{gathered} \tag{1.3}$$

Problem (1.2) is discussed, and exact multiplicity and structure of positive solutions are obtained in section 2.

Exact multiplicity results are usually difficult to obtain for the p-Laplacian Dirichlet problem

$$\begin{gathered} -\Delta_p u = \lambda f(u) \text{ in } \Omega, \\ u = 0 \text{ on } \partial\Omega, \end{gathered} \tag{1.4}$$

where $\Delta_p u = div\left(|\nabla u|^{p-2}\nabla u\right)$ and Ω is a unit ball in R^n $(n \ge 1)$; see, e.g. [15, 2] for $p = 2$. Also see [6-13, 16-19, 22-24] for some many previous studies for (1.4) (or (1.1) when $n = 1$) involving mostly $p = 2$ (the Laplacian case) and only convex or concave nonlinearities, concave-convex nonlinearities, or cubic polynomials; i.e., they need to assume that $p = 2$ and $f''(u)$ changes sign *at most once* on $(0, \infty)$. In particular, we note that a systematic approach by applying a bifurcation theorem of Crandall and Ranibowitz [**3**] was developed in [6, 9-10, 18-19]. When $p \neq 2$, the problem becomes more difficult since certain nice feature inherent to the case $p = 2$ seems to be lost or at least difficult to verify. When $p \neq 2$, to the authors' knowledge, there exists no known exact multiplicity result for general nonlinearities f.

For (1.1) and when $p = 2$, in 1970, Laetsch [**14**] first gave an exact multiplicity result for general nonlinearities f under the assumptions that f satisfies $f(u) > 0$ on $[0, \infty)$, $f(u)$ is a convex function of u, and $f(u)/u$ is eventually strictly increasing. He proved that there exists $\lambda^* > 0$ such that (1.1) has exactly two positive solutions for $0 < \lambda < \lambda^*$, exactly one positive solution for $\lambda = \lambda^*$, and no positive solution for $\lambda > \lambda^*$. His method is based on a quadrature method (called the time map) and on the convexity assumption on the nonlinearity f.

In this paper, we study (1.1) and assume that $f \in C[0,\infty) \cap C^2(0,\infty)$ satisfies (F1)-(F4) stated below. Let $F(u) := \int_0^u f(t)dt$ and $\theta(u) := pF(u) - uf(u)$. The time map formula which we apply to study the p-Laplacian Dirichlet problem (1.1) takes the form as follows:

$$\lambda^{1/p} = \left(\frac{p-1}{p}\right)^{1/p} \int_0^{\alpha} \left[F(\alpha) - F(u)\right]^{-1/p} du := T(\alpha) \text{ for } 0 < \alpha < \infty, \tag{1.5}$$

cf. [**14**, Eq. (2.4)] for $p = 2$. Positive solutions u of (1.1) correspond to $\|u\|_\infty = \alpha$ and $T(\alpha) = \lambda^{1/p}$. Thus to study the exact number of positive solutions of (1.1) is equivalent to studying the shape of the time map $T(\alpha)$ on $(0, \infty)$.

Our main result for exact multiplicity of positive solutions for (1.1) is the next theorem which extends Wang and Long [**23**, Theorem 1.2] from $p = 2$ to any $p > 1$.

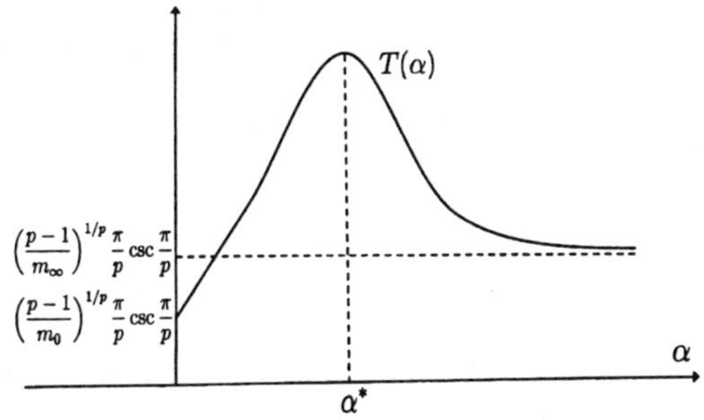

Fig. 1. Graph of $T(\alpha)$. $0 < m_0 \le \infty$, $0 < m_\infty \le \infty$

THEOREM 1.1. *(See Fig. 1) Suppose that* $f \in C[0, \infty) \cap C^2(0, \infty)$ *satisfies*
(F1) $f(0) \ge 0$ *and* $0 < \lim_{u \to 0^+} f(u)/u^{p-1} := m_0 \le \infty$,
(F2) $f(u) > 0$ *for* $u > 0$, *and* $0 < \lim_{u \to \infty} f(u)/u^{p-1} := m_\infty \le \infty$,
(F3) there exist positive numbers $A < B$ *such that*

$$\begin{aligned} &\theta'(u) = (p-1)f(u) - uf'(u) > 0 \text{ on } (0, A), \\ &\theta'(A) = (p-1)f(A) - Af'(A) = 0, \\ &\theta'(u) = (p-1)f(u) - uf'(u) < 0 \text{ on } (A, \infty), \end{aligned}$$

and

$$\begin{aligned} &\theta(u) = pF(u) - uf(u) > 0 \text{ on } (0, B), \\ &\theta(B) = pF(B) - Bf(B) = 0, \\ &\theta(u) = pF(u) - uf(u) < 0 \text{ on } (B, \infty), \end{aligned}$$

(F4) $\dfrac{uf'(u)}{f(u)} \ge \dfrac{-1}{p+1}$ *on* $(0, A)$ *and* $\dfrac{uf'(u)}{f(u)}$ *is increasing on* (A, B).

Then

$$\lim_{\alpha \to 0^+} T(\alpha) = \left(\frac{p-1}{m_0}\right)^{1/p} \frac{\pi}{p} \csc \frac{\pi}{p} \ge 0, \quad \lim_{\alpha \to \infty} T(\alpha) = \left(\frac{p-1}{m_\infty}\right)^{1/p} \frac{\pi}{p} \csc \frac{\pi}{p} \ge 0,$$

and $T(\alpha)$ *has exactly one critical point, a maximum, on* $(0, \infty)$. *Let* α^* *be the critical point for* $T(\alpha)$, *then* $A < \alpha^* < B$.

The exact multiplicity result of positive solutions for (1.1) follows immediately by Theorem 1.1 depending on the values of

$$\lambda_0 := \left(\lim_{\alpha \to 0^+} T(\alpha)\right)^p = \frac{p-1}{m_0} \left(\frac{\pi}{p} \csc \frac{\pi}{p}\right)^p \ge 0,$$

$$\lambda_\infty := \left(\lim_{\alpha \to \infty} T(\alpha)\right)^p = \frac{p-1}{m_\infty} \left(\frac{\pi}{p} \csc \frac{\pi}{p}\right)^p \ge 0.$$

Theorem 1.1 can be proved by modifying the proof of [**23**, Theorem 1.2]; we omit it.

REMARK 1.2. Theorem 1.1 allows that f'' (the convexity of f) changes sign more than once on $(0,\infty)$.

2. A NONLINEAR DIRICHLET PROBLEM OF POLYNOMIAL NONLINEARITIES WITH POSITIVE COEFFICIENTS

In this section we study the exact multiplicity and structure of positive solutions of (1.2) with $f=\sum_{i=1}^m a_i u^{q_i}+\sum_{j=1}^n b_j u^{r_j}$ satisfying (1.3). Recall the Beta function as follows (see e.g. [**20**, p. 18]):

$$B(x,y)=\int_0^1 t^{x-1}(1-t)^{y-1}dt,\ (x,y>0).$$

The next theorem extends Liu and Zhang [**17**, Theorem 1] and Wang and Yeh [**24**, Lemma 3.2] for (1.2) from $p=2$ to any $p>1$.

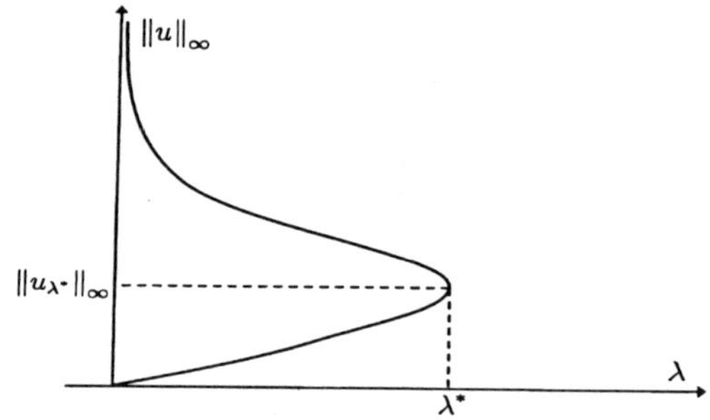

Fig. 2. Bifurcation digram of Problem (1.2).

THEOREM 2.1. *(See Fig. 2) Consider (1.2) where* $f=\sum_{i=1}^m a_i u^{q_i}+\sum_{j=1}^n b_j u^{r_j}$ *satisfies (1.3). Then*

(i) *There exists* $\lambda^*>0$ *such that (1.2) has exactly two positive solutions* u_λ, v_λ *with* $u_\lambda<v_\lambda$ *for* $0<\lambda<\lambda^*$, *exactly one positive solution* u_{λ^*} *for* $\lambda=\lambda^*$, *and no positive solution for* $\lambda>\lambda^*$.

(ii) *For* $0<\lambda<\lambda^*$, *let* u_λ, v_λ *be the two positive solutions of (1.2) with* $u_\lambda<v_\lambda$. *Then* $\|u_\lambda\|_\infty<\|u_{\lambda^*}\|_\infty<\|v_\lambda\|_\infty$, $\lim_{\lambda\to0^+}\|u_\lambda\|_\infty=0$, *and* $\lim_{\lambda\to0^+}\|v_\lambda\|_\infty=\infty$. *More precisely,*

$$\|u_\lambda\|_\infty\sim\left[\frac{pa_1(q_1+1)^{p-1}}{(p-1)\left(B\left(\frac{1}{q_1+1},\frac{p-1}{p}\right)\right)^p}\right]^{\frac{1}{p-1-q_1}}\lambda^{\frac{1}{p-1-q_1}}\ \text{as } \lambda\to0^+, \tag{2.1}$$

$$\|v_\lambda\|_\infty\sim\left[\frac{pb_n(r_n+1)^{p-1}}{(p-1)\left(B\left(\frac{1}{r_n+1},\frac{p-1}{p}\right)\right)^p}\right]^{\frac{1}{p-1-r_n}}\lambda^{\frac{1}{p-1-r_n}}\ \text{as } \lambda\to0^+. \tag{2.2}$$

(III) *The number* λ^* *satisfies*

$$\hat{\lambda} < \lambda^* < \check{\lambda}, \tag{2.3}$$

where

$$\hat{\lambda} := \max_{z>0} \frac{\left(\frac{p}{p-1}z\right)^{p-1}}{\sum_{i=1}^m a_i z^{q_i} + \sum_{j=1}^n b_j z^{r_j}} \geq \frac{\left(\frac{p}{p-1}\right)^{p-1}}{\sum_{i=1}^m a_i + \sum_{j=1}^n b_j}$$

and

$$\check{\lambda} := \max_{z>0} \frac{\left(\frac{p}{p-1}z\right)^{p-1}}{\sum_{i=1}^m \frac{a_i}{q_i+1} z^{q_i} + \sum_{j=1}^n \frac{b_j}{r_j+1} z^{r_j}}.$$

In the next Theorem 2.2, we apply Theorem 2.1 to study the exact multiplicity and structure of positive solutions of

$$\begin{gathered} (\varphi_p(u'(x)))' + \lambda \sum_{i=1}^m a_i u^{q_i} + \sum_{j=1}^n b_j u^{r_j} = 0, \ -1 < x < 1, \\ u(-1) = u(1) = 0, \end{gathered} \tag{2.4}$$

where $f_\lambda(u) := \lambda \sum_{i=1}^m a_i u^{q_i} + \sum_{j=1}^n b_j u^{r_j}$ satisfies (1.3) and (either $r_1 > p-1$ or $b_1 < (p-1)\left(\frac{\pi}{p}\csc\frac{\pi}{p}\right)^p$). Note that, if $r_1 = p-1$ and $b_1 \geq (p-1)\left(\frac{\pi}{p}\csc\frac{\pi}{p}\right)^p$, then it can be easily proved that (2.4) has no positive solution for any $\lambda > 0$.

The next Theorem 2.2 extends Wang and Yeh [**24**, Theorem 2.2] for p-Laplacian problem (2.4) from $p = 2$ to any $p > 1$, and it generalizes Addou *et al.* [**1**, Theorem 2.1(C)] and Sanchez and Ubilla [**21**, Theorem 1.1] in which they studied (2.4) with $m = 1 = n$ and $a_1 = 1 = b_1$. In Theorem 2.2(iv), the proof of the assertion that $\lambda^* \to 0$ as $b_1 \to \left((p-1)\left(\frac{\pi}{p}\csc\frac{\pi}{p}\right)^p\right)^-$ is similar to that of the assertion that $B^* \to 0$ as $b_1 \to \left((p-1)\left(\frac{\pi}{p}\csc\frac{\pi}{p}\right)^p\right)^-$; the proofs of these two assertions are easy but tedious and we omit them. The proofs of the rest of Theorem 2.2 are very similar to those of [**24**, Theorem 2.2 (i)(a),(c), (ii)-(iii), (v)]; we omit the proofs.

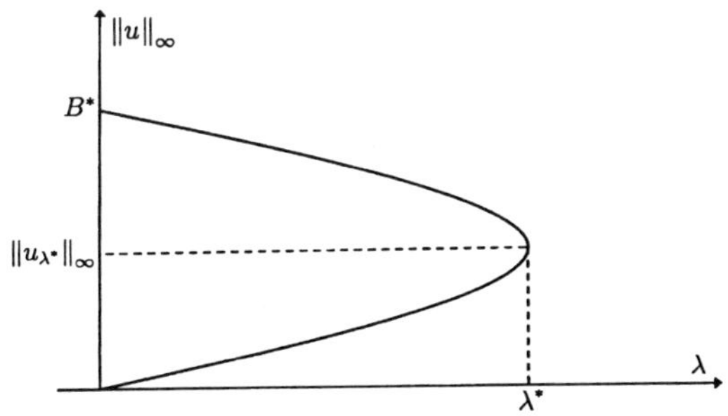

Fig. 3. Bifurcation digram of problem (2.4).

THEOREM 2.2. *(See Fig. 3) Consider (2.4) where $f_\lambda = \lambda \sum_{i=1}^m a_i u^{q_i} + \sum_{j=1}^n b_j u^{r_j}$ satisfies (1.3) and (either $r_1 > p-1$ or $b_1 < (p-1)\left(\frac{\pi}{p}\csc\frac{\pi}{p}\right)^p$). Then*

(i) *There exists $\lambda^* > 0$ such that (2.4) has exactly two positive solutions u_λ, v_λ with $u_\lambda < v_\lambda$ for $0 < \lambda < \lambda^*$, exactly one positive solution u_{λ^*} for $\lambda = \lambda^*$, and no positive solution for $\lambda > \lambda^*$.*

(ii) *Let u be a positive solution of (2.4). Then there exists a unique positive number B^* satisfying*

$$\left(\frac{p-1}{p}\right)^{1/p} \int_0^{B^*} \left[\int_u^{B^*} \sum_{j=1}^n b_j t^{r_j}\,dt\right]^{-1/p} du = 1$$

such that $\|u\|_\infty < B^*$. In particular, if $n = 1$,

$$B^* = \left[\left(\frac{p-1}{pb_1(r_1+1)^{p-1}}\right)^{1/p} B\left(\frac{p-1}{p}, \frac{1}{r_1+1}\right)\right]^{\frac{p}{r_1-p+1}}.$$

(iii) *For $0 < \lambda < \lambda^*$, let u_λ, v_λ be the two positive solutions of (2.4) with $u_\lambda < v_\lambda$. Then $\|u_\lambda\|_\infty < \|u_{\lambda^*}\|_\infty < \|v_\lambda\|_\infty$, $\lim_{\lambda\to 0^+}\|u_\lambda\|_\infty = 0$, and $\lim_{\lambda\to 0^+}\|v_\lambda\|_\infty = B^*$.*

(iv) *If $r_1 = p-1$, then for fixed a_i, b_j, q_i, r_j ($1 \le i \le m$ and $2 \le j \le n$),*

$$\lambda^* \to 0 \text{ and } B^* \to 0 \text{ as } b_1 \to \left((p-1)\left(\frac{\pi}{p}\csc\frac{\pi}{p}\right)^p\right)^-.$$

We now prove Theorem 2.1 as follows.

Proof of Theorem 2.1. (i) Suppose that

$$f = f_{m,n}(u) := \sum_{i=1}^m a_i u^{q_i} + \sum_{j=1}^n b_j u^{r_j},\ m,n \ge 1,$$

satisfies (1.3).

It is easy to see that, for $m,n \ge 1$, $f_{m,n} \in C[0,\infty)\cap C^2(0,\infty)$ satisfies (F1)-(F3) for some positive numbers $A < B$ and $m_0 = \infty = m_\infty$; we omit the proofs.

For (F4),

$$\frac{uf'_{m,n}(u)}{f_{m,n}(u)} > 0 > \frac{-1}{p+1} \text{ on } (0,A)$$

since $f'_{m,n}(u) = \sum_{i=1}^m a_i q_i u^{q_i-1} + \sum_{j=1}^n b_j r_j u^{r_j-1} > 0$ on $(0,\infty)$. So, to complete the proof of (F4), it suffices to prove that $uf'_{m,n}(u)/f_{m,n}(u)$ is increasing on (A,B). Actually, we prove that

$$\frac{uf'_{m,n}(u)}{f_{m,n}(u)} \text{ is increasing on } (0,\infty)$$

by the principle of double induction on positive integers m, n. This proof is exactly the same as that of [**24**, Lemma 3.2]. Hence $f_{m,n}$ satisfies (F4).

So, by Theorem 1.1, $\lim_{\alpha\to 0^+} T(\alpha) = 0 = \lim_{\alpha\to\infty} T(\alpha)$ and $T(\alpha)$ has exactly one critical point, a maximum, on $(0,\infty)$. Thus there exists $\lambda^* := (T(\alpha^*))^p = (\max_{\alpha\in(0,\infty)} T(\alpha))^p > 0$ for some $\alpha^* \in (A,B)$ such that (1.2) has exactly two positive solutions u_λ, v_λ with $u_\lambda < v_\lambda$ for $0 < \lambda < \lambda^*$ (the ordering of u_λ, v_λ can be proved easily), exactly one positive solution u_{λ^*} for $\lambda = \lambda^*$, and no positive solution for $\lambda > \lambda^*$.

(ii) It is easy to see that $\|u_\lambda\|_\infty < \|u_{\lambda^*}\|_\infty < \|v_\lambda\|_\infty$ for $0 < \lambda < \lambda^*$, $\lim_{\lambda\to 0^+} \|u_\lambda\|_\infty = 0$, and $\lim_{\lambda\to 0^+} \|v_\lambda\|_\infty = \infty$. Eqns. (2.1)-(2.2) can be proved by modifying the proofs of [**14**, Theorems 2.5-2.7 and 2.9-2.10]; we omit them.

(iii) For (2.3), we first prove the inequality $\hat{\lambda} < \lambda^*$. For any $z > 0$, by (1.5), we compute that

$$\begin{aligned}
& (\lambda^*)^{1/p} \\
= \ & T(\alpha^*) \\
\geq \ & T(z) \\
= \ & \left(\frac{p-1}{p}\right)^{1/p} \int_0^z [F(z) - F(u)]^{-1/p}\, du \\
= \ & \left(\frac{p-1}{p}\right)^{1/p} \int_0^z [f(d)(z-u)]^{-1/p}\, du \ (\text{for some } d \in (u,z)) \\
> \ & \left(\frac{p-1}{p}\right)^{1/p} \int_0^z [f(z)(z-u)]^{-1/p}\, du \ (f(u) \text{ is strictly increasing on } (0,z)) \\
= \ & \left[\frac{\left(\frac{p}{p-1}z\right)^{p-1}}{f(z)}\right]^{1/p} \\
= \ & \left[\frac{\left(\frac{p}{p-1}z\right)^{p-1}}{\sum_{i=1}^m a_i z^{q_i} + \sum_{j=1}^n b_j z^{r_j}}\right]^{1/p}.
\end{aligned}$$

It follows immediately that

$$\lambda^* > \hat{\lambda} = \max_{z>0} \frac{\left(\frac{p}{p-1}z\right)^{p-1}}{\sum_{i=1}^m a_i z^{q_i} + \sum_{j=1}^n b_j z^{r_j}} \geq \frac{\left(\frac{p}{p-1}\right)^{p-1}}{\sum_{i=1}^m a_i + \sum_{j=1}^n b_j}.$$

We next prove the inequality $\lambda^* < \check{\lambda}$. For $0 < u < \alpha$, we compute that

$$\begin{aligned}
F(\alpha) - F(u) &= \int_0^\alpha f(t)dt - \int_0^u f(t)dt \\
&= \alpha \int_0^1 f(\alpha s)ds - u \int_0^1 f(us)ds
\end{aligned}$$

$$\begin{aligned} &> \left(\int_0^1 f(\alpha\theta)d\theta\right)(\alpha - u) \\ &= \frac{F(\alpha)}{\alpha}(\alpha - u). \end{aligned}$$

So, by (1.5), we obtain

$$\begin{aligned} (\lambda^*)^{1/p} &= \left(\frac{p-1}{p}\right)^{1/p} \int_0^{\alpha^*} [F(\alpha^*) - F(u)]^{-1/p}\, du \\ &< \left(\frac{p-1}{p}\right)^{1/p} \left(\frac{\alpha^*}{F(\alpha^*)}\right)^{1/p} \int_0^{\alpha^*} (\alpha^* - u)^{-1/p} du \\ &= \left(\frac{p}{p-1}\right)^{\frac{p-1}{p}} \frac{\alpha^*}{F(\alpha^*)^{1/p}} \end{aligned}$$

after a simple computation. Then

$$\begin{aligned} \lambda^* &< \left(\frac{p}{p-1}\right)^{p-1} \frac{(\alpha^*)^p}{F(\alpha^*)} \\ &= \frac{\left(\frac{p}{p-1}\alpha^*\right)^{p-1}}{\sum_{i=1}^m \frac{a_i}{q_i+1} (\alpha^*)^{q_i} + \sum_{j=1}^n \frac{b_j}{r_j+1} (\alpha^*)^{r_j}} \\ &\le \max_{z>0} \frac{\left(\frac{p}{p-1}z\right)^{p-1}}{\sum_{i=1}^m \frac{a_i}{q_i+1} z^{q_i} + \sum_{j=1}^n \frac{b_j}{r_j+1} z^{r_j}} \ (= \check{\lambda}). \end{aligned}$$

The proof of Theorem 2.1 is complete.

Acknowledgment. The authors thank Professors Stuart Hastings and Zhaoli Liu for valuable comments and illuminating discussions.

References

[1] I. Addou, A. Benmezaï, S. M. Bouguima, and M. Derhab, Exactness results for generalized Ambrosetti-Brezis-Cerami problem and related one-dimensional elliptic equations, *Electron. J. Diff. Eqns.* **2000** (2000), 1-34.

[2] A. Ambrosetti, H. Brezis, and G. Cerami, Combined effects of concave and convex nonlinearities in some elliptic problems, *J. Funct. Anal.* **122** (1994), 519-543.

[3] M. G. Crandall and P. H. Ranibowitz, Bifurcation, perturbation of simple eigenvalues and linearized stability, *Arch. Rational Mech. Anal.* **52** (1973), 161-180.

[4] J. I. Diaz, Nonlinear partial differential equations and free boundaries. Vol. I. Elliptic equations. *Research Notes in Mathematics,* **106,** Pitman, Boston, MA, 1985.

[5] D. Jiang and W. Gao, Upper and lower solution method and a singular boundary value problem for the one-dimensional p-Laplacian, *J. Math. Anal. Appl.* **252** (2000), 631–648.

[6] P. Korman, Solutions curves for semilinear equations on a ball, *Proc. Amer. Math. Soc.* **125** (1997), 1997-2005.

[7] P. Korman, Uniqueness and exact multiplicity results for two classes of semilinear problems, *Nonlinear Analysis* **31** (1998), 849–865.

[8] P. Korman, Exact multiplicity of positive solutions for a class of semilinear equations on a ball, *E. J. Qualitative Theory of Diff. Equ.*, No. 8. (1999), pp. 1-15.

[9] P. Korman, Y. Li, and T. Ouyang, Exact multiplicity results for boundary value problems with nonlinearities genealising cubic, *Proc. Roy. Soc. Edinburgh Sect.* A **126** (1996), 599-616.

[10] P. Korman, Y. Li, and T. Ouyang, An exact multiplicity result for a class of semilinear equations, *Comm. Partial Differential Equations* **22** (1997), 661–684.

[11] P. Korman and T. Ouyang, Exact multiplicity results for a class of boundary-value problems with cubic nonlinearities, *J. Math. Anal. Appl.* **194** (1995), 328–341.

[12] P. Korman and T. Ouyang, Exact multiplicity results for two classes of boundary value problems, *Differential Integral Equations* **6** (1993), 1507–1517.

[13] P. Korman and J. Shi, Instability and exact multiplicity of solutions of semilinear equations. *Electron. J. Differ. Equ. Conf.* **5** (2000), 311–322.

[14] T. Laetsch, The number of solutions of a nonlinear two point boundary value problem, *Indiana Univ. Math. J.* **20** (1970), 1-13.

[15] P. L. Lions, On the existence of positive solutions of semilinear elliptic equations, *SIAM Review* **24** (1982), 441-467.

[16] Z. Liu, Exact number of solutions of a class of two-point boundary value problems involving concave and convex nonlinearities, *Nonlinear Analysis* **46** (2001), 181-197.

[17] Z. Liu and X. Zhang, A class of two-point boundary value problems, *J. Math. Anal. Appl.* **254** (2001), 599–617.

[18] T. Ouyang and J. Shi, Exact multiplicity of positive solutions for a class of semilinear problems, *J. Differential Equations* **146** (1998), 121–156.

[19] T. Ouyang and J. Shi, Exact multiplicity of positive solutions for a class of semilinear problems: II, *J. Differential Equations* **158** (1999), 94-151.

[20] E. D. Rainville, *Special Functions*, Macmillan, New York, 1960.

[21] J. Sanchez and P. Ubilla, One-dimensional elliptic equation with concave and convex nonlinearities, *Electron. J. Diff. Eqns.* **2000** (2000), 1-9.

[22] J. Wei, Exact multiplicity for some nonlinear elliptic equations in balls, *Proc. Amer. Math. Soc.* **125** (1997), 3235–3242.

[23] S.-H. Wang and D.-M. Long, An exact multiplicity theorem involving concave-convex nonlinearities and its application to stationary solutions of a singular diffusion problem, *Nonlinear Analysis* **44** (2001), 469-486.

[24] S.-H. Wang and T.-S. Yeh, On the exact structure of positive solutions of an Ambrosetti-Brezis-Cerami problem and its generalization in one space variable, submitted.

Symmetry and weighted Sobolev inequalities

Zhi-Qiang Wang

Department of Mathematics and Statistics, Utah State University
Logan, UT 84322, USA

ABSTRACT. We survey the recent results about the symmetry property of extremal functions for a family of weighted Sobolev inequalities due to Caffarelli, Kohn and Nirenberg.

1. Introduction

This paper is concerned with the study of weighted Hardy-Sobolev type inequalities and the study of qualitative property of solutions for related nonlinear partial differential equations. We shall focus on the following weighted Hardy-Sobolev type inequalities due to Caffarelli, Kohn and Nirenberg ([**8**], see also generalizations by Lin in [**24**]). Let $p > 1$ be fixed. There is $C_{a,b} > 0$ such that for all $u \in C_0^\infty(\mathbb{R}^N)$, it holds

$$\left(\int_{\mathbb{R}^N} |x|^{-bq}|u|^q \, dx\right)^{p/q} \leq C_{a,b} \int_{\mathbb{R}^N} |x|^{-pa}|\nabla u|^p \, dx \tag{1}$$

where for $N > p$:

$$-\infty < a < \frac{N-p}{p}, \quad 0 \leq b - a \leq 1, \quad \text{and } q = \frac{pN}{N-p+p(b-a)},$$

and for $N \leq p$:

$$-\infty < a < \frac{N-p}{p}, \quad \frac{p-N}{p} < b - a \leq 1, \quad \text{and } q = \frac{pN}{N-p+p(b-a)}.$$

Let $\mathcal{D}_a^{1,p}(\mathbb{R}^N)$ be the completion of $C_0^\infty(\mathbb{R}^N)$, with respect to the norm

$$||u||_a = \left(\int_{\mathbb{R}^N} |x|^{-pa}|\nabla u|^p dx\right)^{\frac{1}{p}}.$$

Then we see that (1) holds for $u \in \mathcal{D}_a^{1,p}(\mathbb{R}^N)$ and this implies that $\mathcal{D}_a^{1,p}(\mathbb{R}^N)$ is embedded continuously into weighted $L^q(\mathbb{R}^N)$ with weight $|x|^{-bq}$. We define

$$S_p(a,b) = \inf_{u \in \mathcal{D}_a^{1,p}(\mathbb{R}^N)\setminus\{0\}} E_{a,b}(u), \tag{2}$$

to be the best embedding constants, where

$$E_{a,b}(u) = \frac{\displaystyle\int_{\mathbb{R}^N} |x|^{-pa}|\nabla u|^p \, dx}{\left(\displaystyle\int_{\mathbb{R}^N} |x|^{-bq}|u|^q \, dx\right)^{p/q}}. \tag{3}$$

The Euler-Lagrange equation associated to the variational problem (2) and (3) is

$$-div(|x|^{-ap}|\nabla u|^{p-2}\nabla u) = |x|^{-bq}u^{q-1}, \quad u \geq 0, \quad \text{in } \mathbb{R}^N. \tag{4}$$

Extremal functions for $S_p(a,b)$ are ground state solutions of equation (4). This equation is a prototype for nonlinear elliptic PDEs from many branches of physics.

The classical Sobolev ($a = b = 0$) and Hardy ($a = 0, b = 1$) inequalities have played important roles in many applications by virtue of the complete knowledge about the best embedding constants and the extremal functions (e.g., Yamabe problem, prescribed scalar curvature problem, critical exponent problems due to Brezis-Nirenberg, spectral properties of Dirac operators, just mention a few here).

Fundamental questions about weighted Hardy-Sobolev type inequalities like (1) include the best embedding constants, existence and nonexistence of extremal functions, symmetry and geometric property of extremal functions. Great interest in this area has revived in recent years (e.g., **[1]** **[4]** **[5]** **[6]** **[9]** **[10]** **[11]** **[13]** **[14]** **[15]** **[20]** **[21]** **[23]** **[26]** **[27]** **[28]** **[29]** **[30]**, and references therein).

In this note, we shall survey some recent results concerning the above inequalities. In particular, we present the recent results concerning symmetry property of the extremal functions and the symmetry breaking phenomenon for extremal functions. Some open questions will be raised concerning these topics.

2. Symmetry breaking of extremal functions

2.1. The case $p = 2$. For the classical Sobolev inequality ($a = b = 0$) the best constants and the extremal functions were given by Aubin, and Talenti in **[1]**, **[28]**. In **[23]**, Lieb considered the case $a = 0$, $0 < b < 1$ and gave the best constants and explicit minimizers. In **[13]**, Chou and Chu considered the cases $a \geq 0$ and gave the best constants and explicit minimizers. The symmetry of the extremal functions has also been studied in **[23]** and **[13]**. In summary, for $a \geq 0$, *all* nonnegative solutions in $\mathcal{D}_a^{1,2}(\mathbb{R}^N)$ for the corresponding Euler equation (4) are *radial solutions* (in the case $a = b = 0$, they are radial with respect to some point) and explicitly given ([**1**] [**28**] [**23**] [**13**]). This was established in [**13**], using a generalization of the moving plane method (e.g.,[**16, 17**], [**7**])

Recently, Catrina and Wang have discovered the symmetry breaking phenomenon of the extremal functions for a sub-region of parameters when $a < 0$ occurs (partial result was given independently in [**31**]). More precisely, they proved in [**9**] [**10**] that there is a function $h(a)$ defined for $a \leq 0$, satisfying $h(0) = 0$, $a < h(a) < a+1$ for $a < 0$, and $a + 1 - h(a) \to 0$ as $-a \to \infty$, such that for any (a, b) satisfying $a < 0$ and $a < b < h(a)$, the extremal function for $S_2(a, b)$ is non-radial. In a recent preprint [**15**] of Felli and Schneider, it is observed that this curve $h(a)$ can be sharpened to the following

$$h(a) = 1 + a - \frac{N}{2}\left(1 - \frac{N-2-2a}{\sqrt{(N-2-2a)^2 + 4(N-1)}}\right). \tag{5}$$

In summary, we have the following.

THEOREM 2.1. ([**9**], [**10**], [**15**]) *(Symmetry breaking of extremal functions) Let $N \geq 2$. Then for any (a, b) satisfying $a < 0$ and $a < b < h(a)$, the extremal function for $S_2(a, b)$ is non-radial.*

We remark that for $N = 1$ the problem is completely solved in [**10**] and it was proved there that all extremal functions are non-radial.

We recall a reformulation of the inequalities (1) for the case $p = 2$ which was given in [**9, 10**] and the proof of the above result is based on this new formulation. We use notation $x = (\theta, t) \in \mathcal{C} = \mathbf{S}^{N-1} \times \mathbb{R}$. To u a smooth function with compact support in $\mathbb{R}^N \setminus \{0\}$, we associate v a smooth function with compact support on $\mathcal{C}$, by the transformation

$$u(y) = |y|^{-\frac{N-2-2a}{2}} v(\frac{y}{|y|}, -\ln|y|). \tag{6}$$

Here for $y \in \mathbb{R}^N \setminus \{0\}$, with $t = -\ln|y|$ and $\theta = \frac{y}{|y|}$ we have $(\theta, t) \in \mathcal{C}$. It was proved that the mapping (6) is a Hilbert space isomorphism from $\mathcal{D}_a^{1,2}(\mathbb{R}^N)$ to $H^1(\mathcal{C})$, where the inner product on $H^1(\mathcal{C})$ is

$$(v, w) = \int_{\mathcal{C}} \nabla v \cdot \nabla w + \left(\frac{N-2-2a}{2}\right)^2 vw \, d\mu.$$

This enables up to define an energy functional on $H^1(\mathcal{C})$

$$F_{a,b}(v) = \frac{\displaystyle\int_{\mathcal{C}} |\nabla_\theta v|^2 + v_t^2 + \left(\frac{N-2-2a}{2}\right)^2 v^2 \, d\mu}{\left(\displaystyle\int_{\mathcal{C}} v^q \, d\mu\right)^{2/q}}. \tag{7}$$

If $u \in \mathcal{D}_a^{1,2}(\mathbb{R}^N)$ and $v \in H^1(\mathcal{C})$ are related through (6) then

$$E_{a,b}(u) = F_{a,b}(v).$$

Moreover, if u is a solution of (4) then v satisfies

$$-v_{tt} - \Delta_\theta v + \left(\frac{N-2-2a}{2}\right)^2 v = v^{q-1},\ v > 0, \qquad \text{on } \mathcal{C} \tag{8}$$

where $t = -\ln|y|$, and Δ_θ is the Laplace operator on the $N-1$ sphere. In summary we have the following.

PROPOSITION 2.2. *([9, 10]) (i) The transformation* (6) *gives a Hilbert space isomorphism between $\mathcal{D}_a^{1,2}(\mathbb{R}^N)$ and $H^1(\mathcal{C})$.*

(ii) If $u \in \mathcal{D}_a^{1,2}(\mathbb{R}^N)$ and $v \in H^1(\mathcal{C})$ are related through (6) then $E_{a,b}(u) = F_{a,b}(v)$. Therefore, for $S_2(a,b)$ defined in (2), it holds $S_2(a,b) = \inf_{H^1(\mathcal{C})\setminus\{0\}} F_{a,b}(v)$.

(iii) Solutions of (4) and (8) are in one to one correspondence, being related through (6).

Based on this new formulation we may use some standard bifurcation techniques for the problem, and we sketch the proof of Theorem 2.1 below.

The idea is to consider bifurcation by using techniques with linearizations and to show that for certain (a,b), by perturbing the radial solution v_a of equation (8), there are directions in which the energy decreases. Since we are looking at the minimizer on a co-dimension one manifold, we need to find a two dimensional spaces on which the energy decreases. Using the fact that $S_2(a,b)$ is achieved for $a < b < a+1$ ([10]), the minimizer cannot be radial. We work in $H^1(\mathcal{C})$ here. We recall from [10]

$$v_a(t) = \left(\frac{\lambda^2 p}{2}\right)^{1/(p-2)} \left(\cosh\left(\frac{p-2}{2}\lambda t\right)\right)^{-2/(p-2)}. \tag{9}$$

where $\lambda = \lambda_a = (\frac{N-2-2a}{2})^2$. The linearization of (8) at the radial solution v_a decomposes by separation of variables into infinitely many ode's. Denote by $\alpha_k = k(N-2+k)$ the k-th eigenvalue of $-\Delta_\theta$ on $\mathbf{S}^{N-1}$. For $k=0$ or 1 denote by μ_k and f_k the first eigenvalue and the corresponding (positive) eigenfunction in the ODE

$$-f_{k,tt} + \lambda^2 f_k + \alpha_k f_k - (p-1)v_a^{p-2} f_k = \mu_k f_k.$$

This eigenvalue problem is well defined since $v_a(t) \to 0$ as $|t| \to \infty$. Clearly $\mu_0 < 0$. We claim for $a < b < h(a)$ with $a < 0$, $\mu_1 < 0$. It follows from the (9) that

$$-f_{1,tt} + \lambda^2 f_1 + (N-1)f_1 - (p-1)v_a^{p-2} f_1 = \mu_1 f_1,$$

and

$$-f_{1,tt} - \beta\cosh^{-2}(\gamma t) f_1 = (\mu_1 - \lambda^2 - (N-1))f_1,$$

where $\beta = \frac{N(N+2-2(b-a))(N-2-2a)^2}{4(N-2-2(b-a))^2}$ and $\gamma = \frac{(1+a-b)(N-2-2a)}{N-2-2(b-a)}$. As proved in [15], using [18][22] one concludes that the above eigenvalue problem has some discrete spectrum which are simple eigenvalues, and that μ_1 is negative if and only if $a < b < h(a)$ with $a < 0$. Thus, there are at least two negative eigenvulues for the linearized problem. From here one may follow the argument in [10] to show that the energy functional is negative definite in the two dimensional space spanned by the eigenfunctions corresponding to μ_0 and μ_1. This gives the proof for symmetry breaking.

2.2. The general case $p > 1$. For $p > 1$, the general case, we have established symmetry breaking results in [**6**]. The results are stated in the following two theorems.

THEOREM 2.3. *([**6**]) Let $N > p > 1$ be fixed. Then there is a function $h(a)$ defined for $a \leq 0$ satisfying $h(0) = 0$, $h(a) \in (a, a+1)$ for $a < 0$ and $h(a) - (a+1) \to 0$ as $a \to -\infty$, such that for any (a, b) with $a \leq b < h(a)$ the extremal function for $S_p(a, b)$, if it exists, is non-radial.*

THEOREM 2.4. *([**6**]) Let $p > 1$, $N \leq p$, and $b - a \in (\frac{p-N}{p}, 1)$ be fixed. Then for sufficiently large $-a$, the extremal function for $S_p(a, b)$ is nonradial.*

For $p \neq 2$, the approach in [**10**] does not seem to work. The proofs of the above theorems for the general cases make use of a different idea. The idea is to use the radial extremal functions in the radial class to construct suitable test functions to distinguish the least energy in the whole space from the least energy from the radially symmetric class. This approach is in contrast with that used in [**2**] in which instead of the extremal functions in the radial class they used a certain asymptotic property of the extremal functions in the whole functional space. This new method for showing symmetry breaking is rather robust and applies to many other situations where the radial extremal functions develop energy concentrations bounded away from 0. Therefore it should be useful in other symmetry breaking problems (see e.g., [**3**] and references therein).

We describe the idea in more detail and then we sketch the proof below. First we use suitable dilations to make the radial minimizers to develop a sharp layer near $|x| = 1$; so the energy of the dilated minimizers concentrate near $|x| = 1$. Then we take a portion of the radial minimizer as a test function, and the energy can be estimated proportional to the radial energy accordingly, which implies that the best constant $S_p(a, b)$ is less than the least energy for the class of radial functions (denoted by $S_p^R(a, b)$), leading to symmetry breaking. More precisely, we work in $\mathcal{D}_a^{1,p}(\mathbb{R}^N)$ and denote $b = a + c$ with $c \in (0, 1)$ fixed. Let the radial solution be given by $u = u_a$. It is known that the energy functional is invariant under the dilations

$$u \to u^\tau(x) = \tau^{\frac{N-p-pa}{p}} u(\tau x)$$

for $\tau > 0$. For each a we choose $\tau_a = (N - p - pa)^{-\frac{c_0}{N-p-pa}}$ where c_0 depends only on p and c. Then a direct computation as in [**6**] shows that if confined to any tubular neighborhood of the sphere $\{x \mid |x| = 1\}$ both the numerator and the denominator of the energy functional tend to zero as $-a \to \infty$. This means the energy concentrates near $\{x \mid |x| = 1\}$ as $-a \to \infty$. We choose a point x_0 with $|x_0| = 1$ and $\delta \in (0, \frac{1}{2})$ and make a cut-off of $u_a^{\tau_a}$ by defining $w_a = \eta(\frac{|x - x_0|}{\delta}) u_a^{\tau_a}$ where η is a usual cut-off function. Then the energy of w_a should be a fraction of the energy of $u_a^{\tau_a}$ as $-a \to \infty$, depending on δ. This is made precise in **[6]** so we proved there that

$$\limsup_{a \to -\infty} \frac{E_{a,b}(w_a)}{E_{a,b}(u_a^{\tau_a})} \leq C\delta^{(N-1)(1-\frac{p}{q})},$$

where C is a constant independent of δ. Since $E_{a,b}(u_a^{\tau_a}) = S_p^R(a, a+c)$ we get the desired result.

We remark that for $p \neq 2$ the existence of extremal functions for the critical case $b = a$ with $a < 0$ is still open, while for the case $p = 2$ a non-existence result of extremal functions has been obtained for $b = a < 0$ (**[10]** **[20]**). This may be related to the symmetry breaking phenomena.

3. Symmetry property of extremal functions

Following the results from last section, a natural question along the line of study is what are the symmetry of these non-radial extremal functions. First we have the following observation in **[10]** which applies to all bound state solutions of (4).

THEOREM 3.1. *(**[10]**) Let $p = 2$ and $N \geq 2$. For $a < b < a+1$, the minimizer of $S_2(a,b)$, possibly after a dilation $u(x) \to \tau^{\frac{N-2-2a}{2}} u(\tau x)$, satisfies the inversion symmetry:*

$$u\left(\frac{x}{|x|^2}\right) = |x|^{N-2-2a} u(x).$$

Moreover, writing $|x| = e^{-t}$ and $\theta = \frac{x}{|x|}$ we have for fixed θ, $e^{-\frac{N-2-2a}{2}t} u(e^{-t}\theta)$ is even in t and monotonically decreasing in t for $t > 0$.

The result is equivalent to that for any positive solution v of (8), up to a translation in t, v is even with respect to $t = 0$ and monotonic decreasing for $t > 0$. For equation (8) we can use the standard moving plane method to obtain the result.

In **[11]**, we have the following result for the extremal functions in the case $p = 2$.

THEOREM 3.2. *(**[11]**) Let $p = 2$ and $N \geq 2$. Let $c \in (0,1)$ be fixed. Then for sufficiently large $-a$, up to rotations and dilations, the extremal function to $S_2(a, a+c)$ is unique, and has $O(N-1)$ symmetry (i.e., the extremal function is axially symmetric with respect to a line passing through the origin).*

The idea to prove this result is to use asymptotic analysis. Taking advantage of the formulation on the cylinder, as $-a \to \infty$ we may identify a limiting equation of the problem which possesses a radial, non-degenerate solution. More precisely, with $b = a+c$ for $c \in (0,1)$ fixed, when we send $a \to -\infty$, under suitable scaling, equation (8) becomes the following well-known equation defined in $\mathbb{R}^N$: $-\Delta w + w = w^{q-1}$ which possesses a unique radial solution. Since this solution is also non-degenerate in the sense that when linearized the kernel is generated only by the translations invariance in $\mathbb{R}^N$, we may show that problems close to it also have uniqueness. The symmetry result follows from the uniqueness by using arguments like in **[19]**.

Similar results on uniqueness for other type of elliptic problems have been obtained in [**19**].

Very recently in [**26**], we are able to improve the symmetry part of the above result on the extremal functions and to give the exact symmetry for all non-radial extremal functions.

THEOREM 3.3. *([**26**]) Let $p = 2$ and $N \geq 2$. For all (a, b) satisfying $a < 0$ and $a < b < h(a)$, the extremal function u to $S_2(a, b)$ has exact $\mathcal{O}(N-1)$ symmetry. More precisely, up to a rotation, $u(x)$ only depends on the radius r and the angle θ_N between the positive x_N-axis and $\vec{Ox}$, and on each sphere $\{x \in \mathbb{R}^N \mid |x| = r\}$, u is strictly decreasing as the angle θ_N increases, i.e., $u = u(\theta_N, r)$ and $\frac{\partial u}{\partial \theta_N}(\theta_N, r) < 0$ for all $r > 0$ and all $\theta_N \in (0, \pi)$.*

For Theorem 3.3 we work on the alternative formulation on the cylinder. We follow the ideas developed in [**12**] [**25**] recently, which are along the line of research of using the moving plane method via the maximum principle for symmetry of positive solutions of nonlinear elliptic equations ([**16, 17**]). We refer to [**26**] for details.

Another natural question is what is the symmetry of the extremal functions for the parameters (a, b) with $h(a) \leq b < a + 1$ for $a < 0$ (i.e., for parameters above the curve $h(a)$). Partial results are given recently by Smets and Willem.

THEOREM 3.4. *([**29**]) Let $N \geq 3$. Let $b_0 \in (0, 1)$ fixed. Then for (a, b) sufficiently closed to $(0, b_0)$, any extremal function to $S_2(a, b)$ is radially symmetric with respect to the origin.*

The question on the symmetry of extremal functions for $h(a) \leq b < a + 1$, in general, remains open. Finally, for the general case $p \neq 2$, little is known on the symmetry of non-radial extremal functions.

References

[1] T. AUBIN, *Problèmes isopérimétriques de Sobolev,* J. Diff. Geom., **11**, (1976), 573-598.

[2] H. BREZIS, L. NIRENBERG, *Positive solutions of nonlinear elliptic equations involving critical exponents,* Comm. Pure Appl. Math., **36**(1983), 437-477.

[3] H. BREZIS, *Symmetry in nonlinear PDE's*, in Proceedings of Symp. Pure Math., **65** (1999), 1-12.

[4] H. BREZIS, M. MARCUS, *Hardy's inequalities revisited,* Ann. Scuola Norm. Sup. Pisa Cl. Sci., **25** (1997), 217–237 (1998).

[5] H. BREZIS, M. MARCUS, I. SHAFRIR, *Extremal functions for Hardy's inequality with weight,* J. Funct. Anal., **171** (2000), 177–191

[6] J. BYEON, Z.-Q. WANG, *Symmetry Breaking of Extremal Functions for the Caffarelli-Kohn-Nirenberg Inequalities,* Communications in Contemporary Mathematics, **4** (2002), 457-466.

[7] L. A. CAFFARELLI, B. GIDAS, J. SPRUCK, *Asymptotic symmetry and local behavior of semilinear elliptic equations with critical Sobolev growth,* Comm. Pure Appl. Math., **42**, (1989), 271-297.

[8] L. Caffarelli, R. Kohn, L. Nirenberg, *First order interpolation inequalities with weights,* Compositio Mathematica, **53**, (1984), 259 - 275.

[9] F Catrina, Z.-Q. Wang, *On the Caffarelli-Kohn-Nirenberg Inequalities,* Comptes Rendus des Séances de l'Académie des Sciences. Série I. Mathématique, **330** (2000), 437 - 442.

[10] F. Catrina, Z.-Q. Wang, *On the Caffarelli-Kohn-Nirenberg Inequalities: sharp constants, existence (and nonexistence) and symmetry of extremal functions,* Comm. Pure Appl. Math., **54** (2001), 229-258.

[11] F. Catrina, Z.-Q. Wang, *Asymptotic Uniqueness and Exact Symmetry of k-bump Solutions for a Class of Degenerate Elliptic Problems,* Disc. Cont. Dynam. Syst., Added Vol. (2001), 80-88.

[12] J. L. Chern, C. S. Lin, *The symmetry of least-energy solutions for semilinear elliptic equations,* preprint.

[13] K. S. Chou, C. W. Chu, *On the best constant for a weighted Sobolev-Hardy inequality,* J. London Math. Soc., **2**, (1993), 137-151.

[14] E. B. Davies, *A review of Hardy inequalities,* The Mazýa anniversary collection, Vol. 2 (Rostock, 1998), 55–67, Oper. Theory Adv. Appl., 110, Birkhauser, Basel, 1999.

[15] V. Felli, M. Schneider, *Perturbation results of critical elliptic equations of Caffarelli-Kohn-Nirenberg type,* preprint.

[16] B. Gidas, W.-M. Ni, L. Nirenberg, *Symmetry and related properties via the maximum principle,* Comm. Math. Phys., **68** (1979), 209-243.

[17] B. Gidas, W.-M. Ni, L. Nirenberg, *Symmetry of positive solutions of nonlinear elliptic equations in* $\mathbb{R}^N$*,* Adv. Math. Studies, **7A**, (1981), 369 - 402.

[18] F. González-López, N. Kamran, P. J. Olver, *Normalizability of one-dimensional quasi-exactly solvable Schrödinger operators,* Comm. Math. Phys., **153**(1993), 117-146.

[19] M. Grossi, *Uniqueness of the least-energy solution for a semilinear Neumann problem,* Proc. Amer. Math. Soc. **128**(2000), 1665–1672.

[20] T. Horiuchi, *Best constant in weighted Sobolev inequality with weights being powers of distance from the origin,* J. Inequal. and Appl., **1** (1997), 275 - 292.

[21] A. Kufner, A. Sändig, *Some applications of weighted Sobolev spaces.* Teubner-Texte zur Mathematik, 100. Teubner Verlagsgesellschaft, Leipzig, 1987.

[22] L. D. Laudau, E. M. Lifshitz, *Quantum mechanics: non-relativistic theory. Theoretical Physics,* Vol. 3. Pergamon Press Ltd., London-Paris (1958).

[23] E. H. Lieb, *Sharp constants in the Hardy-Littlewood-Sobolev and related inequalities,* Annals of Math., **118**, (1983), 349-374.

[24] C. S. Lin, *Interpolation inequalities with weights,* Comm. PDEs, **11**, (1986), 1515–1538.

[25] C. S. Lin, *Locating the peaks of solutions via the maximum principle: I. The Neumann problem,* Comm. Pure Appl. Math., **54** (2001), 1065-1095.

[26] C. S. Lin, Z.-Q. Wang, *Symmetry of extremal functions for the Caffarelli-Kohn-Nirenberg inequalities,* preprint.

[27] P. L. Lions, *The concentration-compactness principle in the calculus of variations, The limit case,* Revista matematica Ibero Americana, 1 (1985), 145-201, 45-120.

[28] G. Talenti, *Best constant in Sobolev inequality,* Ann. Mat. Pura Appl., **110**, (1976), 353-372.

[29] D. Smets, M. Willem, *Partial symmetry and asymptotic behavior for some elliptic variational problems,* Calc. Var. PDEs, to appear.

[30] Z.-Q. Wang, M. Willem, *Singular minimization problems,* Journal of Differential Equations, **161** (2000), 307 - 320.

[31] M. Willem, *A decomposition lemma and critical minimization problems,* preprint, (1999).

Sign changing solutions of superlinear Schrödinger equations

Tobias Weth

Mathematisches Institut, Universität Giessen Arndtstr. 2, 35392 Giessen Germany

1. Introduction

In this paper I report results on nodal properties of solutions to superlinear (time-independent) Schrödinger equations. These results were obtained in a joint work with T. Bartsch and Z.L. Liu, see **[3]**. We consider equations of the form

$$\begin{cases} -\Delta u + a(x)u &= f(x,u), \quad x \in \mathbb{R}^N, \\ u &\in H^1(\mathbb{R}^N). \end{cases} \tag{1.1}$$

In recent years many existence results have been obtained for (1.1) under various conditions on a and f. In **[5]** (cf. also **[11]**) it has been proved that there are infinitely many solutions of (1.1), e. g. if $a \in \mathcal{C}(\mathbb{R}^N, \mathbb{R}^+)$, $a(x) \to \infty$ as $|x| \to \infty$, and $f(x,u) = |u|^{p-2}u$ with $2 < p < 2^* = 2N/(N-2)$. In this situation the existence of a positive and a negative solution can be found in **[19]**. Moreover, in **[6]** the existence of a sign changing solution has been obtained.

The results from **[1]** for Dirichlet problems on bounded domains suggest that (1.1) should have an unbounded sequence of sign changing solutions. We shall prove such a result here. In addition we improve on the result from **[6]** about the existence of one sign changing solution. Unlike the proof in **[6]** we do not approximate (1.1) by the corresponding Dirichlet problem on the ball $B_R = \{x \in \mathbb{R}^N : |x| < R\}$ and let $R \to \infty$. Instead we develop an approach which allows us to work directly on $H^1(\mathbb{R}^N)$. The methods and the abstract critical point theory from **[1, 6]** or **[14, 15]** require to work on a Banach space $X \subset H^1(\Omega)$ whose cone of positive functions has nonempty interior. On a bounded domain $\Omega \subset \mathbb{R}^N$ one can work on $X = \{u \in \mathcal{C}^1(\overline{\Omega}) : u|_{\partial\Omega} = 0\}$ (see **[1, 2, 6, 12, 13, 14, 15]**). We do not have a substitute for this space if Ω is unbounded. Our new approach has the additional advantage that we do not need a and f to be continuous as in most papers on sign changing solutions (even on bounded domains).

In order to state our results we require the following assumptions:

(A_1) $a \in L^\infty_{loc}(\mathbb{R}^N)$, $\operatorname{ess\,inf} a > 0$.

(A_2) $f : \mathbb{R}^N \times \mathbb{R} \to \mathbb{R}$ is a Caratheodory function, and there are constants $C > 0$ and $p \in (2, 2^*)$ such that

$$|f(x,t)| \le C(1 + |t|^{p-1}) \quad \text{for } x \in \mathbb{R}^N,\ t \in \mathbb{R}.$$

Moreover, $f(x,t) = o(|t|)$ as $t \to 0$ uniformly in x.

(A_3) There is $\eta > 2$ such that $0 \le \eta F(x,t) \le f(x,t)t$ for $x \in \mathbb{R}^N$, $t \in \mathbb{R}$, where $F(x,t) := \int_0^t f(x,s)\,ds$.

(A_4) There is an open subset $\Omega \subset \mathbb{R}^N$ such that $tf(x,t) > 0$ for $x \in \Omega$ and $|t|$ sufficiently large.

(A_5) One of the following conditions is satisfied:

$(A_{5.1})$ For any $M > 0$ and any $r > 0$ there holds:

$$\operatorname{mes}(\{x \in B_r(y) : a(x) \le M\}) \to 0 \quad \text{as } |y| \to \infty,$$

$(A_{5.2})$ $\lim\limits_{|x|\to\infty} \sup\limits_{|t|\le\tau} \frac{|f(x,t)|}{|t|} = 0$ for every $\tau > 0$.

Here 2^* denotes the critical Sobolev exponent, i.e., $2^* = \frac{2N}{N-2}$ for $N \ge 3$ and $2^* = \infty$ for $N = 1, 2$. Moreover, $B_r(y)$ denotes the ball of radius r centered at $y \in \mathbb{R}^N$, and for a measurable set $G \subset \mathbb{R}^N$ we denote by $\operatorname{mes}(G)$ its Lebesgue measure. For some of our results we will also need

(A_6) For every $x \in \mathbb{R}^N$ the function $t \mapsto \frac{f(x,t)}{|t|}$ is nondecreasing on $\mathbb{R}$.

It is easily seen that $(A_2) - (A_4)$ and (A_6) hold for nonlinearities of the form $f(x,t) = \sum\limits_{i=1}^{k} a_i(x)|t|^{p_i-2}$, with $2 < p_i < 2^*$ and nonnegative functions $a_i \in L^\infty(\mathbb{R}^N)$, $i = 1, \ldots, k$, provided that at least one function a_i is positive when restricted to some open subset $\Omega \subset \mathbb{R}^N$. Assumption $(A_{5.2})$ is satisfied if $a_i(x) \to 0$ for $|x| \to \infty$, $i = 1, \ldots, k$, whereas $(A_{5.1})$ holds if

$$a(x) \to \infty \qquad \text{for } |x| \to \infty. \tag{1.2}$$

Note however that one can easily construct potentials a which satisfy $(A_{5.1})$ but not (1.2); see [**4**, p. 551]). Our main results are the following two theorems.

THEOREM 1.1. *Suppose $(A_1) - (A_5)$ hold. Then (1.1) has at least three nontrivial solutions $u_+, u_-, \bar{u}$, where u_+ is positive, u_- is negative and $\bar{u}$ changes sign. If in addition (A_6) holds then $\bar{u}$ has precisely two nodal domains.*

Theorem 1.1 improves on results from [**6, 19**]. That there exists a solution with precisely two nodal domains has so far only been proved for the corresponding Dirichlet problem on a bounded domain; cf. [**7, 9**].

THEOREM 1.2. *Suppose* $(A_1) - (A_5)$ *hold and* f *is odd in* t. *Then* (1.1) *has an unbounded sequence* $(\pm u_k)_{k\geq 1}$ *of sign changing solutions. If in addition* (A_6) *holds then* u_k *has at most* $k+1$ *nodal domains.*

Even on a bounded domain such a result has been proved only quite recently; cf. [**1, 7**]. The bound on the number of nodal domains follows from a nonlinear version of Courant's nodal domain theorem; see Proposition 3.4 below.

2. Invariant subsets of the descending flow and three nontrivial solutions

We first fix some notations. For $1 \leq s \leq \infty$ we denote by $|\cdot|_s$ the norm in $L^s(\mathbb{R}^N)$. Consider the Hilbert space $E := \{u \in H^1(\mathbb{R}^N) : \int_{\mathbb{R}^N} a(x)u^2 < \infty\}$ endowed with the scalar product $(u,v) := \int_{\mathbb{R}^N}(\nabla u \nabla v + a(x)uv)$, $u, v \in E$ and induced norm $\|\cdot\|$. Moreover, we write E^* for the topological dual of E, and $\langle\cdot,\cdot\rangle : E^* \times E \to \mathbb{R}$ for the dual pairing. Finally, for $u \in E$ and $R > 0$ we set $B_R(u) := \{v \in E : \|v-u\| < R\}$ and $S_R := \partial B_R(0)$.

By (A_1) we have a continuous embedding $E \hookrightarrow H^1(\mathbb{R}^N)$, hence

$$E \hookrightarrow L^s(\mathbb{R}^N) \quad \text{for } 2 \leq s \leq 2^*. \tag{2.1}$$

These embeddings are compact for $2 \leq s < 2^*$ in case that $(A_{5.1})$ is satisfied (see [**4**, p. 553]). Note also that by (A_2) for every $\varepsilon > 0$ there exists $K(\varepsilon) > 0$ such that

$$|f(x,t)| \leq \varepsilon|t| + K(\varepsilon)|t|^{p-1} \quad \text{for } x \in \mathbb{R}^N,\ t \in \mathbb{R}. \tag{2.2}$$

Standard arguments using this estimate ensure that the functional

$$\Phi(u) = \frac{1}{2}\|u\|^2 - \int_{\mathbb{R}^N} F(x, u(x))\,dx$$

is of class C^1 on E, and that critical points of Φ are distributional solutions of (1.1). The gradient of Φ has the form $\nabla\Phi : E \to E$, $\nabla\Phi = Id_E - A$ with $A : E \to E$ given by $A(u) := (-\Delta + a)^{-1}[f(\cdot, u(\cdot))]$ for $u \in E$. In other words, $A(u)$ is uniquely determined by the relation

$$(A(u), \varphi) = \int_{\mathbb{R}^N} f(x,u)\varphi \quad \text{for all } \varphi \in E. \tag{2.3}$$

Using (A_5) one can show that Φ satisfies the Palais-Smale condition. Moreover, a standard estimate using (A_3) and (A_4) yields

LEMMA 2.1. (i) *For every finite dimensional subspace* W *of* $C_0^\infty(\Omega) \subset E$ *there holds* $\sup \Phi(W) < \infty$.

(ii) *If* $\sup \Phi(C) < \infty$, *where* $C := \{tu : t \geq 0,\ u \in S\}$ *and* S *is a closed subset of the unit sphere of some finite dimensional subspace* W *of* E, *then there is a number* $R > 0$ *such that* $\Phi(u) \leq -1$ *for every* $u \in C \setminus B_R(0)$.

In the following we consider the convex cones $E^+ := \{u \in E : u \geq 0\}$ and $E^- := \{u \in E : u \leq 0\}$. Moreover, for $u \in E$ we denote by u^+ resp. u^- the positive resp. negative part of u, that is $u^+ = \max\{u, 0\}$, $u^- = \min\{u, 0\}$. Note that $u \in E$ implies $u^\pm \in E^\pm$. Consider the sets

$$D_\varepsilon^+ := \{u \in E : \mathrm{dist}(u, E^+) < \varepsilon\}, \qquad D_\varepsilon^- := \{u \in E : \mathrm{dist}(u, E^-) < \varepsilon\}$$

LEMMA 2.2. *There is an $\varepsilon_0 > 0$ such that for $0 < \varepsilon \leq \varepsilon_0$ there holds*

(i) *$A(\partial D_\varepsilon^-) \subset D_\varepsilon^-$, and every nontrivial solution $u \in D_\varepsilon^-$ of (1.1) is negative;*
(ii) *$A(\partial D_\varepsilon^+) \subset D_\varepsilon^+$, and every nontrivial solution $u \in D_\varepsilon^+$ of (1.1) is positive.*

PROOF. (i) Put $d := \frac{1}{2} \inf_{x \in \mathbb{R}^N} a(x) > 0$, and let $u \in E$, $v = A(u)$. Note that

$$|u^+|_2 = \min_{w \in E^-} |u - w|_2 \leq \frac{1}{\sqrt{2d}} \min_{w \in E^-} \|u - w\| = \frac{1}{\sqrt{2d}} \mathrm{dist}(u, E^-).$$

Similarly, using (2.1) we find for every $s \in [2, 2^*]$ a constant $C_s > 0$ with

$$|u^\pm|_s \leq C_s \,\mathrm{dist}(u, E^\mp) \tag{2.4}$$

Since moreover $\mathrm{dist}(v, E^-) \leq \|v^+\|$, (2.2) and (2.3) yield

$$\begin{aligned}
\mathrm{dist}(v, E^-)\|v^+\| &\leq \|v^+\|^2 \\
&= (v, v^+) \\
&= \int_{\mathbb{R}^N} f(x, u) v^+ \\
&\leq \int_{\mathbb{R}^N} f(x, u^+) v^+ \\
&\leq \int_{\mathbb{R}^N} (d|u^+| + K(d)|u^+|^{p-1}) v^+ \\
&\leq d|u^+|_2 |v^+|_2 + K(d)|u^+|_p^{p-1} |v^+|_p \\
&\leq \left(\frac{1}{2} \mathrm{dist}(u, E^-) + \tilde{K} \,\mathrm{dist}(u, E^-)^{p-1}\right) \|v^+\|,
\end{aligned}$$

with a constant $\tilde{K} > 0$. Hence

$$\mathrm{dist}(A(u), E^-) \leq \frac{1}{2} \mathrm{dist}(u, E^-) + \tilde{K} \,\mathrm{dist}(u, E^-)^{p-1}.$$

So, there exists $\varepsilon_0 > 0$ such that $\mathrm{dist}(A(u), E^-) \leq \frac{3}{4} \mathrm{dist}(u, E^-)$ for every $u \in D_\varepsilon^-$ with $0 < \varepsilon \leq \varepsilon_0$. In particular we have $A(\partial D_\varepsilon^-) \subset D_\varepsilon^-$. If moreover $u \in D_\varepsilon^-$ satisfies $A(u) = u$, then $u \in E^-$. If finally $u \neq 0$, we conclude $u(x) < 0$ for all x by the maximum principle. This completes the proof of (i). Assertion (ii) can be proved analogously. ∎

Denote $E_0 = \{u \in E | \Phi'(u) \neq 0\}$. We recall that a continuous map $\mathcal{V} : E \to E$ is said to be a pseudogradient vector field for Φ if $\mathcal{V}|_{E_0} : E_0 \to E$ is locally Lipschitz continuous and if the following two conditions are satisfied:

(i) $\langle \Phi'(u), \mathcal{V}(u) \rangle \geq \frac{1}{2} \|\Phi'(u)\|^2$ for all $u \in E$;
(ii) $\|\mathcal{V}(u)\| \leq 2\|\Phi'(u)\|$ for all $u \in E$.

If $\mathcal{V}$ is a pseudogradient vector field for Φ, we can integrate $-\mathcal{V}$ and obtain a flow $\varphi : \mathcal{G} \to E$ satisfying

$$\frac{d}{dt}\,\varphi(t,u) \;=\; -\mathcal{V}(\varphi(t,u)), \quad t \geq 0, \qquad\qquad \varphi(0,u) \;=\; u$$

for all $(t,u) \in \mathcal{G}$. Here $\mathcal{G} = \{(t,u) : u \in E,\ 0 \leq t < T(u)\}$, where $T(u) \in (0,\infty]$ is the maximal existence time for the trajectory $\varphi(t,u)$. We call φ the descending flow associated with $\mathcal{V}$. A subset $D \subset E$ is positive invariant for the flow φ if $\varphi(t,u) \in D$ for every $u \in D$, $t \in [0, T(u))$. If D is a positive invariant subset of E, we also consider

$$\mathcal{A}(D) := \{u \in E :\ \varphi(t,u) \in D \ \text{ for some } t \in (0, T(u))\},$$

and in addition we put $\mathcal{A}_0 := \{u \in E \mid \varphi(t,u) \to 0 \text{ as } t \to T(u)\}$. Note that $\mathcal{A}_0$ is open since 0 is a stable solution of (1.1). In view of Lemma 2.2 the next proposition just follows from [**16**, Lemma 3.2].

PROPOSITION 2.3. *Let $0 < \varepsilon \leq \varepsilon_0$. Then there is a pseudogradient vector field $\mathcal{V}$ for Φ such that D_ε^+ and D_ε^- are positive invariant for the associated descending flow φ. Moreover, $\mathcal{A}(D_\varepsilon^+) \supset \partial D_\varepsilon^+$ and $\mathcal{A}(D_\varepsilon^-) \supset \partial D_\varepsilon^-$.*

We also need the following.

LEMMA 2.4. *Let $0 < \varepsilon \leq \varepsilon_0$. Then $\overline{D_\varepsilon^+} \cap \overline{D_\varepsilon^-} \subset \mathcal{A}_0$. In particular there holds $\Phi(u) > 0$ for every $u \in \overline{D_\varepsilon^+} \cap \overline{D_\varepsilon^-} \setminus \{0\}$.*

We close this section with the

PROOF OF THEOREM 1.1. As a consequence of the preceding discussion the existence and the signs of the solutions follow essentially from [**16**, Theorem 3.2]. In order to find a solution with precisely two nodal domains we need to apply this theorem carefully using an idea from [**7**].

We introduce the set

$$M := \{u \in E :\ u^+ \neq 0,\ u^- \neq 0,\ \Phi'(u)u^+ = \Phi'(u)u^- = 0\}.$$

Obviously the set M contains all sign changing solutions of (1.1). Now set $\beta := \inf_{u \in M} \Phi(u)$,and let $(v_n)_n$ be a minimizing sequence in M for β. It is clear that v_n^+ and v_n^- are linearly independent for every n. As in [**7**, Lemma 2.1], we infer from (A_6) that

$$\Phi(v_n^\pm) = \max_{0 \leq t < +\infty} \Phi(tv_n^\pm). \tag{2.5}$$

Setting $C_n := \{tv_n^+ + sv_n^- : t \geq 0,\ s \geq 0\}$, $n \in \mathbb{N}$, (2.5) implies that

$$\sup \Phi(C_n) = \Phi(v_n^+) + \Phi(v_n^-) = \Phi(v_n). \tag{2.6}$$

By Lemma 2.1, there exists $R_n > 0$ such that $\Phi(u) \leq -1, \quad u \in C_n \setminus B_{R_n}(0)$. Now we define the path

$$h_n : [0,1] \to E, \quad h_n(t) = t\frac{2R_n}{\|v_n^+\|}v_n^+ + (1-t)\frac{2R_n}{\|v_n^-\|}v_n^-$$

connecting $h_n(0) \in C_n \cap D_\varepsilon^-$ and $h_n(1) \in C_n \cap D_\varepsilon^+$ in $C_n \setminus B_{R_n}(0) \subset \Phi^{-1}$. [**16**, Theorem 3.2] implies the existence of critical points $u_+ \in D_\varepsilon^+ \setminus D_\varepsilon^-$ and $u_- \in D_\varepsilon^- \setminus D_\varepsilon^+$. In fact, these critical points in principle might depend on n but we just need the existence of one such pair $u_\pm$. By Lemma 2.2, $u_+(x) > 0 > u_-(x)$ for all $x \in \mathbb{R}^N$.

It remains to prove the existence of $\bar{u}$. Consider the map

$$H_n : Q := [0,1] \times [0,1] \to E, \quad H_n(t,s) = sh_n(t)$$

and the sets

$$\mathcal{O}_n := H_n^{-1}(\mathcal{A}_0), \quad \mathcal{O}_n^+ := H_n^{-1}(\mathcal{A}(D_\varepsilon^+)), \quad \mathcal{O}_n^- := H_n^{-1}(\mathcal{A}(D_\varepsilon^-)).$$

Clearly $\mathcal{O}_n$ and $\mathcal{O}_n^\pm$ are open subsets of Q satisfying $\{(t,0) \mid 0 \le t \le 1\} \subset \mathcal{O}_n$ and $\{(t,1) \mid 0 \le t \le 1\} \cap \mathcal{O}_n = \emptyset$. Therefore[**16**, Lemma 3.1] implies the existence of a connected component Γ_n of $\partial_Q \mathcal{O}_n$ intersecting both $\{(0,s) : 0 \le s \le 1\} \subset \mathcal{O}_n^-$ and $\{(1,s) : 0 \le s \le 1\} \subset \mathcal{O}_n^+$. Note that $\Gamma_n \cap \mathcal{O}_n^+ \cap \mathcal{O}_n^- = \emptyset$, since $\mathcal{A}(D_\varepsilon^+) \cap \mathcal{A}(D_\varepsilon^-) \subset \mathcal{A}_0$ by Lemma 2.4. So, we can pick $(t_n, s_n) \in \Gamma_n \setminus (\mathcal{O}_n^- \cup \mathcal{O}_n^+)$. Then $w_n^* := H_n(t_n, s_n)$ is a suitable choice because $w_n^* \in \partial\mathcal{A}_0 \setminus (\mathcal{A}(D_\varepsilon^+) \cup \mathcal{A}(D_\varepsilon^-))$, hence $\Phi(\varphi(t, w_n^*))$ stays positive and bounded away from zero. It follows that the half orbit $\{\varphi(t, w_n^*) : 0 \le t < T(w_n^*)\}$ is relatively compact by the Palais-Smale condition, and that every u_n contained in the omega limit set

$$\omega(w_n^*) \subset \partial\mathcal{A}_0 \setminus (\mathcal{A}(D_\varepsilon^+) \cup \mathcal{A}(D_\varepsilon^-))$$

is a sign changing solution. In order to find a solution with precisely two nodal domains we now prove the existence of a minimizer $\bar{u}$ of Φ on M as a limit of these u_n. By (2.2.6) we have

$$\Phi(u_n) \le \Phi(w_n^*) \le \sup \Phi(C_n) = \Phi(v_n) \to \beta$$

Thus $(u_n)_n$ is a $(\mathrm{PS})_\beta$ sequence in M. Hence a subsequence converges to a critical point $\bar{u}$. Clearly we have

$$\bar{u} \in \partial\mathcal{A}_0 \setminus (\mathcal{A}(D_\varepsilon^+) \cup \mathcal{A}(D_\varepsilon^-)),$$

This implies that $\bar{u}$ is a sign changing solution, so $\bar{u} \in M$ is a minimizer of Φ on M. To finish the proof we only need to show that $\bar{u}$ has precisely two nodal domains. Suppose in contradiction that $\bar{u}$ has three nodal domains Ω_1, Ω_2, Ω_3 such that $\bar{u} > 0$ on Ω_1 and $\bar{u} < 0$ on Ω_2. Since $\bar{u}$ is continuous, it has been shown in [**17**, Lemma 1] that $\bar{u}\chi_{\Omega^*}$ defines an element of E provided that Ω^* is a nodal domain of $\bar{u}$. Then $\bar{u}\chi_{\Omega_1 \cup \Omega_2} \in M$ and $\beta \le \Phi(\bar{u}\chi_{\Omega_1 \cup \Omega_2})$. For any nodal domain Ω^* of $\bar{u}$, we have

$$\langle \Phi'(\bar{u}\chi_{\Omega^*}), \bar{u}\chi_{\Omega^*} \rangle = \langle \Phi'(\bar{u}), \bar{u}\chi_{\Omega^*} \rangle = 0,$$

which implies

$$\Phi(\bar{u}\chi_{\Omega^*}) \ge \left(\frac{1}{2} - \frac{1}{\eta}\right) \|\bar{u}\chi_{\Omega^*}\|^2 > 0$$

by virtue of (A_3). Therefore, we come to a contradiction as

$$\beta \le \Phi(\bar{u}\chi_{\Omega_1 \cup \Omega_2}) < \Phi(\bar{u}\chi_{\Omega_1 \cup \Omega_2}) + \Phi(\bar{u}\chi_{\Omega_3}) \le \Phi(\bar{u}) = \beta.$$

∎

3. Infinitely many sign-changing solutions

This section is devoted to the proof of Theorem 1.2. So, for the remainder of this section we will assume that f is odd in t, hence Φ is an even functional. By Proposition 2.3, one sees that for $0 < \varepsilon \leq \varepsilon_0$ the set

$$\mathcal{E} := \partial\mathcal{A}_0 \setminus (\mathcal{A}(D_\varepsilon^+) \cup \mathcal{A}(D_\varepsilon^-))$$

is positive invariant for the flow φ. For a real number c, we consider the closed and symmetric subsets

$$\mathcal{E}^c := \{u \in \mathcal{E} \mid \Phi(u) \leq c\}, \qquad K_c^* := \{u \in \mathcal{E} : J(u) = c,\ J'(u) = 0\}.$$

Clearly K_c^* is compact and contains only sign changing critical points. We need the following deformation lemma.

LEMMA 3.1. *Let $c \in \mathbb{R}$ and $\mathcal{N}$ be any odd neighborhood of K_c^* in $\mathcal{E}$. Then there exist $\delta > 0$ and an odd and continuous map $\sigma^* : (\mathcal{E}^{c+\delta} \setminus \mathcal{N}) \cup \mathcal{E}^{c-\delta} \to \mathcal{E}^{c-\delta}$ such that $\sigma^*|_{\mathcal{E}^{c-\delta}} = id$.*

The proof is a routine, since $\mathcal{E}$ is a positive invariant set. We now consider the nondecreasing sequence of values

$$c_k := \inf\{c \in \mathbb{R} \mid \gamma(\mathcal{E}^c) \geq k\}, \quad k \in \mathbb{N},$$

which will be shown to be critical values of Φ. Here γ denotes the Krasnoselski genus as defined e.g. in [**18**, p. 94]. To estimate c_k from above, we compare it with the min-max value

$$\beta_k := \inf_{W \subset E, \dim W \geq k} \sup \Phi(W).$$

Lemma 2.1(i) yields $\beta_k < \infty$ for all k. Moreover we have

LEMMA 3.2. $c_k \leq \beta_{k+1}$ *for every* $k \in \mathbb{N}$.

PROOF. Pick an arbitrary $(k+1)$-dimensional subspace W with $\beta := \sup \Phi(W) < \infty$. Moreover, choose $R > 0$ according to Lemma 2.1(ii), so that

$$\Phi(u) \leq -1 \qquad \text{for } u \in W \setminus B_R(0).$$

This implies that $U := W \cap \mathcal{A}_0$ is an open bounded and symmetric neighborhood of 0 in W. Hence $\gamma(\partial_W U) = k+1$. Noting moreover that $\partial_W U \subset W \cap \partial\mathcal{A}_0$ and that

$$\gamma\big((\mathcal{A}(D_\varepsilon^+) \cup \mathcal{A}(D_\varepsilon^-)) \cap \partial\mathcal{A}_0\big) \leq 1$$

by Lemma 2.4, we obtain $\gamma(W \cap \mathcal{E}) \geq k$ using the subadditivity of the genus. Hence $c_k \leq \beta$, as needed. ∎

Now the next proposition follows from Lemma 3.1 in a standard way.

PROPOSITION 3.3. *If $c_k = c_{k+1} = \cdots = c_{k+l} = c$ for some $k \in \mathbb{N}$ and $l \geq 0$, then $\gamma(K_c^*) \geq l+1$. Moreover, $c_k \to +\infty$ as $k \to \infty$.*

If (A_6) is satisfied, we can estimate the number of nodal domains of critical points of Φ by their energy value.

If (A_6) is satisfied, we can estimate the number of nodal domains of critical points of Φ by their energy value.

PROPOSITION 3.4. *Suppose that* (A_6) *is satisfied. Then every weak solution* $u \in E$ *of* (1.1) *with* $\Phi(u) \le \beta_k$ *has at most* k *nodal domains.*

On a bounded domain such a result has been proved in **[7]**, and the proof carries over to the case of an unbounded domain. Proposition 3.4 can be considered as a nonlinear version of Courant's nodal domain theorem (see **[10]**). Note that we do not require the critical points to be isolated, nor do we need to control the critical groups or the Morse indices. We believe that this approach to estimate the number of nodal domains is more useful and more elementary than the one using the Morse index (as in **[8]** or **[2]**).

We may now complete the proof of Theorem 1.2.

PROOF OF THEOREM 1.2. Note that Proposition 3.3 in particular yields a sequence $(\pm u_k)_{k \ge 1}$ of distinct solutions with $\Phi(u_k) \to \infty$ and

$$u_k \in K^*_{c_k} \subset E \setminus (D^+_\varepsilon \cup D^-_\varepsilon). \tag{3.1}$$

Hence $\|u_k\| \to \infty$, and u_k changes sign for every k, proving the first part of Theorem 1.2. The second statement follows immediately from a combination of Lemma 3.2, (3.1) and Proposition 3.4. ∎

References

[1] Bartsch, T., Critical point theory on partially ordered Hilbert spaces. *J. Funct. Anal.* **186** (2001), 117-152

[2] Bartsch, T., Chang, K.-C., and Wang, Z.-Q., On the Morse indices of sign changing solutions of nonlinear elliptic problems. *Math. Z.* **233** (2000), 655–677

[3] Bartsch, T., Liu, Z.L. and Weth, T., Sign changing solutions of superlinear Schrödinger equations. Preprint

[4] Bartsch, T., Pankov, A., and Wang, Z.-Q., Nonlinear Schrödinger equations with steep potential well. *Comm. Contemp. Math.* **4** (2001), 549-569

[5] Bartsch, T. and Wang, Z.-Q., Existence and multiplicity results for some superlinear elliptic problems on $\mathbb{R}^N$. *Comm. Partial Differential Equations* **20** (1995), 1725-1741

[6] Bartsch, T., and Wang, Z.-Q., Sign changing solutions of nonlinear Schrödinger equations. *Topol. Methods Nonlinear Anal.* **13** (1999), 191-198

[7] Bartsch, T., and Weth, T., A note on additional properties of sign changing solutions to superlinear elliptic equations. Preprint

[8] Benci, V., and Fortunato, D., A remark on the nodal regions of the solutions of some superlinear equations. *Proc. Roy. Soc. Edinburgh* **111A** (1989), 123-128

[9] Castro, A., Cossio, J., and Neuberger, J., A sign changing solution for a superlinear Dirichlet problem. *Rocky Mountain J. Math.* **27** (1997), 1041-1053

[10] Courant, R., and Hilbert, D., *Methoden der mathematischen Physik I*, Springer-Verlag Berlin 1968

[11] Costa, D. G., On a class of elliptic systems in $\mathbb{R}^n$. *Electron. J. Differ. Equ.* **1994/07** (1994),

[12] Dancer, E., and Du, Y., Existence of sign-changing solutions for some semilinear problems with jumping nonlinearities. *Proc. Roy. Soc. Edinburgh* **124A** (1994), 1165-1176

[13] Dancer, E., and Du, Y., On sign changing solutions of certain semi-linear elliptic problems. *Appl. Anal.* **56** (1995), 193-206

[14] Li, S.J., and Wang, Z.-Q., Mountain pass theorem in order intervals and multiple solutions for semilinear elliptic Dirichlet problems. *J. d'Analyse Math.* **81** (2000), [illegible]

[15] Li, S.J., and Wang, Z.-Q., Ljusternik-Schnirelmann theory on partially ordered Hilbert spaces. Preprint

[16] Liu, Z.L., and Sun, J.X., Invariant sets of descending flow in critical point theory with applications to nonlinear differential equations. *J. Differential Equations* **172** (2001) 257-299

[17] Müller-Pfeiffer, E., On the number of nodal domains for elliptic differential operators. *J. London Math. Soc. (2)* **31** (1985), 91-100

[18] Struwe, M., *Variational Methods*, Second Edition, Springer-Verlag Berlin 1996

[19] Rabinowitz, P., On a class of nonlinear Schrödinger equations. *Z. Angew. Math. Phys.* **43** (1992), 1019-1029

Nonplanar periodic solutions with a fixed energy for newtonian-like $N \geq 4$-body problems in $R^k (k \geq 3)$

Shiqing Zhang[1] and Qing Zhou[2]

[1] *Department of Mathematics, Yang Zhou University, Yangzhou 225002, P. R. China*

[2] *Department of Mathematics, East China Normal University, Shanghai, 200062, P. R. China*

ABSTRACT. Using variational methods and Moeckel's theorem on the saddle point property of planar central configurations in $R^k(k \geq 3)$, we prove the existence of a noncollision and nonplanar periodic solution for Newtonian-like $N \geq 4$ body problems in $R^k(k \geq 3)$.

1. Introduction

The N-body problems describe the motion of N particles attracting each other in accordance with Newton's law of gravitation. So we have

$$m_i \ddot{q}_i + \nabla_{q_i} V(q) = 0, \quad i = 1, \ldots, N, \tag{1.1}$$

where $q_i \in \mathbb{R}^k$ $(k \geqslant 3)$ is the position and $m_i > 0$ is the mass of the ith particle respectively, and the potential V is given by

$$V(q) = V(q_1, \ldots, q_N) = -\sum_{1 \leqslant i < j \leqslant N} \frac{m_i m_j}{|q_i - q_j|^\alpha}, \alpha > 0 \tag{1.2}$$

In [9], Gordon introduced the strong force condition ($\alpha > 2$) and then applied variational method to study the two body problems. Later he ([10]) also studied the two body problems with Newtonian potential ($\alpha = 1$), he proved that the Keplerian orbit realized the minimun of the Lagrangian action. In the last 20 years,

Authors are partially supported by grants of MOST, NSFC and QSSTF..

many researchers applied the variational methods to study periodic solutions for certain classes of N-body type problems.

In this paper, using the implicit symmetry ([18],[20],[22]) of the equation (1.1)–(1.2) and the constrained variational principle of Ambrosetti-Coti Zelati ([2]) and the saddle point property of planar central configurations in $R^k (k \geq 3)$ (Pacella-Moeckel ([13]–[14])) we will study the existence of noncollision and nonplanar periodic solutions with a fixed energy for the N-body problems with Newtonian-like potentials in $\mathbb{R}^k\,(k \geqslant 3)$.

For any given $m_i > 0 (i = 1, \ldots, N)$ and $h < 0$, we seek for noncollision and nonplanar periodic solutions for

$$\begin{cases} m_i\ddot{q}_i + \nabla_{q_i} V(q_1, \ldots, q_N) = 0 \\ \frac{1}{2}\sum_{i=1}^N m_i|\dot{q}_i(t)|^2 + V(q_1(t), \ldots, q_N(t)) = h \end{cases} \tag{1.3}$$

Definition 1.1 ([3]) $q(t) = (q_1(t), \ldots, q_N(t))$ is called a generalized solution of (1.3), if $q(t)$ is periodic with some period $T > 0$ and for all $i, j = 1, \ldots, N$ there holds

(i) $q_i \in W^{1,2}([0,T]; \mathbb{R}^k)$.
(ii) the collision set $\mathcal{S} = \{t \in [0,T]; x_i(t) = x_j(t)\}$ has measure 0.
(iii) q_i is C^2 on $[0,T]\backslash\mathcal{S}$ and satisfies (1.3) therein.

A generalized solution q with $\mathcal{S} = \varnothing$ is called a noncollision solution. The main result in this paper is the following theorem.

Theorem 1.1. *Suppose that* $0 < \alpha < 2$, $h < 0$ *and* $m_i > 0 (i = 1, \ldots, N)$,

let $M = \sum\limits_{i=1}^N m_i$ and $\mathcal{S}_N$ be the group of permutations on $\{1, \ldots, N\}$, let

$$A_1 = \min_{s\in\mathcal{S}_N}\left\{m_{s(1)}m_{s(2)}2^{\frac{2}{3}} + \sum_{3\leqslant i<j\leqslant N} m_{s(i)}m_{s(j)} + \sum_{1\leqslant i\leqslant 2, 3\leqslant j\leqslant N} m_{s(i)}m_{s(j)}\right\} \tag{1.4}$$

$$A_2 = \left(\frac{1}{2} + \frac{1}{\alpha}\right)\left(\frac{1}{2} + \frac{\alpha}{4}\right)^{2/\alpha} M^{-1}A_1^{\frac{2+\alpha}{\alpha}}, A_3 = A_4^{2/\alpha} \tag{1.5}$$

where A_4 is the minimum value of

$$\phi(x) = \left(\sum_{i<j}\frac{m_i m_j}{|x_i - x_j|}\right)\left(\sum_{i=1}^N m_i|x_i|^2\right)^{\frac{1}{2}}, x \in S \tag{1.6}$$

where
$$S = \left\{x = (x_1, \ldots, x_N) \in \mathbb{R}^{2N}, \sum_{i=1}^N m_i x_i = 0, x_i \neq x_j, i \neq j\right\} \tag{1.7}$$

If $A_2 > A_3$, *then the Lagrangian action functional (2.11) has a minimum* $u = (u_1, \cdots, u_N)$ *on* $\overline{\Lambda}_0$ *which corresponds to a noncollision periodic solution to the equation (1.3). Furthermore, if* $N \geq 4$ *then* u *is nonplanar.*

Remark *We conjecture that $A_2 > A_3$ holds for any $N \geq 4$ and $m_1, \cdots, m_N > 0$.*

2. Existence of Generalized Solutions to (1.3)

Let $O(k)(k \geq 3)$ be the rotational group in $\mathbb{R}^k$ and

$$A(\theta) = \begin{pmatrix} B(\theta) & 0 \\ 0 & -I_{k-2} \end{pmatrix} \in O(k) \tag{2.1}$$

where I_{k-2} is the unit matrix of order $k-2$ and

$$B(\theta) = \begin{pmatrix} \cos\theta & -\sin\theta \\ \sin\theta & \cos\theta \end{pmatrix}, \theta \in (0, 2\pi) \tag{2.2}$$

We introduce the following notations

$$H = \{u \in W^{1,2}(R/Z; \mathbb{R}^k); u(t+1) = u(t)\} \tag{2.3}$$

$$H_{\#} = \{u \in H; u(t + \frac{1}{r}) = A(\frac{2\pi}{r})u(t), r \geq 2 \ \ is \ \ even\} \tag{2.4}$$

Let

$$E = \{u = (u_1, \ldots, u_N); u_i \in H_{\#}, (i = 1, \ldots, N)\} \tag{2.5}$$

$$M_h = \{u \in E | \int_0^1 [V(u) + \frac{1}{2}V'(u)u]dt = h\} \tag{2.6}$$

$$\Lambda_0 = \{u \in M_h; u_i(t) \neq u_j(t), \forall t, i \neq j\}. \tag{2.7}$$

$$\Gamma = \{u \in \bar{\Lambda}_0; \exists t_0 \in [0,1] \text{ and } i_0 \neq j_0 \text{ with } u_{i_0}(t_0) = u_{j_0}(t_0)\}. \tag{2.8}$$

Furthermore, for $u, v \in H$, we let

$$(u, v) = \int \dot{u} \cdot \dot{v}, ||u||^2 = (u, u), \tag{2.9}$$

$$||u||_E^2 = \sum_i m_i ||u_i||^2, u \in E. \tag{2.10}$$

Define the following action functional on Λ_0

$$f(u) = \frac{1}{2}||u||_E^2 \int [h - V(u)]. \tag{2.11}$$

Ambrosetti-Coti Zelati [2] proved that non-constant critical points of f on Λ_0 give rise to periodic solutions of (1.3) after a rescaling of time.

$$T = [\frac{\int_0^1 \frac{1}{2}\sum_{i=1}^N m_i |\dot{u}_i|^2 dt}{\int_0^1 (h - V(u))dt}]^{\frac{1}{2}} \tag{2.12}$$

Lemma 2.1 For $\forall\ u = (u_1, \ldots, u_N) \in \Lambda_0$,

$$||u||_E = \left(\sum_{i=1}^N m_i ||u_i||^2\right)^{\frac{1}{2}} = \left(\sum_{i=1}^N m_i \int_0^1 |\dot{u}_i|^2\right)^{\frac{1}{2}} \tag{2.13}$$

is equivalent to the standard norm on E.

Proof. *We notice that* $\int_0^1 u_i(t)dt = 0$, *so Poincare's inequality implies the Lamma 2.1.*

Lemma 2.2 *f(u) is coercive on the closure* $\bar{\Lambda}_0$ *of* Λ_0,*that is* $f(u^n) \to +\infty$ as $||u^n||_E \to +\infty$.

Obviously,$f(u) > 0$ *and is weakly lower semi-continuous. Similar to [2], we have*

Lemma 2.3 If V is a Newtonian-like potential with $0 < \alpha < 2$, then $\forall\, h < 0$, the minimizer of $f(u)$ on $\overline{\Lambda}_0$ corresponds a generalized periodic solution $q(t) = u(t/T)$ with period $T > 0$.

3. Lower Bound of Lagrangian Action on Collision Solutions

Lemma 3.1 *Let* $q(t) = u(t/T)$ *be a generalized solution to (1.3). Then* $u(t)$ *and* $q(t)$ *satisfy the following equality*

$$[4f(u)]^{1/2} = \int_0^T \left[\frac{1}{2}\sum_{i=1}^N m_i|\dot{q}_i|^2 + \frac{1}{2}\langle \nabla V(q), q\rangle\right] \tag{3.1}$$

In the following, we will estimate the infimum of the functional values on the collision solutions. Let

$$\begin{aligned}
\tilde{H} &= W^{1,2}(\mathbb{R}/T\mathbb{Z}, \mathbb{R}^k),\\
\tilde{H}_\# &= \{x \in \tilde{H}; x\left(t + \frac{T}{r}\right) = A\left(\frac{2\pi}{r}\right)x(t)\},\\
\tilde{E} &= \{q = (q_1, \dots, q_N); q_i \in \tilde{H}_\#, i = 1, \dots, N\},\\
\tilde{\Lambda} &= \{q \in \tilde{E}; q_i(t) \neq q_j(t), \forall t \in \mathbb{R}, i \neq j\},\\
\partial\tilde{\Lambda} &= \{q = (q_1, \dots, q_N); q_i \in \tilde{H}, q_i \in \tilde{H}_\#,\\
&\quad \exists i_0 \neq j_0, \exists \bar{t} \in [0, T] \text{ such that } q_{i_0}(\bar{t}) = q_{j_0}(\bar{t})\}.
\end{aligned} \tag{3.2}$$

For $u \in \partial\Lambda_0$, *we have*

$$q(t) = u(t/T) \in \partial\tilde{\Lambda}. \tag{3.3}$$

Suppose that $q \in \partial\tilde{\Lambda}$, *then there exists* $\bar{t} \in [0, T]$ *and* $i_0 \neq j_0$, *such that* $q_{i_0}(\bar{t}) = q_{j_0}(\bar{t})$. *For simplicity, we can assume that* $\bar{t} = 0$ *and* $i_0 = 1, j_0 = 2$. *Using Lemma 3.1 and the symmetry property* $q_i\left(t + \frac{T}{r}\right) = A\left(\frac{2\pi}{r}\right)q_i(t)$, *we know that*

$$[4f(u)]^{1/2} = \int_0^T \left[\frac{1}{2}\sum_{i=1}^N m_i|\dot{q}_i|^2 + \frac{\alpha}{4}\sum_{1\leqslant i\neq j\leqslant N}\frac{m_i m_j}{|q_i - q_j|^\alpha}\right], \tag{3.4}$$

and denote $[4f(u)]^{1/2}$ *by* $\tilde{f}(q)$.

By the Lagrangian identity, we have

$$\tilde{f}(q) \geq g(q) + h(q) + \tilde{h}(q), \tag{3.5}$$

where

$$g(q) = \frac{r m_1 m_2}{M}\left[\frac{1}{2}\int_0^{T/r}|\dot{q}_1 - \dot{q}_2|^2 + \frac{M\alpha}{2}\int_0^{T/r}\frac{1}{|q_1 - q_2|^\alpha}\right], \tag{3.6}$$

$$h(q) = \frac{1}{M} \sum_{3 \leqslant i < j \leqslant N} m_i m_j \left[\frac{1}{2} \int_0^T |\dot{q}_i - \dot{q}_j|^2 + \frac{M\alpha}{2} \int_0^T \frac{1}{|q_i - q_j|^\alpha} \right], \tag{3.7}$$

and

$$\tilde{h}(q) = \frac{1}{M} \sum_{1 \leqslant i \leqslant 2, 3 \leqslant j \leqslant N} m_i m_j \left[\frac{1}{2} \int_0^T |\dot{q}_i - \dot{q}_j|^2 + \frac{M\alpha}{2} \int_0^T \frac{1}{|q_i - q_j|^\alpha} \right]. \tag{3.8}$$

Notice that the right sides of (3.6)–(3.8) are all sums of the Lagrangian actions for 2-body problems. Using the estimates of the Lagrangian action integral on collision solutions of 2-body problems [8,10], we can give an explicit estimate of $g(q)$.

Let $R_{ij}(t) = |q_i(t) - q_j(t)|$, *then* $R_{ij}(0) = R_{ij}\left(\frac{T}{r}\right) = 0$, *and* $R_{ij}(t) \geqslant 0$. *By Sundman's inequality we have*

$$\begin{aligned} g(q) &= \frac{r m_1 m_2}{M} \left\{ \int_0^{T/r} \frac{1}{2} \dot{R}_{12}^2 + \frac{M\alpha}{2} \int_0^{T/r} R_{12}^{-\alpha} \right\} \\ &\geq \frac{m_1 m_2}{M} \inf \left\{ \varphi(x), x \in W^{1,2}, x(0) = x\left(\frac{T}{r}\right) = 0, x \geqslant 0 \right\} \end{aligned} \tag{3.9}$$

where

$$\varphi(x) = r \int_0^{T/r} \left[\frac{1}{2} |\dot{x}|^2 + \frac{M\alpha}{2} x^{-\alpha} \right] dt \tag{3.10}$$

By a result of Degiovanni and Giannoni ([8]), we have

$$\begin{aligned} &\inf \left\{ \varphi(x); x(0) = x\left(\frac{T}{r}\right) = 0, x(t) \geqslant 0, x \in W^{1,2} \right\} \\ &\quad = T \min_{R \geqslant 0} \left\{ \frac{1}{2} r^2 \left(\frac{2\pi}{T}\right)^2 R^2 + \frac{M\alpha}{2} R^{-\alpha} \right\} \\ &\quad = \left(\frac{1}{2} + \frac{1}{\alpha}\right) \left(\frac{M\alpha^2}{2}\right)^{2/(2+\alpha)} (2\pi r)^{2\alpha/(2+\alpha)} T^{(2-\alpha)/(2+\alpha)}. \end{aligned} \tag{3.11}$$

Let

$$C = \left(\frac{1}{2} + \frac{1}{\alpha}\right) \left(\frac{\alpha^2}{2}\right)^{2/(2+\alpha)} \left(\frac{4\pi^2}{M}\right)^{\alpha/(2+\alpha)}. \tag{3.12}$$

Then

$$g(q) \geqslant C m_1 m_2 r^{2\alpha/(2+\alpha)} T^{(2-\alpha)/(2+\alpha)}. \tag{3.13}$$

For $h(q)$ *and* $\tilde{h}(q)$, *we have*

$$h(q) \geqslant C \sum_{3 \leqslant i < j \leqslant N} m_i m_j T^{(2-\alpha)/(2+\alpha)} \text{ and } \tilde{h}(q) \geqslant C \sum_{1 \leqslant i \leqslant 2, 3 \leqslant j \leqslant N} m_i m_j T^{(2-\alpha)/(2+\alpha)}. \tag{3.14}$$

So, we get

$$\tilde{f}(q) \geqslant C\left(m_1 m_2 r^{2\alpha/(2+\alpha)} + \sum_{3\leqslant i<j\leqslant N} m_i m_j + \sum_{1\leqslant i\leqslant 2, 3\leqslant j\leqslant N} m_i m_j\right) T^{(2-\alpha)/(2+\alpha)}. \tag{3.15}$$

This shows that

$$\inf_{u\in\partial\Lambda_0} [4f(u)]^{\frac{1}{2}} = \inf_{q\in\partial\tilde{\Lambda}} \tilde{f}(q) \geqslant C_1 T^{(2-\alpha)/(2+\alpha)}, \tag{3.16}$$

where

$$C_1 = C \min_{s\in S_N}\left\{m_{s(1)} m_{s(2)} r^{2\alpha/(2+\alpha)} + \sum_{3\leqslant i<j\leqslant N} m_{s(i)} m_{s(j)} + \sum_{1\leqslant i\leqslant 2, 3\leqslant j\leqslant N} m_{s(i)} m_{s(j)}\right\}, \tag{3.17}$$

Hence we can obtain the lower bound

$$f(u) \geqslant \frac{1}{4}\left(\frac{2-\alpha}{-2\alpha h}\right)^{(2-\alpha)/\alpha} C_1^{(2+\alpha)/\alpha} \equiv C_2(-h)^{-(2-\alpha)/\alpha}. \tag{3.18}$$

4 Upper Bound of the Minimum for Lagrangian Action

A critical point of the 0*-homogeneous function* $\phi(x)$ *of (1.6) on the configuration space*

$$S = \left\{x = (x_1, \dots, x_N) \in R^{kN}, \sum_{i=1}^{N} m_i x_i = 0, x_i \neq x_j, i \neq j\right\} \tag{4.1}$$

is called a central configuration, and the central configuration which minimizes the homogeneous function (1.6) is called a minimal central configuration.

Similar to the proof of Long–Zhang[12] we have

Lemma 4.1 *If* $k = 2$*, then a critical point* $u(t) = (u_1(t), \dots, u_N(t))$ *is a minimizer of the Lagrangian action on* $\overline{\Lambda}_0$ *if and only if*

$$u_i(t) = a_i \cos(2\pi t) + b_i \sin(2\pi t), a_i, b_i \in R^2 \tag{4.2}$$

$$\sum_{i=1}^{N} m_i u_i(t) = 0 \tag{4.3}$$

$$\sum_{i=1}^{N} m_i |u_i(t)|^2 = \left(\frac{A_4}{-2h}\right)^2 \tag{4.4}$$

$$\sum_{i<j} \frac{m_i m_j}{|u_i(t) - u_j(t)|} = -2h \tag{4.5}$$

and the central configuration is the minimizing central configuration, and

$$\inf\{f(u) | u \in \overline{\Lambda}_0, k = 2\} = 2\pi^2 \alpha(2-\alpha)^{\frac{2-\alpha}{\alpha}} 2^{-2/\alpha} \cdot A_4^{2/\alpha}(-h)^{\frac{\alpha-2}{\alpha}} \equiv C_3(-h)^{\frac{\alpha-2}{\alpha}} \tag{4.6}$$

We notice that $u_i(t)$ *is a circular orbit when* $\langle a_i, b_i\rangle = 0$ *and* $|a_i| = |b_i|$ *.*

Now we can choose the special loops obtained in Lemma 4.1 to estimate the upper bound for the global minimum of $f(u)$ on $\overline{\Lambda}_0$, hence we have

$$\inf\{f(u)|u \in \overline{\Lambda}_0\} \geq C_3(-h)^{\frac{\alpha-2}{\alpha}} \tag{4.7}$$

We notice that if $A_2 > A_3$, then $C_3 < C_2$. and the minimizer of $f(u)$ on $\overline{\Lambda}_0$ is a noncollison critical point.

5. Second Variation and the Existence of Nonplanar Solutions

For $s \in R$ and $\xi = (\xi_1, \cdots, \xi_N) \in (R^k)^N$, we have

$$\frac{d}{ds^2} f(u+s\xi)|_{s=0} = h\int_0^1 \sum_{i=1}^N m_i|\dot{\xi}_i|^2 dt + (\int_0^1 \sum_{i=1}^N m_i|\dot{\xi}_i|^2 dt)(\int_0^1 \sum_{i<j} \frac{m_i m_j}{|u_i(t)-u_j(t)|} dt)$$

$$+2(\int_0^1 \sum_{i=1}^N m_i \dot{u}_i \dot{\xi}_i dt)(\int_0^1 \sum_{i<j} \frac{-m_i m_j < \xi_i - \xi_j, u_i - u_j >}{|u_i - u_j|^3} dt)$$

$$+\frac{1}{2}\int_0^1 \sum_{i=1}^N m_i|\dot{u}_i|^2 dt \Bigg[-\sum_{i<j} m_i m_j \int_0^1 \frac{|\xi_i - \xi_j|^2}{|u_i - u_j|^3} dt$$

$$+3(\sum_{i<j}\int_0^1 \frac{m_i m_j| < \xi_i - \xi_j, u_i - u_j > |^2}{|u_i - u_j|^5} dt)\Bigg] \tag{5.1}$$

In order to compute the second order variation for $f(u)$ constrained on manifold M_h, we introduce the Lagrangian multiplier function:

$$L(u, \lambda) = f(u) + \lambda g(u) \tag{5.2}$$

Then for critical point u of $f(u)$ on M_h, we have

$$L_u(u, \lambda) = f'(u) + \lambda g'(u) = 0 \tag{5.3}$$

$$L_\lambda(u, \lambda) = g(u) = \int_0^1 \sum_{i<j} \frac{m_i m_j}{|u_i - u_j|} dt + 2h = 0 \tag{5.4}$$

Then

$$\lambda = 0, < L_{uu}(u, \lambda)\xi, \xi > = < f''(u)\xi, \xi > \tag{5.5}$$

In order to prove Theorem 1.1, we need a result of Moeckel ([13]):

Lemma 5.1 *Let $\tilde{x} = (\tilde{x}_1, \cdots, \tilde{x}_N)$ be a planar central configuration of $N \geq 4$ body problems in $R^k(k \geq 3)$, then there exists a normal direction $\eta_i (i = 1, 2, \cdots, N)$ of the central configuration such that*

$$\langle \varphi''(\tilde{x})\eta, \eta \rangle < 0, \eta = (\eta_1, \cdots, \eta_N) \tag{5.6}$$

In the following, we compute

$$\langle \varphi''(\tilde{x})\eta, \eta \rangle = \varphi_1''(0), \tag{5.7}$$

where

$$\varphi_1(s) = \varphi(\tilde{x} + s\eta) \tag{5.8}$$

Then we have

$$\langle \varphi''(\tilde{x})\eta, \eta\rangle = \varphi_1''(0) = \left(\sum_{i=1}^N m_i|\tilde{x}_i|^2\right)^{-1/2} \left(\sum_{i=1}^N m_i|\eta_i|^2\right) U(\tilde{x})$$

$$+ \left(\sum_{i=1}^N m_i|\tilde{x}_i|^2\right)^{1/2} \left(- \sum_{1\leq i<j\leq N} \frac{m_i m_j |\eta_i - \eta_j|^2}{|\tilde{x}_i - \tilde{x}_j|^3}\right) \tag{5.9}$$

If the minimizer u of $f(u)$ on $\overline{\Lambda}_0$ is planar, then by Lemma 4.1 we can choose

$$u_i(t) = \tilde{x}_i \cdot e^{\sqrt{-1}2\pi t} \tag{5.10}$$

On the other hand, if $r = 2$ then we let $\xi = \eta \cdot \sin 2\pi t$ and by (5.1) and (5.8) and (4.5) we have

$$\langle f''(u)\xi, \xi\rangle = \pi^2 \left(\sum_{i=1}^N m_i|\tilde{x}_i|^2\right)^{1/2} \cdot \langle \varphi''(\tilde{x})\eta, \eta\rangle < 0 \tag{5.11}$$

This contradicts the fact that u is a minimizer, so u is nonplanar.

Remark *If we let*

$$A = \begin{pmatrix} 1 & 0 & 0 \\ 0 & -1 & 0 \\ 0 & 0 & -1 \end{pmatrix}, B = \begin{pmatrix} -1 & 0 & 0 \\ 0 & 1 & 0 \\ 0 & 0 & -1 \end{pmatrix}$$

$$H_i = \{u_i \in W^{1,2}(\mathbb{R}/\mathbb{Z}, \mathbb{R}^3) | u_i(t + 1/2) = Au_i(t), u_i(-t) = Bu_i(t)\}$$
$$E = \{u = (u_1, \ldots, u_N) \,|u_i \in H_i\}$$
$$\Lambda = \{u \in E|u_i(t) \neq u_j(t), \ \forall t \in \mathbb{R}, i \neq j\}$$

Then we have

Theorem 1.2 *For $N \geq 2$ body problems in R^3, if $A_2 > A_3$, then the Lagrangian action functional (2.11) has a minimum $u = (u_1, \cdots, u_N)$ on $\overline{\Lambda}$ which corresponds to a noncollision T-periodic solution to the equation (1.3). Furthermore, if $N \geq 4$ then either u is in $x-y$ plane and $x_i(t+1/2) = x_i(t), y_i(t+1/2) = -y_i(t), x_i(-t) = -x_i(t), y_i(-t) = y_i(t)$, or u is nonplanar and satisfies the symmetries $u_i(t + \frac{1}{2}) = Au_i(t)$ and $u_i(-t) = Bu_i(t)$.*

Remark *This paper is essentially the talk given by Zhang S.Q. in Warwick's Workshop on classical n-body problems in April of 2002, in the talk, the authors anounced the existence of a nonplanar and noncollision periodic solution for $N \geq 4$ body problems in R^3 for any given masses $m_1, \cdots, m_N > 0$, the authors worked on*

$$\tilde{\Lambda} = \{u \in \tilde{E}| \sum m_i u_i = 0, u_i(t) \neq u_j(t)\}$$

where $$\tilde{E} = \{u = (u_1, \cdots, u_N)|u_i \in \tilde{H}_i\}$$

$$\tilde{H}_i = \{u_i \in W^{1,2}(R/Z, R^3), u_i(t + 1/r) = A\left(\frac{2\pi}{r}\right) u_i(t), r \geq 2 \ \text{ is } \ \text{even}\}$$

$$A\left(\frac{2\pi}{r}\right) = \begin{pmatrix} \cos\frac{2\pi}{r} & -\sin\frac{2\pi}{r} & 0 \\ \sin\frac{2\pi}{r} & \cos\frac{2\pi}{r} & 0 \\ 0 & 0 & -1 \end{pmatrix}.$$

The authors used the same arguments of the present paper to avoid the collision and prove the minimizer is nonplanar by choosing r in order to $A_2 > A_3$. Zhang S.Q. wrongly thought that $\xi(t) = \eta \cdot \sin 2\pi t$ satisfies $\xi(t + 1/r) = A(2\pi/r)\xi(t)$. But Professors Zhou Q., Chenciner and Terracini discovered that the decreasing direction ξ should depond on $r \geq 2$ since $\xi(t) = \eta \cdot \sin 2\pi t$ don't satisfy the symmetry $\xi(t + 1/r) = A(2\pi/r)\xi(t)$ if $r > 2$, and when r is larger, $\langle f''(q)\xi, \xi\rangle > 0$. But our arguments work for $r = 2$ and masses satisfying $A_2 > A_3$ and maybe more.

Acknowledgments

Parts of this paper were done while the first author visited Professor Shih-Liang Wen in Ohio University from September 1996 to March 1997, the author would like to thank him for his kindness and his interesting on N-body problems,the author would also like to thank Ohio University for the financial support.

References

[1] Ambrosetti, A. and Coti Zelati, V., Closed orbits of fixed energy for a class of N-body problems, *Ann. IHP Analyse nonlineaire* **9,** (1992) 187-200; Addendum to closed orbits of fixed energy for a class of N-body problems, *Ann. IHP Analyse nonlineaire* **9,** (1992) 337-338.

[2] Ambrosetti, A. and Coti Zelati, V., Closed orbits of fixed energy for singular Hamiltonian systems, *Arch. Rat. Mech. Anal.* **112,** (1990) 339-362.

[3] A. Bahri & P. Rabinowitz, Periodic solutions of Hamiltonian systems of three-body type, Ann. IHP. Anal. non linéaire.8, (1991) 561-649.

[4] Bessi, U. and Coti Zelati, V., Symmetries and non-collision closed orbits for planar N-body type problems, *J. Nonlinear Anal. TMA.* **16,** (1991) 587-598.

[5] Bessi, U., Multiple closed orbits of fixed energy for gravitational potentials, *J. Diff. Equ.* **104,** (1993) 1-10.

[6] Chenciner A. and Desolneux N., Minima de l'integrale d'action et equilibres relatifs de n corps, C.R. Acad. Sci. Paris, Sr.I. Math. 326(1998). 1209-1212.

[7] Coti Zelati, V., The periodic solutions of N-body type problems, *Ann. IHP Analyse nonlineaire* **7,** (1990) 477-492.

[8] Degovanni, M. and Giannoni, F., Dynamical systems with Newtonian type potentials, *Ann. Scuola Norm. Sup. Pisa CL Sci.* **15,** (1988) 467-497.

[9] Gordon, W., Conservative dynamical systems involving strong forces, *Trans. Amer. Math. Soc.* **204,** (1975) 113-135.

[10] Gordon, W., A minimizing property of Keplerian orbits, *Amer. J. Math.* **99,** (1977) 961-971.

[11] Long Y. & Zhang S., Geometric characterizations for variational minimization solutions of 3-body problems, Preprint, Jan. 1998, Acta Math. Sinica.16(2000) 579-592; Geometric characterizations for variational minimization solutions of 3 and 4–body problems, Preprint, June 1998.

[12] Long Y. & Zhang S., Geometric characterizations for variational minimization solutions of the 3-body problem with a fixed energy.J.D.E 160(2000),422-438.

[13] Moeckel R., On central configurations, Math. Zeit 205(1990), 499-517.

[14] Pacella F., Central configurations and the equivariant Morse theory, Arch. Rational Mech. Anal. 97(1987),59–74.

[15] Zhang S. Q., Multiple closed orbits of fixed energy for N–body–type problems with gravitational potentials, J. M. A.A. 208(1997), 462–475.
[16] Zhang S.Q. and Zhou Q., A minimizing property of Lagrangian solutions, Preprint, Dec. 1999, a talk given by Zhou Q. invited by Brezis in the University of Paris 6, March, 2000. Acta Math. Sinica 17(2001), 497–500.
[17] Zhang S.Q. and Zhou Q., Variational methods for the choreography solution to the three-body problem, Science in China (Series A), 45(2002), 594–597.
[18] Zhang S.Q. and Zhou Q., A noncollision periodic solution for N–body problems, J. Math. Anal. Appl. 257(2001), 332–342.
[19] Zhang S.Q., Periodic solutions for N–body problems, Progress in Nonlinear Analysis, edited by K.C.Chang and Y.Long, World Scientific, 2000, 423–443.
[20] Zhang S.Q., and Zhou Q., Noncollision periodic solutions with a fixed energy for Newtonian like N–body problems, invited talks given by Zhang S.Q. in Ohio Uiv. and Georgia Inst. of Tech., Feb., April, 1997.
[21] Zhang S.Q., The simultaneous collision orbits for N–body problems, Preprint, Feb., 1999.
[22] Zhang S.Q., Periodic solutions for Newtonian 4-body problems with equal masses in R^3, a mini-talk to Venturelli in ICTP,October 1998.

Existence of positive periodic solution for an impulsive delay differential equation

Aimin Zhao, Weiping Yan and Jurang Yan

Department of Mathematics, Shanxi University Taiyuan, Shanxi 030006, People's Republic of China

ABSTRACT. Sufficient conditions are obtained for the existence of periodic positive solution of an impulsive delay differential equation.

1. Introduction and Preliminaries

The theory of impulsive delay differential equations is emerging as an important area of investigation, since it is a lot richer than the corresponding theory of nonimpulsive delay differential equations. Many evolution processes in nature are characterized by the fact that at certain moments of time experience an abrupt change of state. That was the reason for the development of the theory of impulsive differential equations and impulsive delay differential equations, see the monographs [1] and [2]. For the theory of the delay differential equations, we refer to the monographs [3] and [4].

The purpose of this paper is to study the existence of positive periodic solution of the following impulsive delay differential equation

$$y'(t) = -a(t)y(t) + f(t, y(t-\tau_0(t)), y(t-\tau_1(t)), \cdots, y(t-\tau_n(t))), \text{ a.e. } t > 0, t \neq t_k, \tag{1$_a$}$$

$$y(t_k^+) = b_k y(t_k), \quad k = 1, 2, \cdots. \tag{1$_b$}$$

Some special cases of non-impulsive differential equations of Eq.(1) ($(1)_a$ and $(1)_b$) have been investigated. For example, the delay differential equation

$$N'(t) = -\alpha N(t) + \beta e^{-\gamma N(t-\tau)}, \ t \geq 0,$$

1991 *Mathematics Subject Classification.* 34K13; 34K45; 34A37; 92D25.
Key words and phrases. Periodic solution; existence; impulsive delay equation.

where α, β, γ and τ are positive constants, was used by Wazewska–Czyzewska and Lasota [5] as a model for the survial of red blood cells in an animal. For further investigation in this area, for example, the delay differential equations

$$N'(t) = -\mu N(t) + \sum_{i=1}^{m} p_i e^{-r_i N(t-\tau)}, \ t \geq 0,$$

where μ and p_i, r_i are positive constants, and

$$y'(t) = -\alpha(t)y(t) + \beta(t)e^{-y(t-m\omega)}, \ t \geq 0$$

where α and β are positive ω–periodic functions, see Xu and Li [6] and Graef, Qian and Spikes [7]. The existence of positive periodic solution of $(1)_a$ has been studied by Daqing Jiang and Junjie Wei [8].

In Eq.(1), we will use the following hypotheses:

(H_1) $0 < t_1 < t_2 < \cdots$ are fixed impulsive points with $\lim_{k\to\infty} t_k = \infty$;

(H_2) $\{b_k\}$ is a real sequence and $b_k > 0$, $k = 1, 2, \cdots$;

(H_3) $a : [0, \infty) \to (0, \infty)$, $\tau_i : [0, \infty) \to [0, \infty), i = 0, 1, \cdots, n$, are locally summable functions, and for all $t \geq 0$

$$a(t + \omega) = a(t), \quad \tau_i(t + \omega) = \tau_i(t);$$

(H_4) $b(t) := \prod_{0<t_k<t} b_k$ satisfies

$$b(t + \omega) = b(t), \text{ for all } t \geq 0;$$

(H_5) $f \in C([0, \infty)^{n+2}, [0, \infty))$ and for all $(t, u_0, u_1, \cdots, u_n) \in [0, \infty)^{n+2}$

$$f(t + \omega, u_0, u_1, \cdots, u_n) = f(t, u_0, u_1, \cdots, u_n).$$

Here and in the sequel we assume that a product equals unit if the number of factors is equal to zero. We will only consider the solutions of Eq.(1) with condition

$$y(t) = \phi(t) \text{ for } -\overline{\tau} \leq t \leq 0, \phi \in L([-\overline{\tau}, 0], [0, \infty)), \phi(0) > 0, \tag{1.2}$$

where $L([-\overline{\tau}, 0], [0, \infty))$ denotes the set of Lebesgue measurable functions on $[-\overline{\tau}, 0]$, $\overline{\tau} = \max_{0\leq i\leq n} \sup_{t\in[0,\omega]} \tau_i(t)$.

Definition 1. A function $y : [-\overline{\tau}, \infty) \to (0, \infty)$ is said to be a solution of Eq.(1) on $[-\overline{\tau}, \infty)$, if

(i) $y(t)$ is absolutely continuous on each interval $(0, t_1]$ and $(t_k, t_{k+1}]$, $k = 1, 2, \cdots$;

(ii) for any t_k, $k = 1, 2, \cdots$, $y(t_k^+)$ and $y(t_k^-)$ exist and $y(t_k^-) = y(t_k)$;

(iii) $y(t)$ satisfies $(1)_a$ for almost everywhere (a.e.) in $[0, \infty)\backslash\{t_k\}$ and satisfies $(1)_b$ for every $t = t_k, k = 1, 2, \cdots$.

Definition 2. A solution $y(t)$ of Eq.(1) on $[-\overline{\tau}, \infty)$ is said to be ω-periodic, if

$$y(t + \omega) = y(t) \text{ for all } t \geq 0.$$

Under the above hypotheses (H_1) — (H_5) we consider the non-impulsive delay differential equation

$$x'(t) = -a(t)x(t) + g(t, x(t-\tau_0(t)), x(t-\tau_1(t)), \cdots, x(t-\tau_n(t))), \text{a.e.} t \geq 0 \quad (1.3)$$

with initial condition

$$x(t) = \phi(t) \quad \text{for} \quad -\overline{\tau} \leq t \leq 0, \phi \in L([-\overline{\tau}, 0], [0, \infty)), \phi(0) > 0, \quad (1.4)$$

where

$$g(t, u_0, u_1, \cdots, u_n) = b^{-1}(t) f\left(t, b(t-\tau_0(t))u_0, b(t-\tau_1(t))u_1, \cdots, b(t-\tau_n(t))u_n\right).$$

By a solution $x(t)$ of (3) and (4) we mean an absolutely continuous function $x(t)$ defined on $[-\overline{\tau}, \infty)$ satisfies (3) a.e. for $t \geq 0$ and $x(t) = \phi(t)$ on $[-\overline{\tau}, 0]$.

The following lemmas will be used in the proofs of our results. The proof of the first lemma is similar to that of Theorem 1 in [9] and it will be omitted.

Lemma 1. Assume that (H_1)—(H_5) hold. Then

(i) if $x(t)$ is a solution of (3) on $[-\overline{\tau}, \infty)$, then $y(t) = b(t)x(t)$ is a solution of (1) on $[-\overline{\tau}, \infty)$;

(ii) if $y(t)$ is a solution of (1) on $[-\overline{\tau}, \infty)$, then $x(t) = b^{-1}(t)y(t)$ is a solution of (3) on $[-\overline{\tau}, \infty)$.

Lemma 2. Assume that (H_1)—(H_5) hold. Then the solutions of (1) and (2) are defined on $[-\overline{\tau}, \infty)$ and are positive on $[0, \infty)$.

Proof. Clearly, by Lemma 1, we only need to prove that the solution of (3) and (4) are defined and positive on $[-\overline{\tau}, \infty)$. From (3) and (4) we have that for any $\phi \in L$ and $t > 0$

$$x(t) = \phi(0)e^{-\int_0^t \alpha(s)ds} + \int_0^t e^{-\int_s^t \alpha(r)dr} g(s, x(s-\tau(s)), \cdots, x(s-\tau_n(s)))ds.$$

Hence $x(t)$ is defined on $[-\overline{\tau}, \infty)$ and positive on $[0, \infty)$. The proof of Lemma 2 is complete.

Lemma 3 (Krasnoselskii)[10] Assume that X is a Banach space, K is a cone in X, Ω_1 and Ω_2 are bounded open sets of X such that $0 \in \Omega_1$ and $\overline{\Omega}_1 \subset \Omega_2$, $\Phi : K \cap (\overline{\Omega}_2 \backslash \Omega_1) \to K$ is completely continuous and satisfies

(i) $\|\Phi x\| \leq \|x\|, \forall x \in K \cap \partial\Omega_1$; $\|\Phi x\| \geq \|x\|, \forall x \in K \cap \partial\Omega_2$;

or

(ii) $\|\Phi x\| \geq \|x\|, \forall x \in K \cap \partial\Omega_1$; $\|\Phi x\| \leq \|x\|, \forall x \in K \cap \partial\Omega_2$.

Then Φ has fixed point in $K \cap (\overline{\Omega}_2 \backslash \Omega_1)$.

2. Existence of Positive Periodic Solution

From Lemma 1 we know that Eq.(1) has ω-periodic positive solution if and only if Eq.(3) has ω-periodic positive solution.

We will prove that Eq.(1) has unique positive periodic solution. Let $\overline{b} = \sup_{t\in[0,\omega]} b(t)$, $\underline{b} = \inf_{t\in[0,\omega]} b(t)$, $\sigma = \exp\left(-\int_0^\omega a(\xi)d\xi\right)$, $\sigma_1 = \sigma\underline{b}/\overline{b}$. For

$$u = (u_0, u_1, \cdots, u_n) \in [0, \infty)^{n+1},$$

denote $|u| = \max\{u_0, u_1, \cdots, u_n\}$

Theorem 1. Assume that (H_1)—(H_5) hold and

$$\lim_{|u|\to 0} \max_{t\in[0,\omega]} \frac{f(t,u)}{|u|} = 0, \quad \lim_{\substack{|u|\to\infty \\ u_j \geq \sigma_1 |u| \\ 0\leq j\leq n}} \min_{t\in[0,\omega]} \frac{f(t,u)}{|u|} = \infty. \tag{2.1}$$

Then Eq.(1) has ω-periodic positive solution.

Proof. We extend $a(\cdot), \tau_i(\cdot), i = 0, 1, \cdots, n$ and $g(\cdot, u_0, u_1, \cdots, u_n)$ on $(-\infty, \infty)$ as ω-periodic functions and show that Eq.(3) has ω-periodic solution. Let $X = \{x \in C(R,R), x(t+\omega) = x(t)\}$ and for any $x \in X$, $\|x\| = \sup_{t\in[0,\omega]} |x(t)|$. Then X is a Banach space. Denote

$$G(t,s) = \frac{e^{\int_t^s a(\xi)d\xi}}{e^{\int_0^\omega a(\xi)d\xi} - 1} \tag{2.2}$$

and $A = G(t,t) = G(0,0), B = G(t, t+\omega) = G(0,\omega)$. It is easy to see that an ω-periodic solution of Eq.(3) is equivalent to a solution of integral equation

$$x(t) = \int_t^{t+\omega} G(t,s) g(s, x(s-\tau_0(s)), x(s-\tau_1(s)), \cdots, x(s-\tau_n(s)))ds. \tag{2.3}$$

Let $K = \{x : x \in X, x(t) \geq \sigma\|x\|\}$. It is easy to see that K is a cone in X. Denote $\Phi : K \to X$ by

$$(\Phi x)(t) = \int_t^{t+\omega} G(t,s) g(s, x(s-\tau_0(s)), x(s-\tau_1(s)), \cdots, x(s-\tau_n(s)))ds.$$

From [8,Lemma 2.1] we know that $\Phi(K) \subset K$. We will prove that Φ has fixed point $\widetilde{x}$ on K, that is, Eq.(3) has an ω-periodic positive solution $\widetilde{x}(t)$.

1) From the first formula of condition (5) we know that there exists $\rho_1 > 0$ such that

$$f(t,u) \leq \frac{\underline{b}}{2B\omega\bar{b}}|u|, \quad \text{for } |u| \leq \bar{b}\rho_1, t \geq 0. \tag{2.4}$$

Thus, for $|u| \leq \rho_1$, $t \geq 0$,

$$\begin{aligned} g(t,u) &= b^{-1}(t) f(t, b(t-\tau_0(t))u_0, b(t-\tau_1(t))u_1, \cdots, b(t-\tau_n(t))u_n) \\ &\leq b^{-1}(t) \frac{\underline{b}}{2B\omega\bar{b}} \max_{0\leq j\leq n} b(t-\tau_j(t))u_j \\ &\leq \frac{1}{2B\omega}|u|. \end{aligned}$$

Let $\Omega_1 = \{x : x \in X, \|x\| < \rho_1\}$. Then for any $x \in K \cap \partial\Omega_1$, we have

$$\begin{aligned} (\Phi x)(t) &\leq B \int_t^{t+\omega} g(s, x(s-\tau_0(s)), \cdots, x(s-\tau_n(s)))ds \\ &\leq B \frac{1}{2B\omega} \int_t^{t+\omega} \max_{0\leq j\leq n} x(s-\tau_j(s))ds \\ &\leq \frac{1}{2}\|x\|. \end{aligned}$$

Hence, $\|\Phi x\| < \|x\|$ hold for any $x \in K \cap \partial\Omega_1$.

2) From the second formula of condition (5) we know that there exists $\rho_2 > \rho_1$ such that

$$f(t,u) \geq \frac{2}{\omega\sigma_1 A}|u| \text{ for } |u| \geq \underline{b}\sigma\rho_2 \text{ and } u_i \geq \sigma_1|u|, 0 \leq i \leq n, t \geq 0. \tag{2.5}$$

Let $\Omega_2 = \{x : x \in X, \|x\| < \rho_2\}$. It is obvious that $\overline{\Omega}_1 \subset \Omega_2$. For $x \in K \cap \partial\Omega_2$, denote $u_i(t) = b(t-\tau_i(t))x(t-\tau_i(t)), i = 0, 1, \cdots, n$ and $u(t) = (u_0(t), u_1(t), \cdots, u_n(t))$. Then we have

$$|u(t)| = \max_{0\leq i\leq n} b(t-\tau_i(t))x(t-\tau_i(t)) \geq \underline{b}\sigma\|x\| = \underline{b}\sigma\rho_2,$$

and

$$\begin{aligned} u_i(t) &= b(t-\tau_i(t))x(t-\tau_i(t)) \\ &\geq \underline{b}\sigma\|x\| \\ &\geq \frac{\underline{b}}{\sup\limits_{t\in[0,\omega]} b(t)}\sigma \max_{0\leq j\leq n} b(t-\tau_j(t)) \max_{0\leq j\leq n} x(t-\tau_j(t)) \\ &\geq \sigma_1 \max_{0\leq j\leq n} b(t-\tau_j(t))x(t-\tau_j(t)) = \sigma_1|u(t)|. \end{aligned}$$

Hence,

$$\begin{aligned} (\Phi x)(t) &\geq A\int_t^{t+\omega} g(s, x(s-\tau_0(s)), \cdots, x(s-\tau_n(s)))ds \\ &\geq A\int_t^{t+\omega} b^{-1}(s)f(s, u_0(s), \cdots, u_n(s))ds \\ &\geq A\frac{2}{\omega A\sigma_1\overline{b}}\int_t^{t+\omega} \max_{0\leq j\leq n} b(s-\tau_j(s))x(s-\tau_j(s))ds \\ &\geq \frac{2}{\omega\sigma_1\overline{b}}\,\underline{b}\sigma\omega\|x\| = 2\|x\|. \end{aligned}$$

Therefore, $\|\Phi x\| > \|x\|$ holds for any $x \in K \cap \partial\Omega_2$.

3) $\Phi : K \cap (\overline{\Omega}_2\backslash\Omega_1) \to K$ is continuous. In fact, let $x_m, x_0 \in K \cap (\overline{\Omega}_2\backslash\Omega_1)$ such that $\|x_m - x_0\| \to 0 (m \to \infty)$. $\forall\epsilon > 0$, since $f(t, u_0, u_1, \cdots, u_n)$ is uniformly continuous on $[0, 2\omega] \times [0, \overline{b}\rho_2]^{n+1}$, $\exists\delta > 0$, such that $t \in [0, 2\omega]$, $u_i', u_i'' \in [0, \overline{b}\rho_2]$ and $|u_i' - u_i''| \leq \delta$ imply

$$|f(t, u_0', u_1', \cdots, u_n') - f(t, u_0'', u_1'', \cdots, u_n'')| < \frac{\underline{b}}{B\omega}\epsilon.$$

From the property of x_m and x_0 we know that there exists N such that

$$\|x_m - x_0\| < \frac{\delta}{\overline{b}} \text{ for } m > N.$$

Hence for $t \in [0, 2\omega]$ and $m > N$

$$|b(t-\tau_i(t))x_m(t-\tau_i(t)) - b(t-\tau_i(t))x_0(t-\tau_i(t))| < \delta, i = 0, 1, \cdots, n.$$

Thus we have that

$$\begin{aligned}
&|(\Phi x_m)(t) - (\Phi x_0)(t)| \\
\leq\ & \int_t^{t+\omega} G(t,s)b^{-1}(s)\times \\
&\Big|f(s, b(s-\tau_0(s))x_m(s-\tau_0(s)), \cdots, b(s-\tau_n(s))x_m(s-\tau_n(s))) \\
&\quad - f(s, b(s-\tau_0(s))x_0(s-\tau_0(s)), \cdots, b(s-\tau_n(s))x_0(s-\tau_n(s)))\Big| ds \\
\leq\ & \frac{B}{\underline{b}} \int_t^{t+\omega} \frac{\underline{b}}{B\omega} \epsilon ds = \epsilon,
\end{aligned}$$

which implies that $\|\Phi x_m - \Phi x_0\| \to 0$ as $m \to \infty$.

4) $\Phi(K \cap (\overline{\Omega}_2 \backslash \Omega_1))$ is relatively compact. In fact, since $f(t, u_0, u_1, \cdots, u_n)$ is continuous on $[0, 2\omega] \times [0, \bar{b}\rho_2]^{n+1}$, there exists $M > 0$ such that for any $(t, u) \in [0, 2\omega] \times [0, \bar{b}\rho_2]^{n+1}$, $|f(t, u)| \leq M$. Hence, for any $x \in K \cap (\overline{\Omega}_2 \backslash \Omega_1)$

$$|(\Phi x)(t)| \leq \frac{BM\omega}{\underline{b}}, \quad t \in [0, \omega],$$

which implies that $\Phi(K \cap (\overline{\Omega}_2 \backslash \Omega_1))$ is uniformly bounded. For any $\epsilon > 0$, from the continuity of $G(t, s)$ on $[0, 2\omega] \times [0, 2\omega]$ we know that there exist $\delta_1 > 0$ such that $(t_1, s), (t_2, s) \in [0, 2\omega] \times [0, 2\omega]$ and $|t_1 - t_2| < \delta_1$ implies that

$$|G(t_1, s) - G(t_2, s)| < \frac{\underline{b}}{2\omega M} \epsilon.$$

Hence for any $x \in K \cap (\overline{\Omega}_2 \backslash \Omega_1)$ and $t_1, t_2 \in [0, \omega]$ with $0 < t_2 - t_1 < \min\{\delta_1, \frac{\underline{b}}{4BM}\epsilon\}$ we have

$$\begin{aligned}
&|(\Phi x)(t_2) - (\Phi x)(t_1)| \\
=\ & \Big| \int_{t_2}^{t_2+\omega} G(t_2, s) g(s, x(s-\tau_0(s)), \cdots, x(s-\tau_n(s))) ds \\
&\quad - \int_{t_1}^{t_1+\omega} G(t_1, s) g(s, x(s-\tau_0(s)), \cdots, x(s-\tau_n(s))) ds \Big| \\
\leq\ & \int_{t_2}^{t_2+\omega} |G(t_2, s) - G(t_1, s)| g(s, x(t-\tau_0(s)), \cdots, x(s-\tau_n(s))) ds \\
&+ \int_{t_1+\omega}^{t_2+\omega} G(t_1, s) g(s, x(s-\tau_0(s)), \cdots, x(s-\tau_n(s))) ds \\
&+ \int_{t_1}^{t_2} G(t_1, s) g(s, x(s-\tau_0(s)), \cdots, x(s-\tau_n(s))) ds \\
\leq\ & \frac{\underline{b}\epsilon}{2\omega M} \frac{M\omega}{\underline{b}} + 2\frac{BM}{\underline{b}} |t_2 - t_1| < \epsilon.
\end{aligned}$$

This shows that $\Phi(K \cap (\overline{\Omega}_2 \backslash \Omega_1))$ is equicontinuous. By the Ascoli-Arzela theorem $\Phi(K \cap (\overline{\Omega}_2 \backslash \Omega_1))$ is relatively compact.

Therefore, all conditions of Lemma 3 are satisfied. Hence, the operator Φ has fixed point in $K \cap (\overline{\Omega}_2 \backslash \Omega_1)$. The proof of theorem is complete.

Similarly we have

Theorem 2. Assume that (H_1)—(H_5) hold and

$$\lim_{|u|\to\infty}\max_{t\in[0,\omega]}\frac{f(t,u)}{|u|}=0,\quad \lim_{\substack{|u|\to 0\\ u_j\geq\sigma_1|u|\\ 0\leq j\leq n}}\min_{t\in[0,\omega]}\frac{f(t,u)}{|u|}=\infty. \tag{2.6}$$

Then Eq.(1) has ω-periodic positive solution.

Remark 1. Our results in this paper indicate that under the appropriate linear periodic impulsive perturbations the impulsive delay differential equations remain the original periodicity of the non-impulsive system $(1.1)_a$. Since the function $a(t)$ need not to be continuous, our results extends and improves the corresponding conclusion of [8].

Consider impulsive delay differential equations

$$\begin{cases} y'(t)=-\alpha(t)y(t)+\sum\limits_{i=1}^{l}\beta_i(t)e^{-\gamma_i(t)y(t-m_i\omega)}, & \text{a.e. } t\geq 0, t\neq t_k,\\ y(t_k^+)=b_k y(t_k), \quad k=1,2,\cdots, \end{cases} \tag{2.7}$$

and

$$\begin{cases} y'(t)=-\alpha(t)y(t)+\dfrac{\beta(t)}{1+y^n(t-\tau(t))}, & \text{a.e. } t\geq 0, t\neq t_k,\\ y(t_k^+)=b_k y(t_k), \quad k=1,2,\cdots. \end{cases} \tag{2.8}$$

Corollary 1. Assume that (H_1), (H_2), (H_4) and

(H_6) $\alpha:[0,\infty)\to(0,\infty)$, $\beta_i,\gamma_i:[0,\infty)\to[0,\infty)$, $i=1,2,\cdots,l$, are locally summable periodic functions with common period ω, $m_i, i=1,2,\cdots,l$ are nonnegative integrals

hold. Then Eq.(11) has positive ω-periodic solution.

Corollary 2. Assume that (H_1), (H_2), (H_4) and

(H_7) $\alpha:[0,\infty)\to(0,\infty)$, $\beta,\tau:[0,\infty)\to[0,\infty)$ are locally summable ω-periodic functions, n is a positive constant

hold. Than Eq.(12) has positive ω-periodic solution.

References

[1] V. Lakshmikantham, D. D. Bainov and P. S. Simeonov, Theory of Impulsive Differential Equations, World Scientific, Singapore, 1989.

[2] K. Gopalsamy, Stability and Oscillation in Delay Differential Equations of Population Dynamics, Kluwer Academic Publishers, Dordrecht, 1992.

[3] I. Gyori and G. Ladas, Oscillation Theory of Delay Differential Equations with Applications, Clarendon, Oxford, 1991.

[4] Y. Kuang,Delay Differential Equations with Applications in Population Dynamics, Academic Press, Boston, 1993.

[5] M. Wazewska–Czyzewska and A. Lasota, Mathematical problems of the dynamics of red blood cells system, Ann. of the Polish Math. Soc. Ser III, Appl. Math. 17(1988), 23–40.

[6] Xu Wangwen and Li Jingwen,Global attractivity of the model for the survial of red blood cells with several delays, Ann. Diff. Eqs. 14(1998), 357–363.

[7] J. R. Graef, C. Qian and P. W. Spikes, Oscillation and global attractivity in a periodic delay equation, Canad. Math. Bull. 38(1996), 275–283.

[8] Daqing Jiang and Junjie Wei, Existence of positive periodic solution of nonautonomous delay differential equations, Chinese Annals of Mathematics, 20A(6), 1999, 715–720.
[9] J. Yan and A. Zhao, Oscillation and stability of linear inpulsive delay differential equation, J. Math. Anal. Appl. 227(1998), 187–194.
[10] Dajun Guo, Nonlinear Functional Analysis, Shandong Science and Techenology Publishing Company, Jinan, 1985.

Computational theory and methods for finding multiple critical points

Jianxin Zhou

Texas A&M University, College Station, Texas, U.S.A.

1. Introduction

Let H be a Hilbert space and $J \in C^1(H, \mathbb{R})$. Denote δJ its Frechet derivative and J' its gradient. The objective of this research is to develop computational theory and methods for finding multiple critical points, i.e., solutions to the Euler-Lagrange equation $J'(u) = 0$. A critical point u^* is *nondegenerate* if $J''(u^*)$ is invertible. Otherwise u^* is *degenerate*. The first candidates for a critical point are the local extrema to which the classical critical point theory was devoted in calculus of variation. Most conventional numerical algorithms focus on finding such stable solutions. Critical points that are not local extrema are unstable and called *saddle points*. In physical systems, saddle points appear as unstable equilibria or transient excited states.

(In)Stability is one of the main concerns in system design and control. However, in many applications, performance or maneuverability is more desirable. An unstable solution can be stable enough to accomplish a short term mission before being excited to decay to another desirable state [14,15,22] and it may have much higher performance or maneuverability index.

Due to unstable nature, those multiple solutions are very elusive to be numerically captured. Little is known in the literature. Minimax method is one of the most popular approaches in critical point theory, But most minimax theorems, such as, the mountain pass, various linking and saddle point theorems, focus on the existence issue. They require one to solve a two-level global optimization problem and therefore are not for algorithm implementation. Inspired by the numerical work of Chio-McKenna and Ding-Costa-Chen in [9,12], the idea of Nehari and Ding-Ni

e-mail: jzhou@math.tamu.edu.

to define a solution manifold and the Morse theory, we present a local minimax method (LMM). Consider a semilinear elliptic BVP as a model problem

$$\Delta u(x) + f(x, u(x)) = 0 \quad x \in \Omega \tag{1.1}$$

for $u \in H = H_0^1(\Omega)$ where $\Omega \subset \mathbb{R}^N$ is bounded and open, $f(x,t)$ is nonlinear satisfying $f(x,0) = 0$ and other standard conditions. The variational functional is the energy

$$J(u) = \int_\Omega \{\frac{1}{2}|\nabla u(x)|^2 - F(x, u(x))\}\, dx \text{ with } F(x,t) = \int_0^t f(x,\tau)d\tau. \tag{1.2}$$

It can be shown that (weak) solutions to (1.1) coincide with critical points of J in H.

2. A Local Minimax Method and Its Convergence

For a subspace $H' \subset H$, let $S_{H'} = \{v | v \in H', \|v\| = 1\}$. Let L be a closed subspace in H, called a support, for $v \in S_{L^\perp}$, denote $\{L, v\} = \{tv + w | w \in L,\ t \in \mathbb{R}\}$.

Definition 2.1. A set-valued mapping $P\colon S_{L^\perp} \to 2^H$ is called the *peak mapping* of J w.r.t. L if for $v \in S_{L^\perp}$, $P(v)$ is the set of all local maximum points of J on $\{L, v\}$. A single-valued mapping $p\colon S_{L^\perp} \to H$ is a *peak selection* of J w.r.t. L if $p(v) \in P(v) \quad \forall v \in S_{L^\perp}$. For $v \in S_{L^\perp}$, we say that J has a local peak selection w.r.t. L at v, if there is a neighborhood $\mathcal{N}(v)$ of v and a mapping $p\colon \mathcal{N}(v) \cap S_{L^\perp} \to H$ s.t. $p(u) \in P(u) \quad \forall u \in \mathcal{N}(v) \cap S_{L^\perp}$.

Lemma 2.2. *[18] Let $v_\delta \in S_{L^\perp}$. If J has a local peak selection p w.r.t. L at v_δ s.t. p is continuous at v_δ and $dis(p(v_\delta), L) > \alpha > 0$ for some $\alpha > 0$, then either $J'(p(v_\delta)) = 0$ or for any $\delta > 0$ with $\|J'(p(v_\delta))\| > \delta$, there exists $s_0 > 0$, s.t.*

$$J(p(v(s))) - J(p(v_\delta)) < -\alpha\delta\|v(s) - v_\delta\|, \ \forall 0 < s < s_0$$

with

$$v(s) = \frac{v_\delta + sd}{\|v_\delta + sd\|}, \quad d = -J'(p(v_\delta)).$$

Theorem 2.3. *[18] If J has a local peak selection p w.r.t. L at $v_0 \in S_{L^\perp}$ s.t. (a) p is continuous at v_0, (b) $dis(p(v_0), L) > 0$ and (c) v_0 is a local minimum point of $J(p(v))$ on $S_{L^\perp}$, then $p(v_0)$ is a critical point of J.* ∎

Define a solution set $\mathcal{M} = \{p(v) : v \in S_{L^\perp}\}$. When $L = \{0\}$, $\mathcal{M}$ corresponds to Nehari or Ding-Ni's solution manifold. Then a local minimum point of J on $\mathcal{M}$ yields a saddle point, which can be approximated by, e.g., a steepest descent method. It leads to the following local minimax algorithm

Step 1:: Given $\varepsilon > 0, \lambda > 0$ and $n-1$ previously found critical points $w_1, w_2, \ldots, w_{n-1}$ of J, of which w_{n-1} has the highest critical value. Set $L = \text{span}\{w_1, w_2, \ldots, w_{n-1}\}$. Let $v^1 \in S_{L^\perp}$ be an ascent direction at w_{n-1}. Let $t_0^0 = 1$, $v_L^0 = w_{n-1}$ and set $k = 0$;

Step 2:: Using the initial guess $w = t_0^k v^k + v_L^k$, solve for

$$w^k \equiv p(v^k) = \arg \max_{u \in [L, v^k]} J(u), \quad \text{and denote} \quad t_0^k v^k + v_L^k = w^k \equiv p(v^k);$$

Step 3:: Compute the steepest descent vector $d^k = -J'(w^k)$;

Step 4:: If $\|d^k\| \leq \varepsilon$ then output $w_n = w^k$, stop; else goto Step 5;

Step 5:: Set $v^k(s) = \dfrac{v^k + sd^k}{\|v^k + sd^k\|}$ and use initial guess $u = t_0^k v^k(\frac{\lambda}{2^m}) + v_L^k$ to find $p(v^k(\frac{\lambda}{2^m}))$ in $[L, v^k(\frac{\lambda}{2^m})]$ where t_0^k and v_L^k are found in **Step 2** and, find

$$s^k = \max \left\{ \frac{\lambda}{2^m} \;\middle|\; m \in \mathbb{N}, 2^m > \|d^k\|, \right.$$
$$\left. J(p(v^k(\frac{\lambda}{2^m}))) - J(w^k) \leq -\frac{t_0^k}{2} \|d^k\| \, \|v^k(\frac{\lambda}{2^m}) - v^k\| \right\}.$$

Step 6:: Set $v^{k+1} = v^k(s^k)$ and update $k = k+1$ then goto **Step 2.** ∎

Initial guesses in Steps 2 and 5 are used to closely and consistently trace the position of the previous point $w^k = t_0^k v^k + v_L^k$. This strategy is also to avoid the algorithm from possible oscillating between different branches of the peak mapping P. In Step 3, the steepest descent direction $d^k = -J'(w^k)$ (or the negative gradient) at a known point w^k, is usually computed by solving a corresponding *linearized* problem (ODE or PDE).

DEFINITION 2.4. At $v \in S_{L^\perp}$, if $d = -J'(p(v)) \neq 0$, denote $v(s) = \dfrac{v + sd}{\|v + sd\|}$ and define the stepsize at v by

$$s(v) = \max_{\lambda \geq s \geq 0} \left\{ s \;\middle|\; \lambda > s\|d\|, J(p(v(s))) - J(p(v)) \leq -\frac{1}{2} dis(L, p(v)) \|d\| \|v(s) - v\| \right\}. \tag{2.1}$$

THEOREM 2.5. *Let v^k and $w^k = p(v^k), k = 0, 1, 2, \ldots$ be the sequence generated by the algorithm s.t. p is continuous at v^k and $w^k \notin L$, then the algorithm is strictly decreasing and thus stable.*

THEOREM 2.6. *[19] Let p be a peak selection of J w.r.t. L and J satisfy the (PS) condition. If (a) p is continuous, (b) $dis(L, w^k) > \alpha > 0$, $\forall k = 0, 1, 2, \ldots$ and (c) $\inf_{v \in S_{L^\perp}} J(p(v)) > -\infty$, then (1) $\{w^k\}_{k=0}^\infty$ has a subsequence $\{w^{k_i} = p(v^{k_i})\}$ converges to a critical point; (2) any limit point of $\{w^k\}$ is a critical point of J. In particular, any convergent subsequence $\{w^{k_i}\}$ of $\{w^k\}$ converges to a critical point.*

Let us define $K = \left\{ w \in H \middle| J'(w) = 0 \right\}$ and $K_c = \left\{ w \in H \middle| J'(w) = 0, \quad J(w) = c \right\}$.

THEOREM 2.7. *[19] Assume that V_1 and V_2 are open sets in H with $\emptyset \neq (V_2 \cap S_{L^\perp}) \subset (V_1 \cap S_{L^\perp})$. Set $V_1' = V_1 \cap S_{L^\perp}$ and $V_2' = V_2 \cap S_{L^\perp}$. If a peak selection*

p of J w.r.t. L determined by the algorithm is continuous in V_1' satisfying (a) $p(V_1') \cap K \subset K_c$, where $c = \inf_{v \in V_1'} J(p(v))$ and (b) there is $d > 0$, s.t.

$$\inf\{J(p(v))|v \in V_1';\quad dis(v, \partial V_1') \leq d\} = a > b = \sup\{J(p(v))|v \in V_2'\} > c.$$

Let $\lambda < d$ be chosen in the stepsize rule of the algorithm and $\{w^k\}_{k=0}^{\infty}$ be generated by the algorithm started from any $v^0 \in V_2'$ and $dis(L, w^k) > \alpha > 0\ \forall k = 0, 1, 2, ..,$ then for any $\varepsilon > 0$, there exists an N s.t. for any $k > N$,

$$dis(w^k, K_c) < \varepsilon.$$

THEOREM 2.8. *Assume the peak selection p determined by the algorithm is continuous. If $J(p(\bar{v})) = loc\min_{v \in S_{L^\perp}} J(p(v))$ and $p(\bar{v})$ is an isolated critical point with $dis(p(\bar{v}), L) > 0$, then there exists an open set V in H, $\bar{v} \in V \cap S_{L^\perp}$, s.t. starting from any $v_0 \in V \cap S_{L^\perp}$, the sequence $\{w^k\}_{k=0}^{\infty}$ generated by the algorithm converges to $p(\bar{v})$.*

The above convergence results are based on functional analysis. It is assumed that at each step of the algorithm the computation is *exact.* Convergence results allowing error to exist have been recently obtained by us.

3. A Local Min-L-$\perp$ Method

In the local minimax theory, we need to check if a local peak selection p is continuous or differentiable at v^*. This is very difficult, since p is not explicitly defined and the graph of p is, in general, not closed, an ill condition in numerical computation. On the other hand, non minimax type saddle points, such as the monkey saddles do exist. minimax principle can not cover them. In this section, we present a more general approach to solve those problems.

Applying Nehari's idea to (1.1), Ding-Ni defined a solution manifold [25]

$$\mathcal{M} = \left\{ v \in H_0^1(\Omega) | v \neq 0, \int_\Omega \left[|\nabla v|^2 - v f(v) \right] dx = 0 \right\}, \tag{3.1}$$

and proved that a global minimizer of J on $\mathcal{M}$ yields a solution (MI= 1). Using Green's identity, we can rewrite (3.1) as

$$\begin{aligned} \mathcal{M} &= \left\{ v \in H_0^1(\Omega) | v \neq 0, \int_\Omega v \left[\Delta v + f(v) \right] dx = 0 \right\} \\ &= \left\{ v \in H_0^1(\Omega) | v \neq 0, \langle J'(v), v \rangle_{H_0^1(\Omega)} = 0 \right\}. \end{aligned}$$

it becomes an orthogonal condition. This observation inspires us to

DEFINITION 3.1. Let L be a closed subspace of H. A set-valued mapping $P: S_{L^\perp} \to 2^H$ is an L-$\perp$ mapping of J if $P(v) = \{u \in \{L, v\} : J'(u) \perp \{L, v\}\}\ \forall v \in S_{L^\perp}$. A mapping $p: S_{L^\perp} \to H$ is an L-$\perp$ selection of J if $p(v) \in P(v)\ \forall v \in S_{L^\perp}$. Let $v \in S_{L^\perp}$ and $\mathcal{N}(v)$ be a neighborhood of v. if P (p) is locally defined in $\mathcal{N}(v) \cap S_{L^\perp}$, then P (p) is a local L-$\perp$ mapping (selection) of J at v.

It is clear that a local peak selection is a local L-⊥ selection. The graph of P can be very complex, it may contain multiple branches, U-turn or bifurcation points. We show that such defined L-⊥ selection has several interesting properties.

LEMMA 3.2. *If J is C^1, then the graph* $G = \{(u, v) : v \in S_{L^\perp}, u \in P(v) \neq \emptyset\}$ *is closed.*

Thus the ill-condition for a local peak selection has been removed.

DEFINITION 3.3. Let $v^* \in S_{L^\perp}$ and p a local L-⊥ selection of J at v^*. Assume $u^* = p(v^*) \in L$, we say that u^* is an *isolated* L-⊥ point of J w.r.t. p if there exist neighborhoods $\mathcal{N}(u^*)$ of u^* and $\mathcal{N}(v^*)$ of v^* s.t.

$$\mathcal{N}(u^*) \cap L \cap p(\mathcal{N}(v^*) \cap S_{L^\perp}) = \{u^*\},$$

i.e., for each $v \in \mathcal{N}(v^*) \cap S_{L^\perp}$ and $v \neq v^*$ either $p(v) \notin L$ or $p(v) = u^*$.

THEOREM 3.4. *Let $v^* \in S_{L^\perp}$ and p be a local L-⊥ selection of J at v^* and continuous at v^*. Assume either $p(v^*) \notin L$ or $p(v^*) \in L$ is an isolated L-⊥ point of J w.r.t. p, then a* necessary and sufficient *condition that $u^* = p(v^*)$ is a critical point of J is that there exists a neighborhood $\mathcal{N}(v^*)$ of v^* s.t.*

$$J'(p(v^*)) \perp p(v) - p(v^*), \quad \forall v \in \mathcal{N}(v^*) \cap S_{L^\perp}. \tag{3.2}$$

REMARK 3.5. The nature of a critical point is about orthogonality. Except for J', J is not involved in the theorem. It implies that the result is still true for non variational problem. Replacing J' by an operator $A : H \to H$ in the definition of a local L-⊥ selection p, Theorem 3.4 provides a necessary and sufficient condition that $u^* = p(v^*)$ solves $A(u) = 0$, a potentially useful result in solving multiple solutions to non variational problems. We have the flowing local characterization of a critical point.

LEMMA 3.6. *[33] Let H be a Hilbert space, L be a closed subspace of H and $J \in C^1(H, \mathbb{R})$. Let $v \in S_{L^\perp}$ and p be a local L-⊥ selection of J at v s.t. (a) p is continuous at v, (b) either $p(v) \notin L$ or $p(v) \in L$ is an isolated L-⊥ point of J, (c) v is a local minimum point of the function $J(p(\cdot))$ on $S_{L^\perp}$, then $u = p(v)$ is a critical point of J.* ∎

Replacing the local maximization by a local orthogonalization in Steps 2 and 5 in the local minimax algorithm, we obtain a local min-⊥ algorithm.

Next we study how to check the differentiability of an L-⊥ selection p. When an L-⊥ selection p is introduced, various implicit function theorems can be used to check its continuity or smoothness. Let $L = \{w_1, w_2, .., w_n\}$ and $v \in S_{L^\perp}$. Following the definition of p, $u = t_0 v + t_1 w_1 + ... + t_n w_n = p(v)$ is solved from (n+1) orthogonal conditions

$$\begin{aligned} F_0(v, t_0, t_1, ..., t_n) &\equiv \langle J'(t_0 v + t_1 w_1 + ... + t_n w_n), v\rangle = 0, \\ F_j(v, t_0, t_1, ..., t_n) &\equiv \langle J'(t_0 v + t_1 w_1 + ... + t_n w_n), w_j\rangle = 0, \quad j = 1, ..., n \end{aligned}$$

for $t_0, t_1, ..., t_n$. We have

$$\begin{aligned}\frac{\partial F_0}{\partial t_0} &= \langle J''(t_0v + t_1w_1 + ... + t_nw_n)v, v\rangle,\\ \frac{\partial F_0}{\partial t_i} &= \langle J''(t_0v + t_1w_1 + ... + t_nw_n)w_i, v\rangle, \quad i = 1,2,...,n.\\ \frac{\partial F_j}{\partial t_0} &= \langle J''(t_0v + t_1w_1 + ... + t_nw_n)v, w_j\rangle, \quad j = 1,2,...,n.\\ \frac{\partial F_j}{\partial t_i} &= \langle J''(t_0v + t_1w_1 + ... + t_nw_n)w_i, w_j\rangle, \quad i,j = 1,2,...,n.\end{aligned}$$

By the implicit function theorem, if the $(n+1)\times(n+1)$ matrix

$$Q'' = \begin{bmatrix} \langle J''(u)v, v\rangle, & \langle J''(u)w_1, v\rangle, & ... & \langle J''(u)w_n, v\rangle \\ \langle J''(u)v, w_1\rangle, & \langle J''(u)w_1, w_1\rangle, & ... & \langle J''(u)w_n, w_1\rangle \\ ... & & ... & \\ \langle J''(u)v, w_n\rangle, & \langle J''(u)w_1, w_n\rangle, & ... & \langle J''(u)w_n, w_n\rangle \end{bmatrix} \tag{3.3}$$

is invertible or $|Q''| \neq 0$, then p is differentiable at and near v. This condition can be easily and numerically checked.

4. Instability Analysis by the Local Minimax Method

DEFINITION 4.1. A vector $v \in H$ is said to be a *decreasing direction* of J at a critical point $u^* \in H$ if there exists $T > 0$ s.t. $J(u^* + tv) < J(u^*) \quad \forall\, T > t > 0$.

In general, the set of all decreasing vectors of J at u^* does not form a linear vector space. The maximum dimension of a subspace of decreasing directions of J at u^* can be used to measure local instability of u^* and is called the *local instability index* of u^*. Since such an index lacks of characterization and is too difficult to compute, let us consider several alternatives.

When $J''(u^*): H \to H$ is a self-adjoint Fredholm operator at a critical point u^*, the Hilbert space H has an orthogonal spectral decomposition $H = H^- \oplus H^0 \oplus H^+$ where H^-, H^0 and H^+ are respectively the maximum negative definite, the null and the maximum positive definite subspaces of $J''(u^*)$ in H with $\dim(H^0) < \infty$. The *Morse index* (MI) of u^* is $\mathrm{MI}(u^*) = \dim(H^-)$. When u^* is nondegenerate, i.e., $H^0 = \{0\}$, $\mathrm{MI}(u^*)$ is the local instability index of J at u^* [**28**]. But MI is very expensive to compute and it is ineffective in measuring local instability of a degenerate critical point.

In our local minimax method, $u^* = p(v^*)$ is a local maximum point of J in the subspace $\{L, v^*\}$. The dimension of $\{L, v^*\}$, which equals $\dim(L) + 1$, is known even before we numerically find the solution u^* and can be used to measure local instability of u^* even u^* is degenerate. We establish some relations between MI and the number $\dim(L) + 1$.

THEOREM 4.2. *[32] Let $v^* \in S_{L^\perp}$. Assume that J has a local peak selection p w.r.t. L at v^* s.t. (a) p is continuous at v^*, (b) $u^* \equiv p(v^*) \notin L$ and (c)*

$v^* = \arg\min_{v \in S_{L^\perp}} J(p(v))$. *If either* $H^- \cap \{L, v^*\}^\perp = \{0\}$ *or* p *is differentiable at* v^*, *then* u^* *is a critical point with*

$$\dim(L) + 1 = MI\ (u^*) + \dim(H^0 \cap \{L, v^*\}). \tag{4.1}$$

REMARK 4.3. The result in the last theorem is multi-fold. First, it provides a method to evaluate MI of a saddle point without actually finding $\dim(H^-)$, a very expensive job. Secondly, it indicates that $\dim(L)+1$ is better than MI in measuring local instability for a degenerate saddle point u^*.

One question still remains: how to determine $\dim(H^0 \cap \{L, v^*\})$? Since it is usually very expensive to find H^0, we point out that one does not have to find H^0 to determine $\dim(H^0 \cap \{L, v^*\}$. To see this, let $L = \{w_1, ..., w_n\}$ where $w_1, ..., w_n$ are linearly independent and define the quadratic function

$$\begin{aligned} Q(t_0, t_1, .., t_n) &\equiv \frac{1}{2}\langle J''(u^*)(t_0 v^* + t_1 w_1 + ... + t_n w_n), (t_0 v^* + t_1 w_1 + ... + t_n w_n)\rangle \\ &= \frac{1}{2}(t_0, t_1, ..., t_n) Q''(t_0, t_1, ..., t_n)^T \end{aligned} \tag{4.2}$$

where Q'' is the $(n+1) \times (n+1)$ matrix given in (3.3). Let us observe that $(t_0 v^* + t_1 w_1 + ... + t_n w_n) \in H^0 \cap \{L, v^*\}$ implies $(t_0, t_1, ..., t_n) \in \ker(Q'')$, i.e.,

$$(H^0 \cap \{L, v^*\}) \subset \Big\{t_0 v^* + t_1 w_1 + ... + t_n w_n : (t_0, t_1, ..., t_n) \in \ker\{Q''\}\Big\}.$$

$\ker\{Q''\} = \{0\}$ or $|Q''| \neq 0$ implies $\dim(H^0 \cap \{L, v^*\}) = 0$ even $H^0 \neq \{0\}$.

THEOREM 4.4. *[32] Let* $v^* \in S_{L^\perp}$. *Assume* J *has a local peak selection* p *w.r.t.* L *at* v^* *s.t.* p *is continuous at* v^*, $u^* \equiv p(v^*) \notin L$ *and* $v^* = \arg\min_{v \in S_{L^\perp}} J(p(v))$. *If* p *is differentiable at* v^*, *then* u^* *is a critical point with* $0 \leq \dim(L)+1-MI\ (u^*) \leq \dim(\ker\{Q''\})$.

5. Selected Numerical Examples

For a semilinear elliptic BVP $\begin{cases} \Delta u(x) + |x|^q u^p(x) = 0, & x \in \Omega, \\ u(x) = 0, & x \in \partial\Omega, \end{cases}$
we consider the Lane-Emden equation ($q = 0, p = 3$) on a dumbbell-shaped domain Ω in $\mathbb{R}^2$ and the Henon equation ($q = 3, p = 3$) on the same dumbbell-shaped domain Ω and ($q = 9, p = 3$) on the unit disk Ω in $\mathbb{R}^2$. In all our numerical examples here, we set $\varepsilon < 10.^{-4}$.

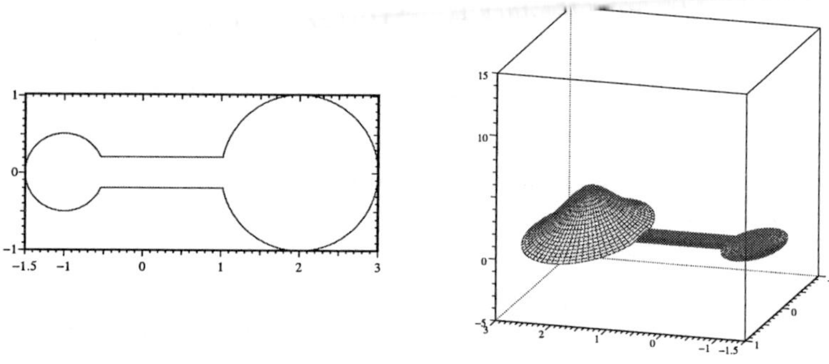

FIGURE 1. Lane-Emden: The domain (left) and the ground state solution (right).

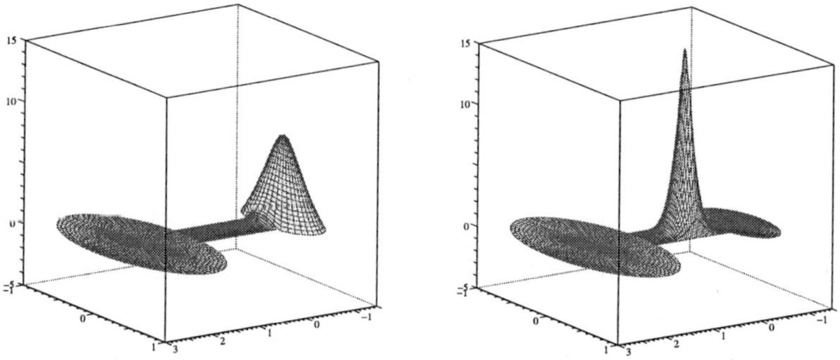

FIGURE 2. Lane-Emden: The 2nd solution MI= 1 (left) and the 3rd solution MI=1 (right).

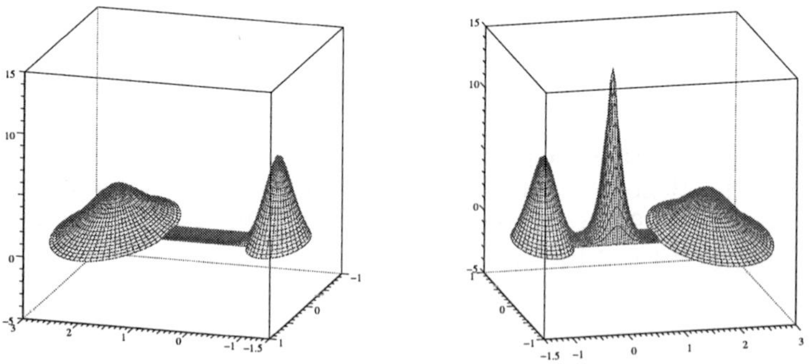

FIGURE 3. Lane-Emden: A solution MI= 2 (left) and a solution MI=3 (right).

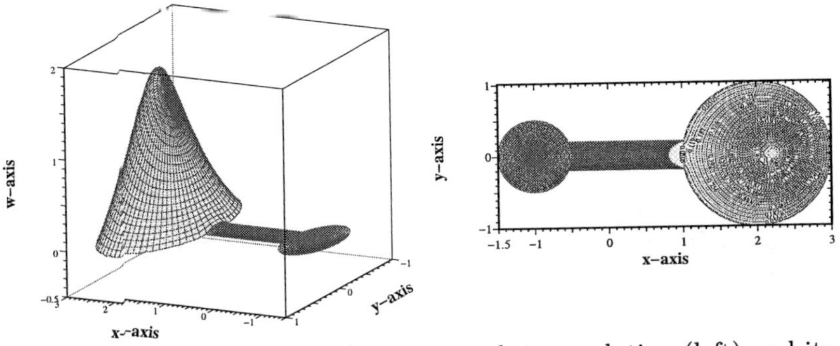

FIGURE 4. Henon: (q=3) The ground state solution (left) and its contours (right).

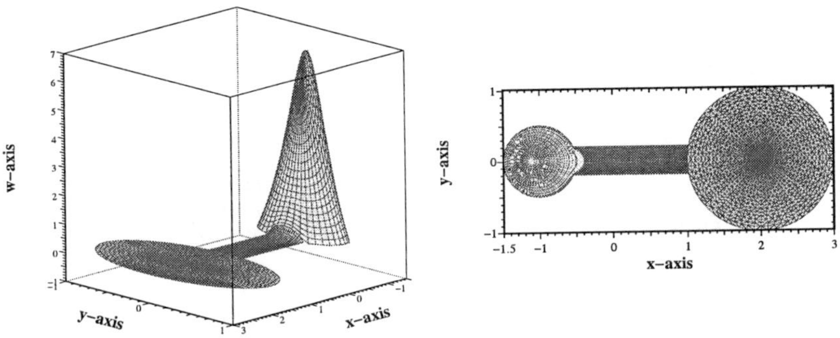

FIGURE 5. Henon: (q=3) The second solution MI= 1 (left) and its contours (right).

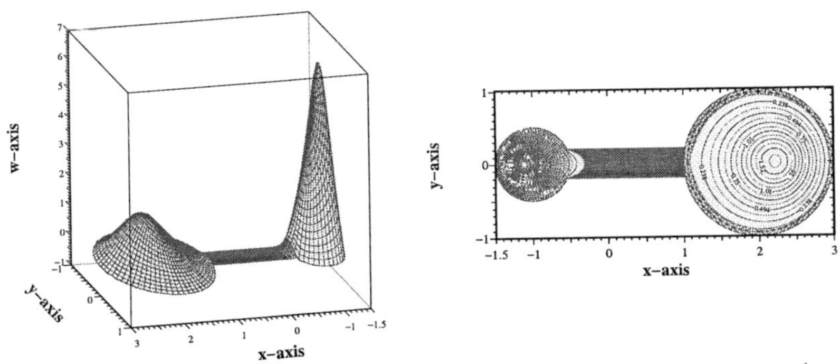

FIGURE 6. Henon: (q=3) A solution MI=2 (left) and its contours (right).

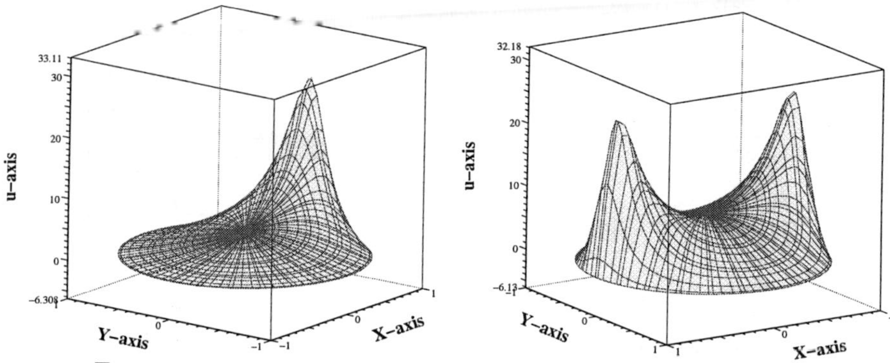

FIGURE 7. Henon: q=9. A 1-peak solution MI= 1 (left) and a 2-peak solution MI=2 (right).

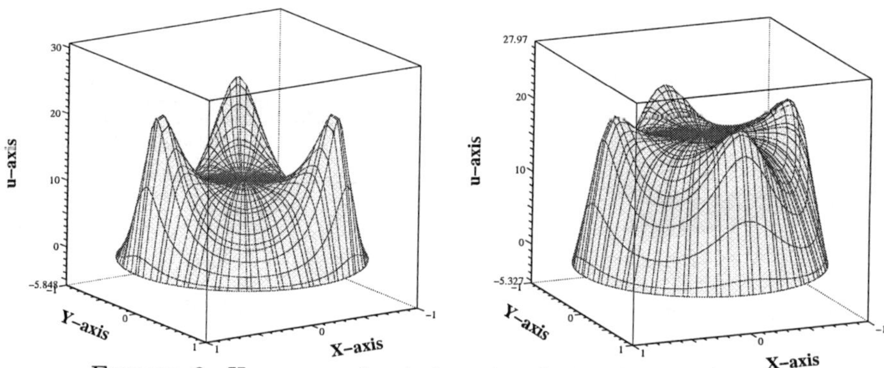

FIGURE 8. Henon: q=9. A 3-peak solution MI= 3 (left) and a 4-peak solution MI=4 (right).

References

[1] A. Ambrosetti and P. Rabinowitz, "Dual variational methods in critical point theory and applications", *J. Funct. Anal. 14(1973)*, 349-381.

[2] A. Bahri and P.L. Lions, "Morse index of some min-max critical points. I. Application to multiplicity results", *Comm. Pure Appl. Math.*, Vol. XLI (1988) 1027-1037.

[3] A. Bahri and P.L. Lions, "Solutions of superlinear elliptic equations and their Morse indices", *Communications on Pure and Applied Math.*, 45(1992), 1205-1215.

[4] T. Bartsch, K. C. Chang and Z. Q. Wang, "On the Morse indicex of sign changing solutions of nonlinear elliptic problems", *Math. Z.* 233 (2000), 655-677.

[5] H. Brezis and L. Nirenberg, "Remarks on Finding Critical Points", *Communications on Pure and Applied Mathematics*, Vol. XLIV, 939-963, 1991.

[6] K.C. Chang, *Infinite Dimensional Morse Theory and Multiple Solution Problems*, Birkhäuser, Boston, 1993.

[7] G. Chen, W.M. Ni and J. Zhou, "Algorithms and visualization for solutions of nonlinear elliptic equation, Part I: Dirichlet Problem", *Int. J. Bifurcation & Chaos*, 7(2000), 1565-1612.

[8] G. Chen, W.M. Ni, A. Porronnet and J. Zhou, "Algorithms and visualization for solutions of nonlinear elliptic equations, Part II: Dirichlet, Neumann and Robin boundary conditions and problems in 3D", Int. J. Bifurcation & Chaos, 7(2001) 1781-1799.

[9] Y. S. Choi and P. J. McKenna, "A mountain pass method for the numerical solution of semi-linear elliptic problems", *Nonlinear Analysis, Theory, Methods and Applications,* 20(1993), 417-437.

[10] Y. Chen and P. J. McKenna, Traveling waves in a nonlinearly suspended beam: Theoretical results and numerical observations, *J. Diff. Equ.,* 136(1997), 325-355.

[11] W.Y. Ding and W.M. Ni, "On the existence of positive entire solutions of a semilinear elliptic equation", *Arch. Rational Mech. Anal.* 91(1986),

[12] Z. Ding, D. Costa and G. Chen, "A high linking method for sign changing solutions for semilinear elliptic equations", *Nonlinear Analysis*, 38(1999) 151-172.

[13] I. Ekeland and N. Ghoussoub, Selected new aspects of the calculus of variations in the large, *Bull. Amer. Math. Soc. (N.S.),* 39(2002), 207-265.

[14] J.J. Garcia-Ripoll, V.M. Perez-Garcia, E.A. Ostrovskaya and Y. S. Kivshar, "Dipole-mode vector solitons", *Phy. Rev. Lett.* 85(2000), 82-85.

[15] J.J. Garcia-Ripoll and V.M. Perez-Garcia, "Optimizing Schrodinger functionals using Sobolev gradients: Applications to quantum mechanics and nonlinear optics", *SIAM Sci. Comp.* 23(2001), 1316-1334.

[16] J. Q. Liu and S. J. Li, "Some existence theorems on multiple critical points and their applications", *Kexue Tongbao* 17 (1984)

[17] S.J. Li and M. Willem, "Applications of local linking to critical point theory", *J. Math. Anal. Appl.* **189** (1995), 6–32.

[18] Y. Li and J. Zhou, "A minimax method for finding multiple critical points and its applications to semilinear PDE", *SIAM J. Sci. Comp.*, 23(2001), 840-865.

[19] Y. Li and J. Zhou, "Convergence results of a local minimax method for finding multiple critical points", *SIAM J. Sci. Comp.*, to appear.

[20] Y. Li and J. Zhou, "Local characterization of saddle points and their Morse indices", *Control of Nonlinear Distributed Parameter Systems*, pp. 233-252, Marcel Dekker, New York, 2001.

[21] J. Mawhin and M. Willem, *Critical Point Theory and Hamiltonian Systems*, Springer-Verlag, New York, 1989.

[22] Z.H. Musslimani, M. Segev, D.N. Christodoulides and M. Soljacic, "Composite Multihump vector solitons carrying topological charge", *Phy. Rev. Lett.* 84(2000) 1164-1167.

[23] Z. Nehari, "On a class of nonlinear second-order differential equations", *Trans. Amer. Math. Soc.* 95 (19960), 101-123.

[24] W.M. Ni, *Some Aspects of Semilinear Elliptic Equations*, Dept. of Math. National Tsing Hua Univ., Hsinchu, Taiwan, Rep. of China, 1987.

[25] W.M. Ni, Recent progress in semilinear elliptic equations, in RIMS Kokyuroku 679, Kyoto University, Kyoto, Japan, 1989, 1-39.

[26] P. Rabinowitz, "Minimax Method in Critical Point Theory with Applications to Differential Equations", *CBMS Regional Conf. Series in Math.,* No. 65, AMS, Providence, 1986.

[27] E. A. Silva, "Multiple critical point for asymtotically quadratic functionals", *Comm. Part. Diff. Euq.* 21 (1996), 1729-1770.

[28] J, Smoller, *Shack Waves and Reaction-Diffusion Equatios*, Springer-Verlag, New York, 1983.

[29] S. Solimini, "Morse index estimates in min-max theorems", *Manuscripta Math.* 63 (1989), 421-453.

[30] M. Struwe, *"Variational Methods"*, Springer, 1996.

[31] M. Willem, *Minimax Theorems*, Birkhauser, Boston, 1996.

[32] J. Zhou, "Local Instability Analysis of Saddle Points by a Local Minimax Method", *in review.*

[33] J. Zhou, "A min-orthogonal method for finding multiple saddle points.", *in review.*